...bers to se...t below

	From Area codes 01923 or 0208:	From the rest of Herts:
vals:	01923 471373	01438 737373
...uiries:	01923 471333	01438 737333
...icom:	01923 471599	01438 737599

BIODIVERSITY IN SCOTLAND: STATUS, TRENDS AND INITIATIVES

THE NATURAL HERITAGE OF SCOTLAND

Each year since it was founded in 1992, Scottish Natural Heritage has organised or jointly organised a conference that has focused attention on a particular aspect of Scotland's natural heritage. The papers read at the conferences, after a process of refereeing and editing, have been brought together as a book. The titles already published in this series are

1. *The Islands of Scotland: a Living Marine Heritage*
    Edited by J. M. Baxter and M. B. Usher (1994), 286pp.

2. *Heaths and Moorland: Cultural Landscapes*
    Edited by D. B. A. Thompson, A. J. Hester and M. B. Usher (1995), 400pp.

3. *Soils, Sustainability and the Natural Heritage*
    Edited by A. G. Taylor, J. E. Gordon and M. B. Usher (1996), 316pp.

4. *Freshwater Quality: Defining the Indefinable?*
    Edited by P. J. Boon and D. L. Howell (1997), 552pp.

This is the fifth book in the series.

BIODIVERSITY IN SCOTLAND: STATUS, TRENDS AND INITIATIVES

Edited by
L. Vincent Fleming, Adrian C. Newton,
Juliet A. Vickery and Michael B. Usher

EDINBURGH: THE STATIONERY OFFICE

First published 1997 by The Stationery Office Limited
South Gyle Crescent, Edinburgh EH12 9EB

Applications for reproduction should be made to the Crown Copyright Unit,
St Clements House, 2–16 Colepate, Norwich NR3 1BQ

British Library Cataloguing in Publication Data
A catalogue record for this book is available from the British Library

ISBN 0 11 495815 7

Cover photography: *Cladonia* lichens by Lorne Gill (SNH), and (inset) the dark-bordered beauty moth (*Epione paralellaria*) by Roy Leverton

PREFACE

The diversity of life forms is astonishing. Over a million species of plants, animals and microbes have been described and named, but this is only a small percentage of all species that inhabit our planet.

A decade ago the word *Bio-Diversity* was conceived, its origin being traced by Wilson (1988). The word rapidly became *Biodiversity*, and although many authors still prefer the term *Biological Diversity* the word gained international credence at the UNCED conference in Rio de Janeiro in 1992. Within five years of that conference, there has been a flurry of activity around the world to consider biodiversity, usually at the species level, but also at the genetic and habitat scales.

Two of the most comprehensive books dealing with global biodiversity issues give an indication of how the subject is being addressed. Groombridge's (1992) book provided an overview. It is instructive to note that only six pages are devoted to genetic diversity, whereas 241 pages are devoted to species diversity and 80 pages to habitat diversity. A further 250 or so pages are then devoted to the use of organisms, the economic value of biodiversity, national and international policies and instruments, and the Biodiversity Convention. Heywood's (1995) book took a more functional approach, including a wealth of information on how ecosystems function, on how to carry out biodiversity inventories and monitoring, on biotechnology, on human influences, and so on. The book does, however, give more prominence to genetic resources and their importance to the economies of many nations.

This present book represents the first attempt to document the biodiversity of Scotland, including the terrestrial, freshwater and marine environments. Although accounts are now available for a number of Scandinavian countries (Sandlund (1992) for Norway, Bernes (1994) for Sweden, and Prip and Wind (1996) for Denmark), there is currently no national account of biodiversity for the United Kingdom. This is despite the fact that the UK was one of the first countries to publish its own Biodiversity Action Plan (Anon., 1994), in which there is a brief account of the biological diversity of the United Kingdom and its dependent territories. A Steering Group set up to take forward the work of the Action Plan published its report in 1995; this included costed action plans for 14 habitats and 116 species (Anon., 1995). The UK Government broadly accepted the recommendations of the Steering Group (Anon., 1996), with its response foreshadowing the conference on which this book is based.

We believe that a focus on Scotland is justified, considering its distinctive biodiversity and its biogeographic affinities with Scandinavian countries. It is our hope that accounts for the other parts of the British Isles will become available in the future so as to permit a more detailed comparison. This book on Scotland is divided into five parts. Part One sets the scene, with overviews of Scotland's biodiversity in both contemporary and historical contexts. Part Two looks at the current position, its nine chapters exploring biodiversity at the genetic, species and habitat levels. Part Three explores concepts of the sustainable management of a range of habitats and species groups for biodiversity conservation and enhancement. These first three parts give a clear indication of the richness of Scotland's biodiversity, as well as demonstrating that the picture is not 'gloomy' for all species. Considerable species richness can, in fact, be found in the most unusual places, as indicated also by Usher (1994).

An innovation at the conference was to include a series of workshops on various aspects of preparing a strategy for biodiversity conservation. All participants had the opportunity of attending one of the workshops, which followed the major themes of *Biodiversity: the UK Action Plan*. The need for inventories was tackled by one workshop; another addressed the data needs and where the information gaps occurred. Not only is research needed to fill the gaps, but it is increasingly realised that effective biodiversity conservation depends on the scientific knowledge base of how ecological and genetical systems function. The whole science of restoration ecology is a growth area in the United States, but needs to be focused onto the particular problems here in Scotland. Both research and restoration were addressed in workshops. The fifth workshop explored aspects of education and public awareness. The accounts of these five workshops form Part Four of the book.

The final part, Part Five, looks at actions for the conservation of biodiversity. Such action will have to include partnerships between the public, private and voluntary sectors. More importantly, a vision of what we want in Scotland will need to be formulated and widely agreed. Only when the goal has been identified can a strategy be developed; that strategy will outline the possible routes between where we are today and where we want to be in the future. Along each of these routes people have their place, either as users of the land or water, or as players in the various action plans. Biodiversity conservation does not just require a partnership between various organisations, but involves a partnership between nature and the human population of Scotland.

It is not always easy to look into the future or to determine what will happen to Scotland's biodiversity in the next millennium. Biodiversity now, be it genetic variability, the number of species, or the range of habitats, is important in securing a biologically diverse world in the future. As editors, we hope that this book will effectively set the scene in the final decade of the twentieth century. It provides the reader with an indication of the current state of knowledge and of trends that are occurring. However, the reader will have to decide how these trends, and the initiatives being taken for the sustainable management of both land and water, will affect the biodiversity of Scotland a decade, or a century, or a millennium from now.

In organising this conference we owe thanks to many people. We particularly thank Sandy Kerr for his contribution to the organisation of the conference, Shiela Wilson for secretarial assistance during the organisation of the meeting, and the Interdisciplinary Symposium Fund of the University of Edinburgh for their generous financial support. Thanks are also due to Jack Matthews, Dave Raffaelli and John Gage for their chapter on marine biodiversity, and to Jonathan Humphrey and Adrian Newton for theirs on forests; both chapters were commissioned after the conference and written rapidly, and we are grateful to these authors for their willingness to help us achieve a better balance of topics in this book. We also thank the many other people who assisted before, during and after the conference; the referees who reviewed the chapters; the Scottish Office, who contributed to the reception; and Lynn Davy for her helpful subediting.

References

Anon. 1994. *Biodiversity: the UK Action Plan*. Cm2428. London, HMSO.

Anon. 1995. *Biodiversity: the UK Steering Group Report*. Vol. 1, *Meeting the Rio Challenge;* Vol. 2, *Action Plans*. London, HMSO.

Anon. 1996. *Government Response to the UK Steering Group Report on Biodiversity*. Cm3260. London, HMSO.

Bernes, C. (Ed.) 1994. *Biological Diversity in Sweden: a Country Study*. Stockholm, Swedish Environmental Protection Agency.

Groombridge, B. (Ed.) 1992. *Global Biodiversity: Status of the Earth's Living Resources*. London, Chapman and Hall.

Heywood, V. H. (Ed.) 1995. *Global Biodiversity Assessment*. Cambridge, Cambridge University Press.

Prip, C. and Wind, P. (Eds.) 1996. *Biological Diversity in Denmark – Status and Strategy*. Copenhagen, Ministry of Environment and Energy.

Sandlund, O. T. (Ed.) 1992. *Biological Diversity in Norway: a Country Study*. Trondheim, Directorate for Nature Management.

Usher, M. B. 1994. Biodiversity: which communities are hiding it? In Leather, S. R., Watt, A. D., Mills, N. J. and Walters, K. F. A. (Eds.) *Individuals, Populations and Patterns in Ecology*. Andover, Intercept, 265–273.

Wilson, E. O. (Ed.) 1988. *Biodiversity*. Washington, National Academy Press.

Vin Fleming and Michael B. Usher
Scottish Natural Heritage
Adrian Newton and Juliet Vickery
University of Edinburgh
Edinburgh, March 1997

CONTENTS

LIST OF CONTRIBUTORS

C. Bain, Royal Society for the Protection of Birds, 17 Regent Terrace, Edinburgh EH7 5BN

J. M. Baxter, Scottish Natural Heritage, 2 Anderson Place, Edinburgh EH6 5NP

H. J. B. Birks, Botanical Institute, University of Bergen, Allégaten 41, N-5007 Bergen, Norway

S. Blake, Scottish Agricultural College, Auchincruive, Ayr KA6 5HW

R. G. H. Bunce, Institute of Terrestrial Ecology, Merlewood Research Station, Grange-over-Sands, Cumbria LA11 6JU

P. D. Carey, Institute of Terrestrial Ecology, Monks Wood, Abbots Ripton, Huntingdon, Cambridgeshire PE17 2LS

R. P. Cummins, Institute of Terrestrial Ecology, Banchory Research Station, Hill of Brathens, Glassel, Banchory, Kincardineshire AB31 4BY

D. M. Davison, Dee Davison Associates, 12 (3F1) Iona Street, Edinburgh EH6 8SF

J. Drewitt, Scottish Natural Heritage, 2 Anderson Place, Edinburgh EH6 5NP

C. Easton, Scottish Office Agriculture, Environment and Fisheries Department, Pentland House, 47 Robb's Loan, Edinburgh EH14 1TY

E. P. Easton, Scottish Conservation Projects, c/o Central Scotland Countryside Trust, Hillhouse Ridge Farm, Shottskirk Road, Shotts ML7 5JJ

R. A. Ennos, Institute of Ecology and Resource Management, The University of Edinburgh, Darwin Building, Mayfield Road, Edinburgh EH9 3JU

B. C. Eversham, Institute of Terrestrial Ecology, Monks Wood, Abbots Ripton, Huntingdon, Cambridgeshire PE17 2LS

L. V. Fleming, Scottish Natural Heritage, 2 Anderson Place, Edinburgh EH6 5NP

G. N. Foster, Scottish Agricultural College, Auchincruive, Ayr KA6 5HW

D. D. French, Institute of Terrestrial Ecology, Banchory Research Station, Hill of Brathens, Glassel, Banchory, Kincardineshire AB31 4BY

J. D. Gage, The Scottish Association for Marine Sciences, Dunstaffnage Marine Laboratory, PO Box 3, Oban, Argyll PA34 4AD

C. A. Galbraith, Joint Nature Conservation Committee, Monkstone House, City Road, Peterborough PE1 1JY (current address: Scottish Natural Heritage, 2 Anderson Place, Edinburgh EH6 5NP)

S. D. Gillings, British Trust for Ornithology, The National Centre for Ornithology, The Nunnery, Thetford, Norfolk IP24 2PU

S. Housden, Royal Society for the Protection of Birds, 17 Regent Terrace, Edinburgh EH7 5BN

J. W. Humphrey, Forestry Commission Research Agency, Northern Research Station, Roslin, Midlothian EH25 9SY

A. J. Kerr, Scottish Natural Heritage, 2 Anderson Place, Edinburgh EH6 5NP (current address: Scottish Natural Heritage, Caspian House, Mariner Court, Clydebank Business Park, Clydebank G81 2NR)

E. C. Mackey, Scottish Natural Heritage, 2 Anderson Place, Edinburgh EH6 5NP

P. S. Maitland, Fish Conservation Centre, Gladshot, Haddington, East Lothian EH41 4NR

A. Manning, Institute of Cell, Animal and Population Biology, University of Edinburgh, Ashworth Laboratories, King's Buildings, Edinburgh EH9 3JT

J. B. L. Matthews, The Scottish Association for Marine Sciences, Dunstaffnage Marine Laboratory, PO Box 3, Oban, Argyll PA34 4AD

D. I. McCracken, Scottish Agricultural College, Auchincruive, Ayr KA6 5HW

J. Miles, Scottish Office Agriculture, Environment and Fisheries Department, Pentland House, 47 Robb's Loan, Edinburgh EH14 1TY

A. C. Newton, Institute of Ecology and Resource Management, The University of Edinburgh, Darwin Building, Mayfield Road, Edinburgh EH9 3JU

P. Racey, Department of Zoology, University of Aberdeen, Tillydrone Avenue, Aberdeen AB24 2TZ

D. G. Raffaelli, Department of Zoology, University of Aberdeen, Culterty Field Station, Newburgh, Ellon, Aberdeenshire AB41 0AA

S. M. Redpath, Institute of Terrestrial Ecology, Banchory Research Station, Hill of Brathens, Glassel, Banchory, Kincardineshire AB31 4BY

I. Ribera, Scottish Agricultural College, Auchincruive, Ayr KA6 5HW

G. E. Rotheray, National Museums of Scotland, Royal Museum of Scotland, Chambers Street, Edinburgh EH1 1JF

M. Scott, Plantlife, Strome House, North Strome, Lochcarron, Ross-shire IV54 8YJ

M. R. Shaw, National Museums of Scotland, Royal Museum of Scotland, Chambers Street, Edinburgh EH1 1JF

C. Sydes, Scottish Natural Heritage, 2 Anderson Place, Edinburgh EH6 5NP

D. B. A. Thompson, Scottish Natural Heritage, 2 Anderson Place, Edinburgh EH6 5NP

G. Tudor, Scottish Natural Heritage, 2 Anderson Place, Edinburgh EH6 5NP

M. B. Usher, Scottish Natural Heritage, 2 Anderson Place, Edinburgh EH6 5NP

J. A. Vickery, Institute of Cell, Animal and Population Biology, University of Edinburgh, Ashworth Laboratories, King's Buildings, Edinburgh EH9 3JT (current address: British Trust for Ornithology, The Nunnery, Nunnery Place, Thetford, Norfolk IP24 2PU)

R. Watling, Royal Botanic Garden Edinburgh, 20A Inverleith Row, Edinburgh EH3 5LR

A. Watt, Institute of Terrestrial Ecology, Edinburgh Research Station, Bush Estate, Penicuik, Midlothian EH26 0QB

M. R. Young, Department of Zoology, University of Aberdeen, Culterty Field Station, Newburgh, Ellon, Aberdeenshire AB41 0AA

FOREWORD

It is a privilege to write this foreword to *Biodiversity in Scotland: Status, Trends and Initiatives*. The book is based on a conference that provided an overview of biodiversity in Scotland. It is important that we understand how best to develop a strategy for biodiversity conservation. Biodiversity is a policy area rooted in science, and if proper progress is to be made scientists, politicians and policy makers must understand each other.

It was therefore very encouraging that there was such an excellent response to the conference. It helped us all to reach an increased understanding and to see more clearly how we could progress. Biodiversity cannot be confined to one country; living organisms do not respect national boundaries. Biodiversity is therefore a multinational issue with international conventions that commit us to take action, not only for our own biodiversity, but also for the biodiversity of other nations that may be affected by our actions. It is awesomely multidimensional in time and space. It involves complex relationships between the needs of different species and habitats, and the contrasting needs of those who have a strong interest in, and commitment to, the countryside or seas. It is right that we use this publication to focus on Scotland's biodiversity and Scotland's contribution internationally.

Scotland has a truly outstanding biodiversity. Biodiversity is most simply and effectively understood as the variety of life, ranging in scale from genes and DNA to whole habitats and ecosystems. We have species and habitats in Scotland that are important, in both the national and international context. Examples are the native pinewoods, the extensive blanket bogs, the bryophyte-rich Atlantic woodlands and the enormous colonies of breeding seabirds.

Scotland's outstanding biodiversity stems from a combination of influences. First, there are the natural features of geology, geography, and climate. Secondly, Scotland's habitats have been moulded by human influence for several thousands of years. Scotland today is a result of the complex interaction of nature with the activities of farmers, farm tenants, crofters, foresters and fishermen and their stewardship of their surroundings. My personal view is that people have not done such a bad job.

I should like to emphasise that Scotland remains a working countryside where many people are dependent for their livelihood on our natural resources. As Minister, my portfolio covers farming, forestry, the environment and rural development. It is a priority to encourage a balanced and integrated approach to all these issues. The Government is fully committed to maintaining Scotland's great wealth of habitats and species. Protection, however, can best be achieved in partnership, through dialogue with, and by taking account of the needs of, those who live and work in Scotland's countryside or around our coasts. Action for biodiversity provides an opportunity to conserve and enhance our environment while still enabling the Government to take forward strategies for economic development in rural Scotland.

In May 1996, I announced the Government's response to the UK Biodiversity Action Plan Steering Group Report. The Government endorsed the main proposals of the Report for follow-up action by Country Groups in Scotland, England, Wales and Northern Ireland. The Scottish Biodiversity Group brings together a partnership of those interests who can make a positive contribution to action on biodiversity. Membership includes representation from The Scottish Office, Scottish Natural Heritage, the Forestry Commission, the Scottish Environment Protection Agency, the Convention of Scottish Local Authorities, the Royal Society for the Protection of Birds, the Scottish Wildlife Trust, Plantlife, the research and scientific community, the Confederation of British Industry, the National Farmers' Union of Scotland, the Scottish Fishermen's Federation, the Scottish Crofters' Union, and the Scottish Landowners' Federation. The Scottish Biodiversity Group has five roles. These are to:

1. oversee the implementation of costed action plans for particular species and habitats of importance to Scotland;
2. prepare the next tranche of costed action plans for other species and habitats of particular interest in Scotland;
3. advise on good practice and promote the preparation of local biodiversity action plans;
4. help to promote public awareness and involvement in biodiversity through the community; and
5. oversee the Scottish aspects of a network of information and data on biodiversity.

The Scottish Group met for the first time in July 1996, ahead of any of the other Country Groups. Three subgroups – Costed Action Plans, Local Biodiversity Action Plans, and Public Awareness and Involvement – have been established. The Costed Action Plan Subgroup will take forward the detailed work on the costed action plans for both species and habitats which are either wholly or substantially Scottish. Of the 116 species for which action plans were published in the UK Steering Group Report, some 50 occur in Scotland; we shall take the lead on 23 of these. The Local Biodiversity Action Plan Subgroup will take forward progress on the development of local biodiversity plans. This work will lead to guidance on best practice to assist local authorities. In this particular area, lessons can be learned from south of the border where local biodiversity action plans are generally further advanced. The vital role of the Public Awareness and Involvement Subgroup is to promote biodiversity interests and education among and throughout all parts of the community. This work will build on our distinctive education institutions, and on the influential report on environmental education in Scotland, *Learning for Life* (Edinburgh, The Scottish Office, 1993).

The Scottish Biodiversity Group will make an important input into the development of an information network on biodiversity. All this talk of Groups and Subgroups may sound excessively bureaucratic. We do not intend it to be so. The organisational structure has been set up to reflect not only the very large programme of work ahead, but also the Government's determination to work in partnership with all the varied interests who have a role in delivering biodiversity objectives.

The Government's commitment to biodiversity cannot be judged simply by the setting up of the Scottish Biodiversity Group. The Government already spends very substantial resources directly and indirectly on biodiversity through, for example, Scottish Natural Heritage (SNH), the Scottish Office Agriculture, Environment and Fisheries Department's agri-environmental and research programmes, the Royal Botanic Garden, the Forestry Commission, universities and higher education institutions, and the Scottish Environment Protection Agency. Let me give you four examples of current programmes that assist biodiversity. First, there is the new Countryside Premium Scheme on which we have recently consulted. This is designed to provide assistance to farmers and crofters for the management, enhancement and creation of particular habitats and features of importance to the natural heritage of Scotland. Second, the Biodiversity Woodland Improvement Grant was recently introduced by the Forestry Commission to assist in the improvement of existing woodlands in relation to the species and habitats plans in the Steering Group Report. Third, SNH's Species Action Programme, launched in the summer of 1995, focused action on 17 species, with an additional 6 species being added in 1996. Finally, the Nitrates Focus Group was established to report on how the prevention of pollution from agricultural activities may be improved. The Government is fully determined to play its part, in partnership with others, to achieve the targets set out in the action plans.

On 25 September 1996 I launched a document entitled *Wild Geese and Agriculture in Scotland: a Discussion Paper* (Edinburgh, The Scottish Office, 1996). This aims to take forward the efforts made by SNH and others to resolve the conflicts, or perceived conflicts, between farming and geese in many parts of Scotland. It is a good example of how the Government is seeking an integrated approach to the resolution of problems, aiming to build a consensus that will protect Scotland's valuable natural heritage and biodiversity without undermining the legitimate interests of Scottish farmers and crofters. The discussion chapter recognises the importance of biodiversity issues, for example by distinguishing clearly between the status of the various species of geese that migrate to Scotland, while looking for solutions that recognise the needs of those who make their living from the land.

We all have an important role to play in conserving our biodiversity. The Government is fully committed and will play its part, but it cannot do everything. Working in partnership with all those who have major interests in the countryside and seas is the way real progress can be made. The conference, and the publication of this book, provide a solid foundation for taking forward the important task of protecting Scotland's biodiversity. I commend this joint initiative of the University of Edinburgh and Scottish Natural Heritage.

The Earl of Lindsay
Minister for Agriculture, Forestry and the Environment, Scottish Office
November 1996

PART ONE
SCOTLAND'S BIODIVERSITY IN CONTEXT

PART ONE

Scotland's Biodiversity in Context

The distribution of organisms is profoundly influenced by environmental conditions, such as variation in geology, soils, topography and climate. The biodiversity of Scotland may therefore be usefully considered by reference to the physiographic characteristics of the landscape. The land area of Scotland covers approximately 78,829 km², including some 800 islands, mostly off the west and north coasts. Previous glacial activity has resulted in a highly indented coastline, and as a result the ratio of coastline to land area is relatively high (approximately 1 km linear coastline to each 7 km² of land area). Apart from the enclosed sea lochs, the extensive marine environment bordering Scotland includes the relatively shallow North Sea and the continental shelf extending into the Atlantic Ocean.

Geologically, Scotland has an ancient and complex history, the oldest rocks being formed some 3,300 million years ago. The geological diversity of Scotland is exceptionally high, with rocks from all of the major geological epochs represented. Much of central and northern Scotland is dominated by a relatively hard, crystalline core of schists and granites, with softer sedimentary strata being more characteristic further south. This geological diversity is associated with highly varied geomorphology. Scotland is characterised by areas of high ground, with a rugged topography resulting from glacial activity. Soils are also complex and diverse, with peats and gleys dominating much of upland western Scotland, podzols being more characteristic of the drier eastern side, and brown forest soils developing in the relatively warm southeast (Anon., 1995).

The climate of Scotland is profoundly influenced by its maritime position on the edge of a continental land mass, with prevailing westerly winds and the warm waters of the North Atlantic Drift, or Gulf Stream, along the west coast, producing relatively mild and wet conditions. The east of Scotland is characterised by a more continental climate, with arctic-alpine conditions associated with the higher mountains inland. Although the Scottish mainland is only 240 km across at its widest point, there is a marked gradient in rainfall across the country. Annual totals range from less than 600 mm on parts of the east coast to more than 3,800 mm on some of the western mountains (Anon., 1995).

Scotland's complex pattern of environmental variation has clearly had a major influence on the distribution of different groups of organisms, and the composition of communities. Usher evaluates the existence of distinct biogeographical zones within Scotland, and highlights the fact that species are not distributed evenly across the landscape, but tend to occur in concentrations or 'hotspots'. Despite its isolated position at the edge of the Eurasian continent, Scotland's biodiversity is considerable, and comparable to that of Scandinavian countries. To an extent this is a consequence of the pronounced altitudinal variation characteristic of Scotland, with the arctic, boreal and temperate zones, as well as Atlantic, alpine and continental zones of the European continent, occurring in juxtaposition in neighbouring areas within Scotland (Anon., 1995).

In a recent analysis of land use in Scotland, around 80% of the land area was classified as open countryside or as mosaics of grassland and moorland. An additional 15% is covered by woodland, with only 2% associated with urban or rural development (Anon., 1995). Although this might suggest that Scotland has a predominantly wild landscape, most of the land has been profoundly influenced by human activity over millennia. For this reason, it is essential that the biodiversity of Scotland is viewed within its historical context, the subject of Birks' chapter. Although forest developed over much of Scotland following the end of the last glaciation, arctic-alpine and Atlantic communities have their own distinctive histories, illustrating the impact of long-term climatic changes on biodiversity. Techniques such as pollen analysis also provide a baseline by which to assess the impact of human activity on Scotland's biodiversity over the past 6,000 years, and enable the possible changes we may expect in the future to be evaluated.

Reference

Anon. 1995. *The Natural Heritage of Scotland: an Overview*. Perth, Scottish Natural Heritage.

1 SCOTLAND'S BIODIVERSITY: AN OVERVIEW

Michael B. Usher

Summary

1. This chapter attempts a synthesis of biodiversity in Scotland. Variation below the species level is still poorly understood. Variation at the species level and above is therefore considered in detail.

2. Scotland, including its surrounding seas, has about 90,000 species of all forms of life, unicellular, fungi, plants and animals.

3. Species are not distributed evenly across Scotland; 'hotspots' occur in the south-west, the Central Belt (perhaps owing to recorder effort), the Inner Hebrides, the Central Highlands and along the Moray Firth, although 'coldspots' tend to be concentrated in the north.

4. Scotland is rich in plant communities, with 64% of those defined by the National Vegetation Classification.

5. Scotland's complex topography results in 'fuzzy' biogeographical zones, but each can be characterised by both its species and its plant communities.

6. Although Scotland is situated on an island at a northern latitude, its biodiversity is considerable, and compares favourably with that of Denmark, Norway and Sweden.

1.1 Introduction

Biodiversity as a concept spans the hierarchy of biological systems (Allen and Starr, 1982). Although conventionally thought of at three levels – genetic, species, ecosystem – biodiversity is perhaps better described in the following way:

1. 'below species' (including genetic and molecular, but also including cultivars, varieties, breeds, subspecies, etc.);
2. 'species' (certainly useful for plants and animals, but possibly of less relevance for bacteria, viruses, etc.); and
3. 'above species' (including assemblages, communities, ecosystems and broadening out into the field of landscape ecology).

The aim of this chapter is to focus on two levels in this hierarchy, leaving the 'below species' level to others (e.g. Ennos and Easton, this volume). It is, however, at this genetic level that there may be a very considerable amount of 'hidden' diversity. The study of soil bacteria in a small quantity of Norwegian beech forest soil (Torsvik *et al.*, 1990b) indicated that 'the genetic diversity corresponds to about 4,000 different genomes of standard soil bacterial size ... this seems to imply that the bacterial population in soil is composed of a large number of genetically separate clones'. Moreover, Torsvik *et al.*'s conclusion was that their results 'indicate that soil contains a vast number of bacteria which are virtually unknown'. This important investigation, together with the associated methodological study (Torsvik *et al.*, 1990a), indicated the complexity of the species composition or clonal structure of bacterial communities, and, more importantly, the possible magnitude of biodiversity that remains to be quantified.

There are many doubts about the number of separate virus or bacterial types in Scotland. Perhaps the species concept is inappropriate for these organisms, and it may be more appropriate to talk of different genomes or individual strains. However, until studies such as those of Torsvik *et al.* have been repeated in Scotland, and on the bacteria from a series of different terrestrial, freshwater and marine habitats, it remains uncertain how wide a range of different genomes or separate clones might exist.

For the purpose of this review of biodiversity, at the 'species' and 'above species' levels in Scotland, the terrestrial, freshwater and marine environments are considered. It is simple to delimit the extent of the terrestrial and freshwater environments (Figure 1.1), but there are more difficulties in defining the marine environment. The 12 mile limit (Figure 1.1) has been used for convenience, although this has had to be interpreted 'flexibly' in attempting to enumerate which species occur in Scottish waters.

1.2   How many species are there in Scotland?

The number of species is recorded precisely for very few groups of plants and animals, and has to be estimated, or sometimes guessed, for the remainder of groups. For example, we know precisely the number of amphibians (one frog, two toads and three newts) and terrestrial reptiles (adder, slow-worm and common lizard). Even for well-known groups such as the mammals there remains doubt as to whether one or two species of pipistrelle bat occur (Racey, this volume), and for the birds there is doubt as to whether the Scottish crossbill is either a distinct species or a subspecies of the common crossbill. Such questions will only affect the total count of species in Scotland by a very small amount. There are other groups, especially the viruses, bacteria (of all types) and protozoa, which are very poorly known, and for which estimates could be out by orders of magnitude. Between these extremes are groups such as the fungi, Insecta and Arachnida, for which a considerable amount of progress has been made but with different families or orders studied to different extents. For example, the lichenised fungi, Lepidoptera (butterflies and moths) and Araneae (spiders) have been well studied

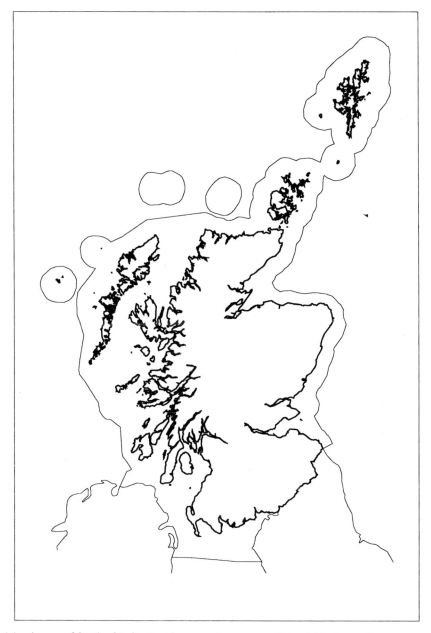

Figure 1.1 A map of Scotland indicating the approximate area within which the species count was made. The area of terrestrial and freshwater habitats is indicated by a thick line. The approximate 12 mile (19.3 km) limit is shown by a thinner line. Portions of England, Ireland and Northern Ireland are also shown as a thin line.

and the number of species in Scotland can be given with reasonable accuracy, whereas many other fungal groups, the Diptera (flies) and Acari (mites) are less well known and the number of species can only really be guessed.

For the flowering plants there is also another difficulty resulting from the apomictic species. These are species where the ovule is diploid and hence they produce swarms of genetically identical individuals, but apparently genetically different from nearby swarms. Thus hundreds of 'microspecies' can be recognised. This is particularly the case for two genera, *Hieracium* (hawkweeds) and *Taraxacum* (dandelions), and for the species group *Rubus fruticosus* (brambles). In order to give a total number of species of vascular plant, 'microspecies' have been ignored, but the taxonomic sections have to be counted, with 11, 9 and 3 sections for *Hieracium*, *Taraxacum* and *Rubus fruticosus,* respectively.

The various counts, estimates and guesses are included in Table 1.1. Given the different levels of precision, it is not appropriate to give an overall total number of species. However, in the terrestrial and freshwater environments there might be slightly more than 50,000 species, whereas in the sea around Scotland there are probably at least 39,000 species (see also Davison and Baxter, this volume). This gives an approximation for Scotland's total species richness of 90,000 species. This is divided into 49% single-celled organisms, 23% plants and fungi, and 28% animals (Figure 1.2).

1.3   Are there 'hotspots' of species diversity?

Considering species, a 'hotspot' is a defined geographical area where there are more species present than in an average, similarly sized area. Similarly, a 'coldspot' is a geographical area that has an unusually low number of species. Because the number of species tends to be linked to the size of an area (see, for example, Usher, 1979; Rafe *et al.*, 1985), it is preferable to analyse equally sized geographical areas if hotspots or coldspots are to be located.

Most biological recording in Scotland is based on the national grid, so that it is possible to use nested squares of $100 \, \text{km} \times 100 \, \text{km}$, $10 \, \text{km} \times 10 \, \text{km}$, $1 \, \text{km} \times 1 \, \text{km}$ or $0.1 \, \text{km} \times 0.1 \, \text{km}$. Because most mapping is undertaken at the 10 km grid square scale, this was used for an analysis of hotspots and coldspots (Carey *et al.*, 1996). Data for the best recorded groups held by the Biological Records Centre – vascular plants, bryophytes (both mosses and liverworts) and invertebrates (terrestrial and freshwater molluscs and diurnal insects, including butterflies, grasshoppers and dragonflies) – and by the British Trust for Ornithology for birds (both wintering and breeding), have been analysed separately, and then the four sets have been combined. Each 10 km grid square was assessed for the percentage of the *i*th taxonomic group (T_i) that occurred in it and then an index of species richness (B) was calculated as the average of the four T_i values. Thus, each taxonomic group had equal weighting, irrespective of the number of species that it contained. The results are shown in Plate 1. Full results, for each of the taxonomic groups, are given by Carey *et al.* (1996).

Despite a scatter of points, it would appear that there are five geographical areas of Scotland that tend to have above-average species richness. These are the Solway Coast, the Edinburgh to East Lothian area, the southern Hebrides and adjacent mainland, the Central Highlands, and the Moray Firth Coast. The Edin-

Table 1.1 Number of species native to Scotland and an assessment of the accuracy with which this is known.

Taxonomic group	Terrestrial and freshwater species	Marine species (as reported in Davison, 1996)	Source (for terrestrial and freshwater species only)
Viruses	1,000**	2,300*	J. I. Cooper
Monera			
Eubacteria	1,500*	1,700*	R. Pickup
Archaebacteria	100*	N/A	
Protista			
Protozoa	10,500**	27,000*	C. R. Curds
Fungi (all groups including lichens)	9,000**	140***	Watling (this volume)
Plantae			
Algae	6,800**	2,200**	B. A. Whitton
Bryophyta	928****	N/A	R. D. Porley
Vascular plants	1,075****	5****	D. J. McCosh
Animalia			
Mesozoa	N/A	10*	
Porifera	8****	275***	Maitland (1977)
Cnidaria	3****	300***	P. Cornelius
Ctenophora	N/A	3****	
Platyhelminthes	500**	400**	D. I. Gibson
Nemertina	2****	70**	Maitland (1977)
Aschelminthes	2,000**	100**	L. May; Maitland (1977); Heath *et al.* (1977)
Acanthocephala	30*	20*	D. I. Gibson
Entoprocta	N/A	50***	
Bryozoa	10**	300**	M. E. Spencer Jones
Mollusca	166****	675****	M. P. Kerney; Maitland (1977)
Sipuncula	N/A	15***	
Echiura	N/A	3****	
Annelida	70***	850***	R. W. Martin; C. R. Curds; Dring (1994); Maitland (1977)
Arthropoda			
Pauropoda	6****	N/A	A. D. Barber
Diplopoda	34****	N/A	R. E. Jones
Chilopoda	25****	N/A	A. D. Barber
Insecta	14,000**	N/A	Rotheray (1996)
Crustacea	300***	2,200***	D. T. Bilton; Maitland (1977)
Arachnida	2,000**	50***	G. Legg; P. Hillyard; D. Nellist; Turk (1953)
Pycnogonidia	N/A	14****	
Tardigrada	30***	16****	Maitland (1977)
Uniramia	N/A	3****	
Chaetognatha	N/A	20****	
Pogonophora	N/A	10***	
Brachiopoda	N/A	11****	
Phoronida	N/A	2****	
Echinodermata	N/A	120***	

continued

Table 1.1—continued

Taxonomic group	Terrestrial and freshwater species	Marine species (as reported in Davison, 1996)	Source (for terrestrial and freshwater species only)
Hemichordata	N/A	10***	
Chordata			
Tunicata	N/A	100****	
Vertebrata			
Lampreys	1****	2****	Maitland (1972); Maitland *et al.* (1994)
Hagfish	N/A	1****	
Cartilaginous fish	N/A	34****	
Bony fish	17****	189****	Maitland (1972); Maitland *et al.* (1994)
Amphibians	6****	N/A	H. R. Arnold
Reptiles	3****	1****	H. R. Arnold
Birds	242****	N/A	Thom (1986)
Mammals	41****	22****	H. R. Arnold

Key: * guess, ** guess, but with some data, *** estimate, based on reasonable data; **** actual count or close to this. Sources are personal communications, except where a date of publication is given.

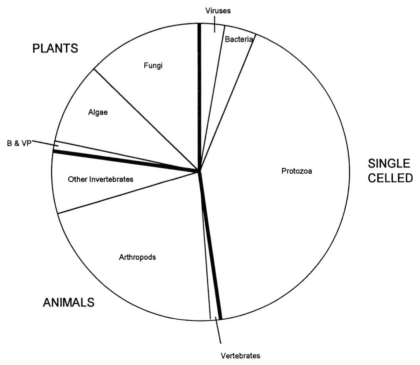

Figure 1.2   A pie chart of species richness in Scotland. The three major divisions of this chart are the single-celled organisms (viruses, bacteria and Protozoa), plants and fungi, and animals. B & VP indicates bryophytes and vascular plants.

burgh and Lothian area may be apparently species-rich owing to recording stimulated by many local naturalists over two centuries, but it is unlikely that the other hotspot areas are due to recorder effort. The Solway Firth is an area that represents the northern limit of many southern species (the 'Solway Line' of Adams (1990)), such as natterjack toad (*Bufo calamita*) and annual rest-harrow (*Ononis reclinata*). The southern Hebridean area is particularly interesting because it is associated with the area of greatest bryophyte species richness, with particular invertebrate species richness in its northern part, and with considerable bird species richness in its southern part, as well as extremely varied topography. The Moray Firth is important for the numbers of birds, invertebrates and vascular plants. The Central Highlands gains its species richness from the invertebrates, bryophytes and, to some extent, the vascular plants associated with the calcareous, schistose rocks.

It is apparent that there are three geographical areas that are, on average, species-poor (Plate 1). These are the northern Grampians, the northwest Highlands, and Lewis. All of these areas are associated with extensive areas of acid rocks, peatlands, and a cold and often wet climate. The birds and vascular plants tend to be species-poor in all three areas, the invertebrates in Lewis and the northwest Highlands, but the bryophytes are species-poor in none of these areas (although there are some coldspots in the northwest Highlands). Perhaps the most remarkable coldspot is the whole of the centre of the Isle of Lewis.

Plate 1 shows areas of greatest and least diversity based on current knowledge of species' distributions, which is good for birds, reasonably good for vascular plants, but less good for invertebrates and bryophytes. The coverage of recording is too poor for other species groups for an analysis to show any more than areas where recorders have been active. Thus care needs to be exercised in interpreting the results, which undoubtedly reflect recording effort. However, determining patterns of species distribution is only a start; the reasons for these patterns need to be determined, and if species have declined in their abundance or distributional range, it may be important to consider their restoration, as has been suggested for several of the rarer species (Anon., 1995).

At present it is not possible to carry out a similar analysis in the marine environment although undoubtedly areas of the west coast, where the warmer waters of the Gulf Stream meet the colder Arctic waters, support relatively large numbers of species.

1.4   How many communities are there in Scotland?

If a count is to be made, the problem is one of the definition of a community. As Ricklefs (1990) has pointed out, the term 'community' has been used in a large number of contexts by ecologists. He says 'it usually is applied to a group of populations that occur together, but there ends any similarity among definitions'. It can be applied to populations of plants, or animals, or microbes, or any combination of them, but it is usually delimited in both space and time (Rosenzweig, 1995). It is, therefore, not possible to define the number of communities occurring in Scotland without first accepting a definition of a community.

For practical purposes one definition that can be used is that of the National Vegetation Classification (NVC) (Rodwell, 1991–1997). The classification is based largely on vascular plants, although bryophytes are also used to characterise some of the communities. The classification aims only to classify areas of semi-natural terrestrial vegetation, and it does not use any animal or microbial data. Although wetland vegetation is included in the classification, the scheme hardly extends to communities in fresh, brackish or salt waters. Approaches to the classification of aquatic habitats tend to rely more on the nature of the substrate than on the plants (or animals) that occur in them (Usher, 1997). In the marine environment, the Marine Nature Conservation Review (Hiscock, 1996) is developing a classification of biotopes. This could be considered as the marine equivalent to the NVC, and will define biotopes in terms of their physical habitats and distinctive assemblages of conspicuous species.

The NVC lists about 220 communities and 560 subcommunities (the actual numbers might change pending publication of the final volume and as more communities are discovered in Scotland). Approximately 64% of the communities recognised in Great Britain occur in Scotland (Table 1.2). Scotland is particularly rich in upland communities, and also rich for maritime and mire communities. Compared with the remainder of Great Britain, the grassland communities are poorly represented in Scotland. Terrestrial habitat diversity is further assessed by Miles *et al.* (this volume).

One hundred and forty-one terrestrial NVC communities can be recognised in Scotland, but the number of freshwater and marine communities remains unas-

Table 1.2 The communities of the National Vegetation Classification in Great Britain and their occurrence in Scotland (extracted from Rodwell, 1991–97).

Vegetation type	Code	No. of communities in GB	No. of communities in Scotland	Percentage of GB communities in Scotland	No. of sub-communities in GB
Calcareous grasslands	CG	14	5	36	43
Heaths	H	22	14	64	72
Mires	M	38	27	71	76
Maritime[a]	MC	12	9	75	34
Mesotrophic grasslands	MG	13	7	54	35
Swamps/fens	S	28	17	61	69
Strandline, etc.[a]	SD	19	13	68	54
Saltmarshes[a]	SM	28	15	54	49
Uplands	U	21	19	90	55
Woodlands	W	25	15	60	76
Total	—	**220**	**141**	**64**	**563**

[a]Possible changes in these statistics may occur when volume 5 is published.

sessed. The Marine Nature Conservation Review (Hiscock, 1996) should give statistics comparable to those of the NVC. General principles for the classification of rivers have been discussed by Naiman *et al.* (1992), and further focus has been given to classifications in Scotland by some of the papers in Maitland *et al.* (1994). The problem will remain one of definition; it is only when there is reasonable agreement on what is a community that the number in Scotland can be counted, and this count can then be compared with the number in Great Britain or in other countries. The NVC has provided one such basis, biased towards plants, and we await general acceptance of freshwater and marine classifications.

1.5   Do groups of species and communities tend to co-occur?

At the level of the community, defined by its species, the question is trivial. However, if one looks at the whole of Scotland, how are species and communities distributed geographically and is there co-occurrence? A spatial analysis of the distribution of the four taxonomic groups analysed previously was described by Carey *et al.* (1994, 1995) and the work has been developed by Usher and Balharry (1996) into a series of twelve biogeographical zones in Scotland. These zones range in area from zone 8 (Barra and Tiree) with 243 km^2 to zones 2 (Grampian Fringe and Southern Uplands) and 1 (Central and Southern Lowlands) with 14,975 and 16,532 km^2, respectively (Plate 2).

The analysis of the species that are characteristic of the zones (Usher and Balharry, 1996) showed that the plants were particularly important. If the ten most characteristic species for each zone are considered, vascular plants accounted for 54 of these species (45%) and bryophytes for 49 species (41%). The molluscs (9 species) and diurnal insects (3 species) contributed only 10% of the characteristic species, and birds only 4%. A similar analysis of the characteristic NVC communities (J. S. Rodwell, pers. comm.) indicated a broader spread, with all vegetation types contributing (Table 1.3). There is a strong correlation, as would be expected, between the geographical location of the zone and the type of NVC communities that are characteristic of it. For example, the two high-altitude zones, 4 and 5, are strongly characterised by heath and upland communities. The East Coast (zone 3) is strongly characterised by its sand-dune communities, and the Western and Northern Isles by their maritime, dune and saltmarsh communities.

For a detailed investigation of a high-altitude zone, the Cairngorm zone (4) can be selected. Usher and Balharry (1996) showed that the characteristic species were the alpine foxtail (*Alopecurus borealis*) and alpine cat's-tail (*Phleum alpinum*) grasses, the downy willow (*Salix lapponum*), and three herbaceous species, alpine speedwell (*Veronica alpina*), chickweed willowherb (*Epilobium alsinifolium*) and sheathed sedge (*Carex vaginata*). There were also three mosses, *Cynodontium tenellum* (a montane species), *Philonotis seriata* (a species of acidic flushes at high altitudes) and *Stegonia latifolia* (a species of calcareous alpine ledges), and finally there was a liverwort, *Tetralophozia setiformis,* a species of dry acid rocks occurring in Scotland above 300 m. The zone is strongly characterised by heathland communities, with *Calluna vulgaris – Cladonia arbuscula* heath (H13), *Calluna vulgaris – Arctostaphylos uva-ursi*

Table 1.3 For each of the twelve biogeographical zones (*sensu* Usher and Balharry, 1996) (see Plate 2), the ten most characteristic plant communities (cf. the National Vegetation Classification communities defined by Rodwell (1990–1997)) are shown. The community codes are given in Table 1.2. Dots indicate that the vegetation types were not among the ten most characteristic of the zone.

Biogeographical zone (and number)	Vegetation types									
	CG	H	M	MC	MG	S	SD	SM	U	W
Central and Southern Lowlands (1)	·	·	2	·	1	4	·	·	·	3
Grampian Fringe and Southern Uplands (2)	·	2	4	·	·	·	·	·	3	1
East Coast (3)	·	1	·	2	1	·	6	·	·	·
Cairngorm (4)	·	5	1	·	·	·	·	·	3	1
Western Highlands (5)	·	4	1	·	·	·	·	·	5	·
Western Highlands Fringe (6)	·	1	3	·	·	1	·	·	3	2
Western Isles (North) and North Mainland (7)	1	2	5	1	·	1	·	·	·	·
Barra and Tiree (8)	·	1	·	3	·	·	3	3	·	·
Western Isles (South) (9)	·	1	1	3	·	·	4	1	·	·
Northern Isles (10)	·	1	·	3	2	·	3	1	·	·
Argyll and Inner Hebrides (11)	1	·	1	1	1	1	·	4	·	1
Galloway Coast (12)	·	·	·	3	·	·	2	4	·	1
Total	**2**	**18**	**18**	**16**	**5**	**7**	**18**	**13**	**14**	**9**

heath (H16), *Vaccinium myrtillus – Deschampsia flexuosa* heath (H18), *V. myrtillus – Cladonia arbuscula* heath (H19) and *V. myrtillus – Rubus chamaemorus* heath (H22), although there are also three upland communities, *Carex bigelowii – Polytrichum alpinum* heath (U8), *Juncus trifidus – Racomitrium lanuginosum* heath (U9) and *Saxifraga aizoides – Alchemilla glabra* banks (U15). The other two communities are *Juniperus communis – Oxalis acetosella* woodland (W19) and *Cratoneuron commutatum – Festuca rubra* spring (M37).

This one example has been given to demonstrate what is obvious in the more uniform eastern part of Scotland: both the characteristic species and the characteristic communities are logically those that would be associated with high, or intermediate, or low altitudes. In the west, where over a horizontal distance of a kilometre or two the altitude can range from sea level to over 1,000 m, the situation is more complex. The use of characteristic species and communities is therefore more problematic.

Biogeographical zonation provides a basis for biodiversity conservation (see also Brown *et al.*, 1993). Within each of the zones, the characteristic species and characteristic communities can be recognised, and areas with large (or small) numbers of species or communities (the hotspots and coldspots) can be recognised. The data on species hotspots (Plate 1) may indicate 10 km grid squares with an unusual species richness, or they may indicate extreme topographical variation

PLATE 1

Percentile

- 91 - 100
- 81 - 90
- 11 - 20
- 1 - 10

Plate 1. A map of Scotland showing 'hotspots' and 'coldspots' of biodiversity. These terms are defined in the text (Usher, this volume), but the dark red and dark blue squares represent the 10% of the 10km grid squares with the greatest and least species richness respectively, and the orange and the pale blue the next 10 percent of squares with the greatest and least species richness respectively.

PLATE 2

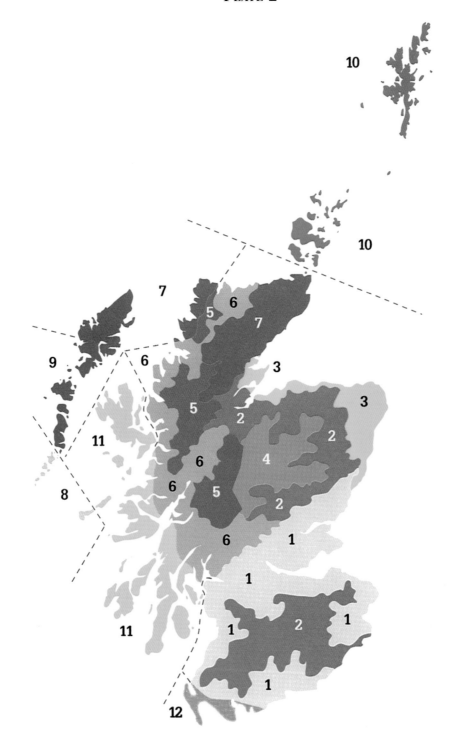

Plate 2. A map showing the approximate boundaries of a division of Scotland into 12 biogeographical zones (see Usher, this volume). The Zones are : 1 Central and Southern Lowlands; 2 Grampian Fringe and Southern Uplands; 3 East Coast; 4 Cairngorm; 5 Western Highlands; 6 Western Highlands Fringe; 7 Western Isles (North) and North Mainland; 8 Barra and Tiree; 9 Western Isles (South); 10 Northern Isles; 11 Argyll and the Inner Hebrides; and 12 Galloway Coast. This map is taken from Usher and Balharry (1996).

providing more habitats for species, or they may indicate aspects of the effort that recorders have expended. In any case, it is those 'red' squares that would repay further consideration, looking at what is characteristic of the biogeographical zone in which they are located and questioning why they are particularly species-rich. Similarly, hotspots for communities can be determined, totalling the number of NVC communities in each 10 km grid square (Figure 1.3) and again questioning why some are community-rich. There is a strong degree of recorder bias as some grid squares apparently lack any NVC communities, especially in the northeast of Scotland, near the Ross-Sutherland border, in the Central Belt and just north of the border with England. However, it is apparent that there are three areas with a large number of vegetation types: Galloway, the Central Highlands just north of the Highland Fault, and the northwest Highlands including the island of Skye. The first two of these areas correlate with species 'hotspots', but the third largely coincides with a species 'coldspot'. Biodiversity priorities will have to be determined: is it better to focus on large numbers of species or on large numbers of communities?

1.6 Discussion

The estimate of about 50,000 species in the terrestrial and freshwater environments of Scotland needs to be compared with those for other parts of the world if Scotland's species richness, or species paucity, is to be evaluated. Scandinavian countries have been selected for comparison because they are geographically close to Scotland, and have similar landscapes and oceanic climates (Table 1.4). The estimate of 500–2,000 viruses (J. I. Cooper, pers. comm.) cannot be compared because there are no estimates for the other countries; the situation for bacteria is similar. However, the taxonomy of these groups is poorly understood, and the species concept may not be appropriate. In addition, for the protozoa there are taxonomic and definitional problems that make comparisons difficult. For all three groups it is noticeable that Scotland is likely to have at least 20% of the world's total number of 'species'. This is either because the 'species' are widely distributed and generally common, or because the knowledge of 'species' is so rudimentary that the true position on this planet is unknown.

Estimates of numbers of algal species are also difficult owing to the taxonomic recognition of species and the way in which marine and freshwater species tend to be combined. Sandlund (1992) stated

> About 300–350 species of algae are probably common in Norwegian coastal waters, and a similar number in fresh water. The total number of species worldwide is extremely uncertain. Estimates of marine diatoms, for example, vary from 5,000 to 100,000, and it has been estimated that 15,000 species are to be found in Norwegian coastal waters. 478 benthic algae have been described from the Norwegian coast. The Chrysophyceae and green algae (Chlorophyceae) are the best known groups of freshwater algae in Norway, and about 2,000 species have been described, ...

For the fungi and lichens, Scotland appears to have of the order of half of the UK species, but more species than Norway (Table 1.4). The similarity of the

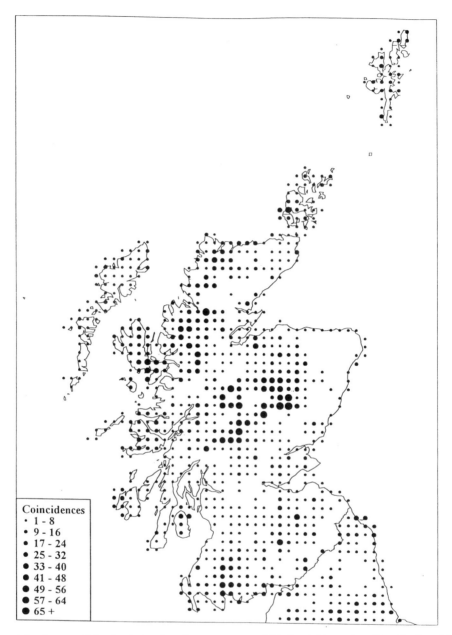

Figure 1.3 A map of Scotland showing the occurrence of National Vegetation Classification (NVC) communities (from Rodwell, 1997). There is a degree of recorder effect in this diagram: not all 10 km grid squares have been worked for their community richness. However, a few squares, indicated by the largest black dots, have a richness that approaches half of all NVC communities recorded in Scotland.

bryophytes is interesting, but the comparative poverty of the British Isles for vascular plants is noticeable.

For the animals, the counts of arthropods for Scandinavian countries and

Table 1.4 A comparison of species richness of some terrestrial and freshwater biota in Scotland, the United Kingdom, Scandinavia and the planet Earth. Data are drawn from Table 1.1 for Scotland, Prip and Wind (1996) for Denmark, Sandlund (1992) for Norway, Bernes (1994) for Sweden, and Anon. (1994) for the UK and the Earth.

Taxonomic group	Scotland	United Kingdom	Denmark	Norway	Sweden	Earth
Viruses	1,000[a]	?	?	?	?	>5,000
Bacteria	1,600	?	?	?	?	>4,000
Protozoa	10,500	>20,000	?	?	c. 2,000	>40,000
Algae	6,800	>20,000	516[b]	?	494[b]	>40,000
Fungi and lichens	9,000	>16,500	5,700	6,800	>8,000	>87,000
Bryophytes	928	1,000	584	1,056	1,050	>14,000
Vascular plants	1,070	1,480	1,447	c. 1,800	1,972	>262,000
Non-arthropod invertebrates	c. 2,800	>3,000	>2,680	?	?	>90,000
Arthropods	c.16,400	>25,500	>18,500	>17,100	28,200	>1,190,000
Freshwater fish	18	38	?	40	?	>8,500
Amphibians	6	6	14	5	13	>4,000
Reptiles	3	6	7	5	6	>6,500
Birds (breeding)	179	210	285	249	250	9,881
Mammals	41	48	49	57	69	4,327

[a]This is the geometric mean of the range 500–2,000.
[b]This figure includes some marine species.

Scotland are reasonably similar. The number of non-arthropod invertebrates in Scotland appears large as it is similar to that for the United Kingdom. Scotland appears to be species-poor for both freshwater fish and mammals, but for other vertebrate groups the figures are similar.

The expectation might have been that Scotland would appear species-poor in these comparisons. There are three reasons for such an hypothesis. First, Scotland is part of an island. Second, it occupies the northern third or so of that island. Third, it has a small surface area, with 78,829 km^2, compared with 244,157 km^2 for the UK, 386,958 km^2 for Norway, and 449,964 km^2 for Sweden, but only 43,093 km^2 for Denmark. Usually studies have indicated that a ten-fold increase in area approximately doubles the number of species. This species-area relationship has been written in the form $S = cA^z$, where S is the number of species, A is the area, c is a constant depending on the species richness of the taxon under investigation (it is actually the number of species per unit area), and z is another constant whose value is usually about 0.3 (MacArthur and Wilson, 1967). From this relationship, and if the first two of the three factors listed above are excluded, it could be expected that the UK has about 40% more species than Scotland, that Norway and Sweden have about 60 and 70% more species, respectively, and that Denmark has about 17% fewer species. The data in Table 1.4 tend to indicate the comparative species richness of Scotland rather than its species poverty, although such a conclusion is based on a comparison with Scandinavian countries and not with territories situated nearer to the equator, which are likely to be more species-rich.

Given the increasing interest in biological resources, the nation's capital of biodiversity, how does one protect, conserve or enhance what exists today? Nature is dynamic. The collection of species in a defined area will be stable over periods of tens to hundreds of years. However, species can move around in space through time, and hence some that were in Scotland a few thousand years ago are no longer so (Birks, this volume). Coope (1986) similarly showed that the distribution patterns of some beetles across the Palaearctic has changed significantly over time. Thus, although the species entity itself is stable, its geographical distribution will be constantly changing in relation to shifts in climate.

The definition of a community is difficult, and as species move spatially the collection of populations at any one location will change. Does this mean that the community changes? Communities are also changed by systems of land husbandry. For example, the dynamic nature of Scottish hill communities was outlined by Miles (1988), who showed that with low herbivore pressure the successional pathways were from grassland to dwarf shrub and scrub communities and eventually (if there was a seed source) to woodland dominated by birch or pine. As the herbivore pressure increased the pathways would be balanced at dwarf shrub or bracken communities, whereas at intense herbivore pressure no trees regenerated, scrub and dwarf shrubs died back, and grasslands were created.

Such dynamism needs to be considered carefully in the preparation of species and habitat action plans (Anon., 1995). Conservation of Scotland's biodiversity needs to consider the whole land area, as well as the fresh water and seas around

Scotland, and not just a handful of protected areas. The latter have a distinct role as 'sources' of species. A wide geographical coverage of protected areas, perhaps located in the 'hotspots', concentrating on what is characteristic of each biogeographical zone, should provide the basis for protecting Scotland's 90,000 or so species, and the genetic variability that each encompasses. However, the rest of the land and water outside protected areas will harbour populations of most of these species, and will represent a significant proportion of their gene bank. Action plans for biodiversity must encompass the biodiversity both within and outwith protected areas, and recognise that both positive and negative changes will be taking place. If the plans do not, in 100 years' time they may be deemed to have been a failure. If a similar conference to this were to be held in 2096, what would be its verdict on our plans (Fleming, this volume; Kerr and Bain, this volume) and aspirations (Manning, this volume) in 1996? The answer lies as much in the net changes of species richness and distribution as in the viability of local communities and local populations of species.

Acknowledgements

I thank Drs John Baxter, Vin Fleming and Des Thompson for commenting on a draft of this chapter. I acknowledge the support of many of my colleagues in SNH, especially Drs John Baxter and Stephen Ward, in the compilation of Table 1.1.

References

Adams, P. 1990. *Saltmarsh Ecology*. Cambridge, Cambridge University Press.

Allen, T. F. H. and Starr, T. B. 1982. *Hierarchy: Perspectives for Ecological Complexity*. Chicago and London, University of Chicago Press.

Anon. 1994. *Biodiversity: the UK Action Plan*. Cm2428. London, HMSO.

Anon. 1995. *Biodiversity: the UK Steering Group Report*. London, HMSO.

Bernes, C. (Ed). 1994. *Biological Diversity in Sweden: a Country Study*. Stockholm, Swedish Environmental Protection Agency.

Brown, A., Birks, H. J. B. and Thompson, D. B. A. 1993. A new biogeographical classification of the Scottish uplands. II. Vegetation-environment relationships. *Journal of Ecology*, **81**, 231–251.

Carey, P. D, Dring, J. C. M., Hill, M. O., Preston, C. D. and Wright, S. M. 1994. Biogeographical zones in Scotland. Scottish Natural Heritage Research, Survey and Monitoring Report No. 26.

Carey, P. D., Hill, M. O. and Fuller, R. J. 1996. Biodiversity hotspots in Scotland. Scottish Natural Heritage Research, Survey and Monitoring Report No. 77.

Carey, P. D., Preston, C. D., Hill, M. O., Usher, M. B. and Wright, S. M. 1995. An environmentally defined biogeographical zonation of Scotland designed to reflect species distribution. *Journal of Ecology*, **83**, 833–845.

Coope, G. R. 1986. The invasion and colonization of the North Atlantic islands: a palaeoecological solution to a biogeographic problem. *Philosophical Transactions of the Royal Society of London*, B **314**, 619–635.

Davison, D. M. 1996. An estimation of the total number of marine species that occur in Scottish coastal waters. Scottish Natural Heritage Review No. 63.

Dring, J. C. M. 1994. Support for the Biological Records Centre 1993/4: first annual report. Part 3. Summaries of species occurrence. Joint Nature Conservation Committee Report No. 187.

Heath, J., Brown, D. J. F. and Boag, B. 1977. *Provisional Atlas of Nematodes of the British Isles*. Dundee, Scottish Horticultural Research Institute and Huntingdon, Institute of Terrestrial Ecology.

Hiscock, K. (Ed.) 1996. *Marine Nature Conservation Review: Rationale and Methods*. Peterborough, Joint Nature Conservation Committee.

MacArthur, R. H. and Wilson, E. O. 1967. *The Theory of Island Biogeography*. Princeton, Princeton University Press.

Maitland, P. S. 1972. *A Key to the Freshwater Fishes of the British Isles with Notes on their Distribution and Ecology*. Ambleside, Freshwater Biological Association.

Maitland, P. S. 1977. *A Coded Checklist of Animals Occurring in Fresh Water in the British Isles*. Edinburgh, Institute of Terrestrial Ecology.

Maitland, P. S., Boon, P. J. and McLusky, D. S. (Eds.) 1994. *The Fresh Waters of Scotland: a National Resource of International Significance*. Chichester, Wiley.

Miles, J. 1988. Vegetation and soil change in the uplands. In Usher, M. B. and Thompson. D. B. A. (Eds.) *Ecological Change in the Uplands*. Oxford, Blackwell Scientific Publications, 57–70.

Naiman, R. J., Lonzarich, D. G., Beechie, T. J. and Ralph, S. C. 1992. General principles of classification and the assessment of conservation potential in rivers. In Boon, P. J., Calow, P. and Petts, G. E. (Eds.) *River Conservation and Management*. Chichester, Wiley, 93–123.

Prip, C. and Wind, P. (Eds.) 1996. *Biological Diversity in Denmark – Status and Strategy*. Copenhagen, Danish Forest and Nature Agency.

Rafe, R. W., Usher, M. B. and Jefferson, R. G. 1985. Birds on reserves: the influence of area and habitat on species richness. *Journal of Applied Ecology*, **22**, 327–335.

Ricklefs, R. E. 1990. *Ecology*, 3rd Edn. New York, Freeman.

Rodwell, J. S. (Ed.) 1991–1997. *British Plant Communities*, Vols 1–5. Cambridge, Cambridge University Press.

Rosenzweig, M. L. 1995. *Species Diversity in Space and Time*. Cambridge, Cambridge University Press.

Rotheray, G. E. 1996. Why conserve Scottish insects? In Rotheray, G. E. and MacGowan, I. (Eds.) *Conserving Scottish Insects*. Edinburgh, Edinburgh Entomological Club, 11–16.

Sandlund, O. T. (Ed.) 1992. *Biological Diversity in Norway: a Country Study*. Trondheim, Directorate for Nature Management.

Thom, V.M. 1986. *Birds in Scotland*. Calton, Poyser.

Torsvik, V., Salte, K., Sørheim, R. and Goksøyr, J. 1990a. Comparison of phenotypic diversity and DNA heterogeneity in a population of soil bacteria. *Applied and Environmental Microbiology*, **56**, 776–781.

Torsvik, V., Goksøyr, J. and Daae, F. L. 1990b. High diversity in DNA of soil bacteria. *Applied and Environmental Microbiology*, **56**, 782–787.

Turk, F. A. 1953. A synonymic catalogue of British Acari. *Annals and Magazine of Natural History*, series 12, **6**, 1–26 and 81–99.

Usher, M. B. 1979. Changes in the species-area relationships of higher plants in nature reserves. *Journal of Applied Ecology*, **16**, 213–215.

Usher, M. B. 1997. Principles of nature conservation evaluation. In Boon, P. J. and Howell, D. L. (Eds.) *Freshwater Quality: Defining the Indefinable?* Edinburgh, Stationery Office, 199–214.

Usher, M. B. and Balharry, D. 1996. *Biogeographical Zonation of Scotland*. Perth, Scottish Natural Heritage.

2 SCOTTISH BIODIVERSITY IN A HISTORICAL CONTEXT

H. J. B. Birks

Summary

1. In a European context, the biodiversity of Scotland's flora today is remarkable for its Atlantic species that also occur along the western seaboard of Europe, on the Azores, and in tropical areas, and its mountain flora of arctic-alpine and arctic species. Nowhere else in the world do these elements co-exist so extensively.

2. Very little is known about the Quaternary history of Atlantic species in Scotland.

3. Abundant fossils of mountain species indicate that local and regional extinction of some arctic and arctic-alpine taxa has occurred in Scotland since the late-glacial. It is likely that some taxa survived the last glaciation around the margins of the ice-sheet and rapidly invaded the recently deglaciated landscapes as the ice-sheets began to retreat. Some of these taxa survive at high altitudes above the potential treeline in the Highlands, whereas others went extinct in Scotland but survive in Norway today.

4. The post-glacial (Holocene) forest history of Scotland has involved the differential colonisation and spread of tree taxa in the first 5,000 years with subsequent regional- and local-scale extinction as a result of climate change and human activity in the late Holocene.

5. Changing patterns of floristic richness revealed from detailed pollen-stratigraphical records reflect the changing mosaic structure of the landscape over the past 12,000 years.

6. Freshwater systems have undergone major changes in recent centuries as a result of acidification and, more recently, eutrophication.

7. The surviving mountain flora in Scotland is threatened with further extinction as a result of climate warming predicted to occur in the coming decades.

2.1   Introduction

When viewed in a European context, the biodiversity of Scotland's vascular plant, bryophyte, and lichen flora is remarkable in two ways. First, Scotland has a unique flora of Atlantic species of ferns, bryophytes, and lichens, many of which occur along the western seaboard of Europe, on the Azores and the Canary Islands, and in parts of the tropics, so-called Macaronesian–Tropical distributions (Figure 2.1). Second, Scotland provides the southernmost and/or westernmost European occurrences for several arctic, arctic–alpine, and boreal species that are common in Scandinavia (Figure 2.1). Scotland has a unique flora because of this combination of Atlantic and arctic–alpine, arctic, and boreal species (Ratcliffe and Thompson, 1988). Scotland's flora is thus not an impoverished version of Norway's flora. Norway, although rich in mountain species, has a very impoverished version of Scotland's Atlantic flora, whereas Scotland has a slightly impoverished version of Norway's mountain flora but a very rich Atlantic flora.

It is this mixture of species of contrasting phytogeographical elements that gives Scotland its unique flora. Nowhere else in the world can one find species with Macaronesian or Tropical affinities growing at their northernmost known world localities (e.g. *Trichomanes speciosum* (Figure 2.1), *Hymenophyllum tunbrigense* (Figure 2.1), *Adelanthus decipiens, Jubula hutchinsiae*) occurring in the same area as arctic species such as *Koenigia islandica* (Figure 2.1), *Cerastium arcticum* (Figure 2.1), *Alopecurus alpinus,* and *Arenaria norvegica* growing in their southernmost European localities.

The present flora of Scotland poses several interesting questions. What are the origins of this diverse flora? Did any plants survive in Scotland during the last glaciation? Have any species gone extinct since the last glaciation? Were forest trees once commoner? Is there evidence for local extinctions of forest trees in the post-glacial?

Answers to these questions can be provided by palaeoecological studies of peats or freshwater-loch sediments, involving detailed pollen and plant macrofossil analyses to provide evidence about past flora and vegetation and radiocarbon-dating to provide an independent chronology for the palaeoecological recon-structions.

In this chapter, I consider what is known about the Quaternary history of the Atlantic flora in Scotland and then discuss the history of Scotland's mountain flora with particular emphasis on questions of glacial survival and post-glacial (Holocene) extinction over the past 10,000 years. I then outline the history of the Scottish tree flora, the development of mid-Holocene forest patterns, and the local extinction of trees from the islands. Changing patterns of floristic and landscape diversity over the past 12,000 years are discussed and recent changes in freshwater systems are briefly considered. I conclude by discussing the possibility of future extinctions in the Scottish flora as a result of climate warming in the coming decades.

PLATE 3 PINEWOODS

Intermediate wintergreen (Pyrola media), a characteristic pinewood herb (Photo: M.B. Usher).

Callicera rufa, a rare hoverfly that breeds in water-filled rot-holes in standing pines (Photo: S.G. Ball).

Pinewood bracken (Pteridium pinetorum var. pinetorum), recently described from Speyside (Photo: M.B. Usher).

Native pinewood, Rothiemurchus (Photo: L. Gill).

PLATE 4 OAKWOODS

A multi-stemmed oak richly covered with epiphytic bryophytes and ferns, Glen Nant (Photo: L. Gill).

Hay-scented buckler-fern (Dryopteris aemula), characteristic of oakwoods on the Atlantic seaboard (Photo: M.B. Usher).

Sessile oak woodland, Ariundle National Nature Reserve (Photo: M.B. Usher).

The Scotch argus butterfly (Erebia aethiops) (Photo: M.B. Usher).

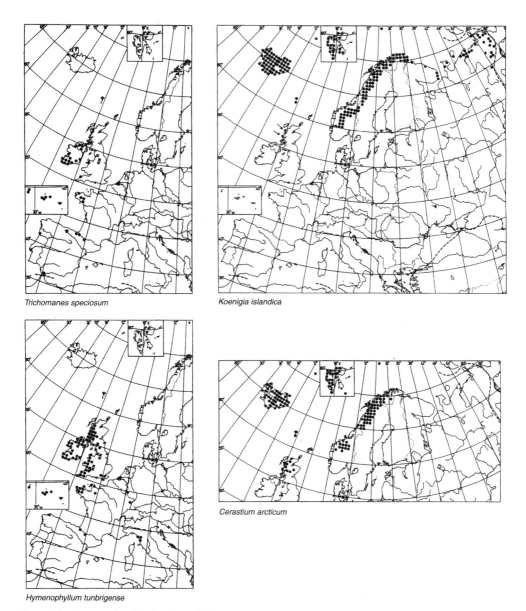

Trichomanes speciosum

Koenigia islandica

Hymenophyllum tunbrigense

Cerastium arcticum

Figure 2.1   Present-day distributions of *Trichomanes speciosum* and *Hymenophyllum tunbrigense* (both species with an Atlantic distribution) and *Koenigia islandica* and *Cerastium arcticum* (both species with an Arctic distribution) in Europe. Note that the map for *T. speciosum,* prepared in 1972, does not include recent records of this plant in Scotland (see Ratcliffe *et al.*, 1993). Modified from Jalas and Suominen (1972, 1979, 1983).

2.2   The Atlantic flora

Scotland, in particular western parts of the Mainland and the Inner Hebrides, has a remarkable flora of ferns, bryophytes, and lichens, many of which have

Macaronesian–Tropical distributions globally and occur in their northernmost world localities on Skye or nearby Wester Ross. Many of these species occur in shaded, permanently humid habitats within the zone of potential woodland, in particular in rocky woodlands, ravines, and gorges (Ratcliffe, 1968). In addition above the potential woodland zone there is a group of bryophytes, mainly large leafy liverworts, with extraordinary disjunct world distributions such as Scotland, Ireland, western N America, Yunnan, and the Himalayas. Examples include *Herbertus aduncus, Mastigophora woodsii* and *Leptodontium recurvifolium* (Ratcliffe, 1968). These distributions are as spectacular as the well-known disjunct distributions of *Eriocaulon aquaticum* (Inner Hebrides, Ireland, eastern N America) and *Spiranthes romanzoffiana* (Scotland, Ireland, SW England, N America, Kamchatka).

There are about 150 or more Atlantic bryophytes and ferns growing in Scotland today and a comparable number of Atlantic lichens. We know almost nothing about their Quaternary history in Scotland or elsewhere because of the almost total absence of fossil remains of all liverworts and lichens and the extreme rarity of fossil remains of Atlantic mosses and ferns. We do, for example, know that *Hymenophyllum wilsonii* and *Eriocaulon aquaticum* have been growing on Skye for much of the Holocene, as shown by the occurrence of their spores and pollen in Skye loch sediments.

Plant macrofossils (seeds, fruits, leaves, etc.) preserved in presumed interglacial peat beds on Shetland (Birks and Ransom, 1969; Birks and Peglar, 1979; cf. Hall *et al.*, 1993; Duller *et al.*, 1995) provide unambiguous evidence for the local growth of the heaths *Daboecia cantabrica, Erica mackaiana* and *Bruckenthalia spiculifolia* in Scotland prior to the last glaciation. *Daboecia* and *E. mackaiana* have Atlantic distributions today in Europe, occurring in NW Spain, Portugal and western Ireland (see Mitchell and Watts, 1970), whereas *B. spiculifolia* is today confined to the Balkan mountains (Birks and Peglar, 1979; Whittington, 1994). It is striking that *Calluna vulgaris* is very rare or absent from pre-Holocene deposits and did not achieve any dominance until the current Holocene postglacial (Stevenson and Birks, 1995).

There is thus a disappointingly meagre factual basis for assessing the history of Atlantic species in Scotland (or elsewhere). It is clear that species like *D. cantabrica* and *E. mackaiana* were once more widespread and it is possible that many Atlantic species may have become increasingly relictual during the Quaternary as a result of the numerous cycles of local extinction, chance re-establishment and expansion, and subsequent extinction associated with the glacial-interglacial climatic cycles of the past 1–2 million years. This model of repeated extinction and re-establishment would explain, in part, the disjunct distributions of many Atlantic species today.

2.3 The mountain flora

Macrofossils of plants that we associate with the Scottish mountains today occur frequently in late-glacial (*c.* 12,000–10,000 ^{14}C years ago) deposits in Scotland. There are several macrofossil records of, for example, *Salix herbacea, Salix reticulata,*

Oxyria digyna, Silene acaulis, Dryas octopetala, Betula nana, Saxifraga oppositifolia, Saxifraga cespitosa, Saxifraga aizoides, Draba incana, Thalictrum alpinum, Minuartia rubella, Cerastium alpinum, Lychnis alpina, Orthothecium rufescens, and *Polytrichum sexangulare,* as well as pollen records of *Koenigia islandica* (see, for example, Birks and Mathewes, 1978; Birks, 1984; Huntley, 1994; Webb and Moore, 1982). These fossils indicate that at the end of the last glaciation the present-day Scottish mountain flora was more widespread than today and occurred down to low elevations over much of Scotland.

Macrofossils have also been found of plants that do not grow in Scotland today and have thus gone extinct since the late-glacial. These include *Papaver* sect. *Scapiflora* (Conolly, 1961; Huntley, 1994; H. H. Birks, 1994), *Silene furcata* (Dickson, 1992; H. H. Birks, unpublished) and *Cassiope* (Huntley, 1994). Table 2.1 summarises the numbers of montane taxa that have gone extinct since the last glacial stage in Ireland, Britain (including Scotland) and Scandinavia.

There has been the greatest number of extinctions of mountain taxa in Ireland, not surprisingly because of the small area of high mountains in Ireland and the highly oceanic mild climate (Ratcliffe, 1991). Arctic and arctic–alpine plants often require climates conducive to the cessation of growth in autumn, a period of quiescence associated with cold winters and stable snow conditions, and a rapid transition to the summer growing season (Moe, 1995). An oceanic climate is generally not favourable for many arctic–alpines, as many gardeners know.

There is little obvious reason for particular extinctions. For example, it is difficult to see why *Luzula spicata, Betula nana,* or *Minuartia stricta* went extinct in Ireland or why *Papaver* sect. *Scapiflora, Salix polaris,* or *Cassiope* were lost from Scotland, whereas other species still occur (e.g. *Koenigia islandica, Saxifraga cernua, Saxifraga cespitosa*) in Scotland. Interestingly nine of the plants that went extinct in Ireland occur in Britain today and 13 of the 19 that have gone extinct in Britain occur in Scandinavia today (Table 2.1).

These patterns of extinctions parallel present-day patterns of richness in mountain floras of Ireland, Scotland, and western Norway, with fewest taxa in the extreme west and increasing richness eastwards with a more continental climate

Table 2.1 Numbers of arctic-alpine and arctic taxa that have gone extinct since the last (Weichselian/Devensian) glacial stage, the total number of arctic-alpine and arctic taxa identified from Weichselian/Devensian deposits (including the late-glacial), the extinction percentage, and the number of taxa that have gone extinct but that grow today in the nearest area to the east. Data from all available sources.

	Ireland	*Britain*	*Scandinavia*
Extinct taxa	11	14	3
Total taxa identified	29	75	148
Extinction (%)	38	19	2
Number of extinct taxa growing in area to east	9	13	—

(Moe, 1995). In general, plants that are absent from the mountains in the extreme west of Norway tend to be absent from the British Isles too (Moe, 1995). Species that have become extinct in Ireland or Britain since the last glaciation do not occur in the westernmost Norwegian mountains but occur further inland in more continental areas, for example on the Hardangervidda.

The documented extinctions of arctic–alpine and arctic taxa in Scotland since the last glaciation were almost certainly a result of the rapid climate change that occurred at the beginning of the post-glacial acting directly on plant growth (Dahl, 1951) and indirectly as a result of competition from more rapidly growing, taller, more robust lowland species and from forest development (Ratcliffe, 1991). Changes in soil and in the extent of permanently open well-drained habitats may also have been important, along with chance factors in extinction and survival (Ratcliffe, 1960). The result of all these factors is the persistence in Scotland of very small isolated populations of some species that were once more common 10,000–12,000 years ago (e.g. *Artemisia norvegica, Saxifraga cespitosa, S. rivularis, Minuartia rubella*). These populations have survived at high altitudes above the mid-Holocene treeline (Birks, 1988) on open soils and in the absence of competition. Many are slow growing and have limited reproductive capacity today. They are under threat of extinction as a result of climate warming in the near future (Ratcliffe, 1991).

In addition to the national-scale extinctions that have occurred since the late-glacial (Table 2.1), local extinctions of many mountain taxa have also occurred in, for example, the Southern Uplands and on Skye and the Outer Hebrides. As the palaeoecological record is inevitably incomplete because of the vagaries of fossilisation, it is likely that more extinctions have occurred than are documented by the Quaternary fossil record.

It is striking that some plants that occur in northern England and Scandinavia today and that were widespread during the late-glacial in England and Wales have never been found in the Scottish late-glacial (e.g. *Polemonium caeruleum*; Pigott, 1958). There is a group of Pennine limestone and Scandinavian plants that are absent from Scotland for no obvious reason (e.g. *Carex ornithopoda, Actaea spicata, Cypripedium calceolus, Polemonium caeruleum*).

It is also important to consider the glacial history of the Scottish mountain flora and the question of plant survival on unglaciated mountain tops and coastal cliffs, so-called nunataks (see, for example, Dahl, 1955, 1987). There is now good geological evidence for unglaciated mountain tops in parts of the Highlands and Islands (see, for example, Dahl, 1955; Ballantyne, 1990, 1994; Ballantyne and McCarroll, 1995; McCarroll *et al.*, 1995). There is, however, no evidence for plant growth on such nunataks but there is abundant evidence for the growth of open-ground, pioneer mountain plants around the southern, western, northern, and eastern edges of the last ice-sheet (see, for example, H. H. Birks, 1994) during the last glaciation. Palaeobotanical, statistical, and biosystematic studies all suggest that there is no need to invoke the nunatak hypothesis of glacial survival of plants (see, for example, Nordal, 1987; H. H. Birks, 1994; H. J. B. Birks, 1994,

1996) to explain the contemporary distributions of northwest European mountain plants.

2.4 Forest history of Scotland

The temporal and spatial patterns of tree colonisation and spread within Scotland during the Holocene can be reconstructed from radiocarbon-dated pollen diagrams (Birks, 1989; see Figure 2.2 for an example). The major Scottish forest trees (*Betula, Corylus avellana, Pinus sylvestris, Ulmus, Quercus, Alnus glutinosa*) spread into Scotland at different times from different directions and migrated at different rates (Figure 2.3). Interestingly, the northwest Highland pine populations that differ genetically from other *Pinus sylvestris* populations (Kinloch *et al.*, 1986) appear to have a Holocene history and source area different from eastern Scottish pine populations (Birks, 1989; Figure 2.3).

Bennett (1989) used all the available pollen-stratigraphical data in conjunction with soil data and present-day ecological preferences of the tree taxa to reconstruct a map of forest types for the British Isles 5,000 years ago. His reconstruction (Figure 2.4) suggested that oak was dominant in the south and west of Scotland, pine was prominent in the eastern and north-west Highlands, and birch was important in the far north and west. Recent work by Bennett and his colleagues (Bennett *et al.*, 1990, 1992; Bunting, 1994; Fossitt, 1996) indicated that Shetland, Orkney and the Western Isles may have supported extensive areas of open woodland and scrub in the early and mid-Holocene, refuting earlier suggestions of 'open scrub or treeless vegetation' in these areas (Birks, 1988). Bennett's (1985) forest reconstructions also suggested that there were large unforested areas even at the time of maximum forest cover, especially at high altitudes (see Figure 2.4). Birks (1988) attempted to reconstruct the maximum extent of treelines in the past, and suggested that the maximum altitude of the treeline in Scotland varied from about 800 m in the Cairngorms to 450 m on Skye.

One of the most striking changes in Scotland's forests was the widespread decline and local extinction of *Pinus sylvestris* about 4,400 years ago in northern and western Scotland (Figure 2.3; Birks, 1975; Bennett, 1984). Bennett (1995a) recently reviewed the possible causes of this apparently rapid decline of pine and discussed the difficulties in disentangling the factors that influence the abundance of pine trees in the landscape from the factors that control the local preservation of pine stumps in peats. It is possible that the pine decline was irregular and patchy in its spatial and temporal distribution and may have been a result of climatic change, particularly of precipitation (Bennett, 1995a).

Bennett (1995b) also discussed evidence for the local extinction of trees from islands around Scotland, in particular of *Pinus, Alnus, Quercus,* and *Betula* about 3,000–4,000 years ago. All these extinctions occurred after the arrival of Neolithic farmers and the introduction of crops and domesticated animals to islands whose ecosystems had developed in the absence of such crops, animals, and land-use practices. These extinctions are some of the earliest examples of tree extinctions

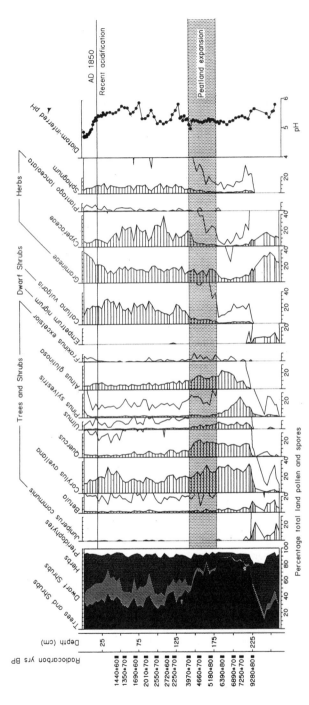

Figure 2.2 Pollen diagram from the Round Loch of Glenhead, Galloway, showing the percentages of the major pollen and spores plotted against depth, along with the radiocarbon dates (in ^{14}C years before present (BP)) and the surface-water pH inferred from the diatom assemblages preserved in the sediments. The times of expansion of blanket bog and of recent acidification are indicated. The original pollen data have been kindly provided by A. C. Stevenson (Jones *et al.*, 1989); the diatom-inferred pH values are from Birks *et al.* (1990).

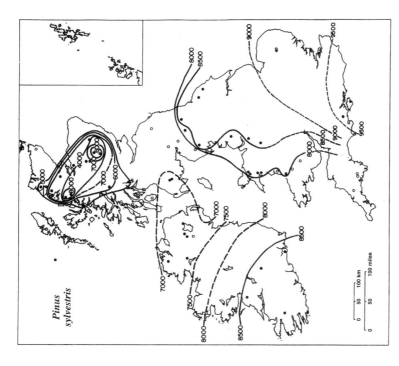

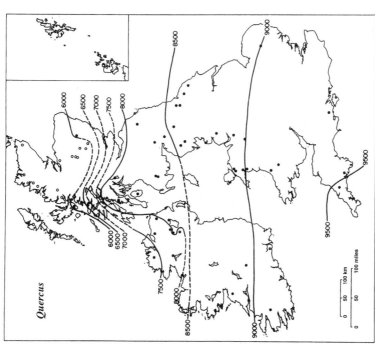

Figure 2.3 Tree-spreading maps of *Quercus* and of *Pinus sylvestris* in the British Isles. The contours show the times (in ^{14}C years before present) by which the tree taxa had reached the sites shown by solid dots. Sites where there is no pollen-analytical evidence for local occurrence are shown as open circles. Modified from Birks (1989).

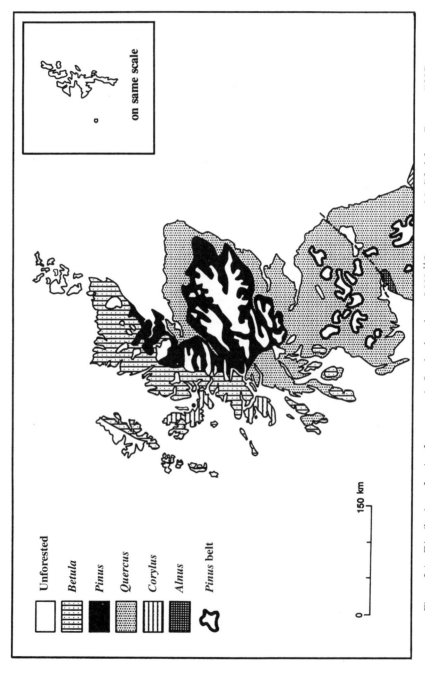

Figure 2.4   Distribution of major forest types in Scotland at about 5,000 ^{14}C years ago. Modified from Bennett (1985).

on islands resulting from human impact and are analogous to the later ecological impacts of Polynesians and Europeans colonising oceanic islands (Bennett, 1995b).

2.5   Diversity changes over the past 12,000 years

Birks and Line (1992) used rarefaction analysis to reconstruct the changing patterns of floristic richness through time from pollen-stratigraphical data. As floristic richness at the spatial scale recorded by pollen data is primarily a function of the mosaic structure of the landscape within the pollen-source area, the reconstructed changing patterns of floristic richness are an indirect reflection of landscape diversity.

The late-glacial was characterised by a high landscape diversity but this quickly decreased as the landscape became forested in the early and mid-Holocene (Birks and Line, 1992). With soil deterioration, deforestation, and the spread of blanket bog in the late Holocene (Figure 2.2), floristic and landscape diversity decreased to very low levels. In contrast in more fertile areas with extensive human activity, such as crofting, floristic diversity has increased in the past 2,000 years, presumably because of the creation of a landscape mosaic of fields, field boundaries, etc. (Birks and Line, 1992).

2.6 Recent changes in freshwater systems

Freshwater systems are particularly sensitive to environmental change, particularly different types of atmospheric and catchment pollution. Many lochs in Scotland on acid-sensitive bedrock such as granite have undergone major changes in their floristic composition as a result of recent acidification with surface-water pH values falling to 5 or lower (see Figure 2.2) (see, for example, Jones *et al.*, 1989, 1993; Battarbee and Allott, 1994; Allott *et al.*, 1995). These changes, due to the introduction of strong acid anions associated with acid deposition since about AD 1800, have resulted in the local extinction of some diatom species and major changes in the relative abundance of other diatom taxa and of aquatic macrophytes. Local extinction of some macrophyte taxa has occurred in naturally acid clear-water lochs of pH 5.5–6.0 that have become chronically acidified and now have a pH of 4.6–4.8 (Raven, 1988). The most affected areas are Galloway, Arran, the Trossachs, and Rannoch but acidified lochs are also known from the Cairngorms, Morvern, Strontian, and the northwest Highlands (Battarbee and Allott, 1994). Liming of chronically acidified lochs (e.g. Loch Fleet in Galloway) has caused major changes in the algal flora leading to the invasion and expansion of new species rather than to the re-establishment of the pre-acidification flora (Flower *et al.*, 1990).

Other changes in Scotland's freshwaters (e.g. Loch Leven) have occurred as a result of eutrophication, especially in the agricultural lowlands (Battarbee and Allott, 1994). An increasing problem is the impact of nutrient inputs to the large naturally oligotrophic lochs in the Scottish Highlands as a result of fish farming, sewage effluent disposal, and fertilisers from forestry (Battarbee and Allott, 1994). There is evidence, for example, for increasing input of nutrients over the past 40 years at Loch Ness leading to recent changes in the algal flora of the loch (V. J. Jones, personal communication).

2.7   Predicted changes of species distributions

The record of plant macrofossils shows that extinctions of some arctic and arctic–alpine species have occurred since the late-glacial. Several species of mountain plant in Scotland today persist as small isolated populations in their westernmost or southernmost European localities, particularly in the Cairngorms.

Huntley (1995) and Huntley *et al.* (1995) considered the possible effects that global warming under different $2 \times CO_2$ scenarios (doubling of pre-industrial atmospheric carbon dioxide concentration) would have on the European distribution of selected species. Their analyses of *Pinus sylvestris, Silene acaulis* and *Hymenophyllum wilsonii* are particularly relevant to Scottish biodiversity. In these analyses, the present-day distributions of selected species were mapped in modern climate space of temperature sum (degree days) above a 5°C threshold, the mean temperature of the coldest month, and an index of moisture availability based on the ratio of actual to potential evapotranspiration. Climate predictions from different general circulation models (GCMs) under $2 \times CO_2$ scenarios were then used to see where the realised climatic niche of the species today would be found in the future.

The potential predicted range of *Pinus sylvestris* (Huntley, 1995) in Europe under a doubled 'greenhouse effect' suggested the extinction of pine in Scotland and over much of southern Scandinavia. Similarly *Silene acaulis*, as a representative of the arctic–alpine flora in Scotland, was predicted to become extinct in the British Isles (Huntley *et al.*, 1995). The Atlantic filmy fern, *Hymenophyllum wilsonii*, was also predicted to go extinct in the British Isles under one GCM simulation or to become much rarer under a different simulation (Huntley *et al.*, 1995).

If these potential predicted ranges are correct (see also Hendry, 1993; Elmes and Free, 1994; Watt *et al.*, this volume), they have important implications for the boreal, arctic, arctic-alpine, and Atlantic components of Scottish plant biodiversity, the very components of greatest importance in a European context. As Ratcliffe (1991) noted, 'many of our mountain plants would be greatly at risk from any appreciable rise in temperatures, and concern for their survival gives particular reason for campaigning vigorously against the most ominous of environmental hazards'.

2.8   Conclusion

The basic conclusion from a historical perspective of Scotland's plant diversity is that the Scottish flora, as we see it today, is a result of progressive extinctions over a series of glacial-interglacial cycles. Extinctions have occurred nationally, regionally, and locally as a result of major climatic changes, competition, and habitat change, and, more recently, human activity and land-use change. If the predictions of Huntley *et al.* (1995) are correct, further but more numerous extinctions of the Atlantic and the mountain flora in Scotland seem likely under conditions of a doubling of pre-industrial atmospheric carbon dioxide concentrations and a mean global temperature increase of 2–5°C in the coming decades.

Acknowledgements

I am indebted to Rick Battarbee, Keith Bennett, Hilary Birks, Viv Jones, Derek Ratcliffe, and Des Thompson for many valuable discussions about Scotland's biodiversity, past and present, and to Hilary Birks and Sylvia Peglar for help in the preparation of the manuscript.

References

Allott, T. E. H., Golding, P. N. E. and Harriman, R. 1995. A palaeolimnological assessment of the impact of acid deposition on surface-waters in north-west Scotland, a region of high sea-salt inputs. *Water, Air and Soil Pollution*, **85**, 2425–2430.

Ballantyne, C. K. 1990. The Late Quaternary glacial history of the Trotternish Escarpment, Isle of Skye, Scotland, and its implications for ice-sheet reconstruction. *Proceedings of the Geologists' Association*, **101**, 171–186.

Ballantyne, C. K. 1994. Gibbsitic soils on former nunataks: implications for ice sheet reconstruction. *Journal of Quaternary Science*, **9**, 73–80.

Ballantyne, C. K. and McCarroll, D. 1995. The vertical dimensions of Late Devensian glaciation on the mountains of Harris and southeast Lewis, Outer Hebrides, Scotland. *Journal of Quaternary Science*, **10**, 211–223.

Battarbee, R. W. and Allott, T. E. H. 1994. Palaeolimnology. In Maitland, P. S., Boon, P. J. and McLusky, D. S. (Eds.) *The Fresh Waters of Scotland: A National Resource of International Significance*. Chichester, Wiley, 113–130.

Bennett, K. D. 1984. The post-glacial history of *Pinus sylvestris* in the British Isles. *Quaternary Science Reviews*, **3**, 133–155.

Bennett, K. D. 1985. A provisional map of forest types for the British Isles 5000 years ago. *Journal of Quaternary Science*, **4**, 141–144.

Bennett, K. D. 1995a. Post-glacial dynamics of pine (*Pinus sylvestris* L.) and pine woods in Scotland. In Aldhous, J. R. (Ed.) *Our Pinewood Heritage*. Farnham, Forestry Commission, 23–39.

Bennett, K. D. 1995b. Insularity and the Quaternary tree and shrub flora of the British Isles. In Preece, R. C. (Ed.) *Island Britain: a Quaternary Perspective*. London, Geological Society, 173–180.

Bennett, K. D., Boreham, S., Sharp, M. J. and Switsur, V. R. 1992. Holocene history of environment, vegetation and human settlement on Catta Ness, Lunnasting, Shetland. *Journal of Ecology*, **80**, 241–273.

Bennett, K. D., Fossitt, J. A., Sharp, M. J. and Switsur, V. R. 1990. Holocene vegetational and environmental history at Loch Lang, South Uist, Western Isles, Scotland. *New Phytologist*, **114**, 281–298.

Birks, H. H. 1975. Studies in the vegetational history of Scotland. IV. Pine stumps in Scottish blanket peats. *Philosophical Transactions of the Royal Society of London*, B **270**, 181–226.

Birks, H. H. 1984. Late-Quaternary pollen and plant macrofossil stratigraphy at Lochan An Druim, north-west Scotland. In Haworth, E. Y. and Lund, J. W. G. (Eds.) *Lake Sediments and Environmental History*. Leicester, University of Leicester Press, 377–405.

Birks, H. H. 1994. Plant macrofossils and the nunatak theory of per-glacial survival. In Lotter, A. F. and Ammann, B. (Eds.) *Festschrift Gerhard Lang. Dissertationes Botanicae*, **243**, 129–143.

Birks, H. H. and Mathewes, R. W. 1978. Studies in the vegetational history of Scotland V. Late Devensian and early Flandrian pollen and macrofossil stratigraphy at Abernethy Forest, Inverness-shire. *New Phytologist*, **80**, 455–484.

Birks, H. J. B. 1988. Long-term ecological change in the British Uplands. In Usher, M. B. and Thompson, D. B. A. (Eds.) *Ecological Change in the Uplands*. Oxford, Blackwell, 37–56.

Birks, H. J. B. 1989. Holocene isochrone maps and patterns of tree-spreading in the British Isles. *Journal of Biogeography*, **16**, 503–540.

Birks, H. J. B. 1994. Is the hypothesis of survival on glacial nunataks necessary to explain the present-day distributions of Norwegian mountain plants? *Phytocoenologia*, **23**, 399–426.

Birks, H. J. B. 1996. Statistical approaches to interpreting diversity patterns in the Norwegian mountain flora. *Ecography*, **19**, 332–340.

Birks, H. J. B., Juggins, S. and Line, J. M. 1990. Lake surface-water chemistry reconstructions from palaeo-limnological data. In Mason, B. J. (Ed.) *The Surface Waters Acidification Programme*. Cambridge, Cambridge University Press, 301–313.

Birks, H. J. B. and Line, J. M. 1992. The use of rarefaction analysis for estimating palynological richness from Quaternary pollen-analytical data. *The Holocene*, **2**, 1–10.

Birks, H. J. B. and Peglar, S. M. 1979. Interglacial pollen spectra from Sel Ayre, Shetland. *New Phytologist*, **83**, 559–575.

Birks, H. J. B. and Ransom, M. E. 1969. An interglacial peat at Fugla Ness, Shetland. *New Phytologist*, **68**, 777–796.

Bunting, M. J. 1994. Vegetation history of Orkney, Scotland: pollen records from two small basins in west Mainland. *New Phytologist*, **128**, 771–792.

Conolly, A. P. 1961. Some climatic and edaphic indications from the late-glacial flora. *Proceedings of the Linnean Society of London*, **172**, 56–62.

Dahl, E. 1951. On the relation between summer temperature and the distribution of alpine vascular plants in the lowlands of Fennoscandia. *Oikos*, **3**, 22–52.

Dahl, E. 1955. Biogeographic and geologic indications of unglaciated areas in Scandinavia during the glacial ages. *Bulletin of the Geological Society of America*, **66**, 1499–1519.

Dahl, E. 1987. The nunatak theory reconsidered. *Ecological Bulletin*, **38**, 77–94.

Dickson, J. H. 1992. Some recent additions to the Quaternary flora of Scotland and their phytogeographical, palaeoclimatic and ethnobotanical significance. A review. *Acta Botanica Fennica*, **144**, 51–57.

Duller, G. A. T., Wintle, A. G. and Hall, A. M. 1995. Luminescence dating and its application to key pre-Late Devensian sites in Scotland. *Quaternary Science Reviews*, **14**, 495–519.

Elmes, G. W. and Free, A. 1994. *Climate Change and Rare Species in Britain*. London, HMSO and Natural Environment Research Council.

Flower, R. J., Cameron, N. G., Rose, N., Fritz, S. C., Harriman, R. and Stevenson, A. C. 1990. Post–1970 water-chemistry changes and palaeolimnology of several acidified upland lakes in the UK. *Philosophical Transactions of the Royal Society of London,* B **327**, 427–433.

Fossitt, J. A. 1996. Late Quaternary vegetation history of the Western Isles of Scotland. *New Phytologist*, **132**, 171–196.

Hall, A. M., Gordon, J. E. and Whittington, G. 1993. Early Devensian Interstadial peat at Sel Ayre, Shetland. In Birnie, J. F., Gordon, J. E., Bennett, K. D. and Hall, A. M. (Eds.) *The Quaternary of Shetland: Field Guide*. Cambridge, Quaternary Research Association, 104–118.

Hendry, G. A. F. 1993. Forecasting the impact of climatic change on natural vegetation with particular reference to the boreal and sub-arctic floras. In Holten, J. I., Paulsen, G. and Oechel, W. C. (Eds.) *Impacts of Climatic Change on Natural Ecosystems, with Emphasis on Boreal and Arctic/Alpine Areas*. Trondheim, Norwegian Institute for Nature Management and the Directorate for Nature Management, 136–150.

Huntley, B. 1994. Late Devensian and Holocene palaeoecology and palaeoenvironments of the Morrone Birkwoods, Aberdeenshire, Scotland. *Journal of Quaternary Science*, **9**, 311–336.

Huntley, B. 1995. Plant species' response to climate change: implications for the conservation of European birds. *Ibis,* **137**, S127–S138.

Huntley, B., Berry, P. M., Cramer, W. and McDonald, A. P. 1995. Modelling present and potential future ranges of some European higher plants using climate response surfaces. *Journal of Biogeography,* **22**, 967–1001.

Jalas, J. and Suominen, J. 1972. *Atlas Florae Europaeae*. Vol. 1, *Pteridophyta (Psilotaceae to Azollaceae)*. Helsinki.

Jalas, J. and Suominen, J. 1979. *Atlas Florae Europaeae*. Vol. 4, *Polygonaceae*. Helsinki.

Jalas, J. and Suominen, J. 1983. *Atlas Florae Europaeae*. Vol. 6, *Caryophyllaceae (Alsinoideae and Paronychioideae)*. Helsinki.

Jones, V. J., Flower, R. J., Appleby, P. G., Natanski, J., Richardson, N., Rippey, B., Stevenson, A. C. and Battarbee, R. W. 1993. Palaeolimnological evidence for the acidification and atmospheric contamination of lochs in the Cairngorm and Lochnagar areas of Scotland. *Journal of Ecology*, **81**, 3–24.

Jones, V. J., Stevenson, A. C. and Battarbee, R. W. 1989. Acidification of lakes in Galloway, south-west Scotland: a diatom and pollen study of the post-glacial history of the Round Loch of Glenhead. *Journal of Ecology*, **77**, 1–23.

Kinloch, B. B., Westfall, R. D. and Forrest, G. I. 1986. Caledonian Scots Pine: origins and genetic structure. *New Phytologist*, **104**, 703–729.

McCarroll, D., Ballantyne, C. K., Nesje, A. and Dahl, S. O. 1995. Nunataks of the last ice sheet in northwest Scotland. *Boreas*, **24**, 305–323.

Mitchell, G. F. and Watts, W. A. 1970. The history of the Ericaceae in Ireland during the Quaternary epoch. In Walker, D. and West, R. G. (Eds.) *Studies in the Vegetational History of the British Isles*. London, Cambridge University Press, 13–21.

Moe, B. 1995. Studies of the alpine flora along an east-west gradient in central Western Norway. *Nordic Journal of Botany*, **15**, 77–89.

Nordal, I. 1987. Tabula rasa after all? Botanical evidence for ice-free refugia in Scandinavia reviewed. *Journal of Biogeography*, **14**, 377–388.

Pigott, C. D. 1958. *Polemonium caeruleum* L. *Journal of Ecology*, **46**, 507–525.

Ratcliffe, D. A. 1960. The mountain flora of Lakeland. *Proceedings of the Botanical Society of the British Isles*, **4**, 1–25.

Ratcliffe, D. A. 1968. An ecological account of Atlantic bryophytes in the British Isles. *New Phytologist*, **67**, 365–439.

Ratcliffe, D. A. 1991. The mountain flora of Britain and Ireland. *British Wildlife*, **3**, 10–21.

Ratcliffe, D. A. and Thompson, D. B. A. 1988. The British uplands: their ecological character and international significance. In Usher, M. B. and Thompson, D. B. A. (Eds.) *Ecological Change in the Uplands*. Oxford, Blackwell, 9–36.

Ratcliffe, D. A., Birks, H. J. B. and Birks, H. H. 1993. The ecology and conservation of the Killarney Fern (*Trichomanes speciosum* Willd.) in Britain and Ireland. *Biological Conservation*, **66**, 231–247.

Raven, P. J. 1988. Occurrence of *Sphagnum* in the sublittoral of several small oligotrophic lakes in Galloway, south west Scotland. *Aquatic Botany*, **30**, 223–230.

Stevenson, A. J. and Birks, H. J. B. 1995. Heaths and moorland: long-term ecological changes, and interactions with climate and people. In Thompson, D. B. A., Hester, A. J. and Usher, M. B. (Eds.) *Heaths and Moorlands: Cultural Landscapes*. Edinburgh, HMSO, 224–239.

Webb, J. A. and Moore, P. D. 1982. The Late Devensian vegetational history of the Whitelaw Mosses, southeast Scotland. *New Phytologist*, **91**, 341–398.

Whittington, G. 1994. *Bruckenthalia spiculifolia* (Salisb.) Reichenb. (Ericaceae) in the late Quaternary of western Europe. *Quaternary Science Reviews*, **13**, 761–768.

PART TWO
CURRENT STATUS AND TRENDS
OF BIODIVERSITY IN SCOTLAND

PART TWO

CURRENT STATUS AND TRENDS OF BIODIVERSITY IN SCOTLAND

Concern over accelerating rates of loss of biodiversity, and the signing of international agreements such as the Convention on Biological Diversity and Agenda 21, have highlighted the need for the world's flora and fauna to be inventoried and monitored. Indeed, parties to the Convention are now committed to identify and monitor biodiversity, for both conservation and sustainable use. The task is an enormous but crucial one: the resulting data provide baseline information for the assessment of change and provide the fundamental biological information required to set priorities and devise viable and effective strategies for biodiversity conservation and management (Stork and Samways, 1995).

In the past much attention has focused on biological variation at the species level; the term biodiversity is often used as a synonym for species diversity, and diversity at other scales, from genes to landscapes, remains poorly described and understood. Even at the species scale our level of understanding is very patchy. Estimates of the total number of named and described species range from 1.4 to 1.8 million; possibly less than 15% of all organisms (Jermy *et al.*, 1995). Recording has not been systematic: some groups, such as birds and mammals, have received a disproportionate amount of attention while others, such as deep-sea invertebrates and microorganisms, have been almost entirely neglected (see also Usher, this volume).

The papers in this section provide a review of Scotland's biodiversity at the genetic, species and landscape level in terrestrial and aquatic environments. Despite increasing human pressure for land, Scotland still retains at least five broad habitat types of international conservation importance: blanket bog, machair, heath, montane vegetation, and native woodland. These are discussed in detail by Miles *et al*. Notable changes in land use are highlighted, for example, the conversion of moorland and blanket bog to conifer woodland. The need for continued political and practical action to stem gradual decline in habitat quality caused by factors such as widespread overgrazing, burning and deposition of atmospheric pollutants is also addressed.

The internationally important marine ecosystems of Scottish waters, which support an estimated 40,000 species, are further considered in chapters by Davison and Baxter and by Matthews *et al.* Both emphasise the paucity of basic information on the composition and geographic range of marine taxa. Sampling in all but a narrow intertidal zone requires specialised and expensive technology, but the need to understand marine food chains in the context of fish stocks and environmental change has provided new impetus for offshore surveys. The overlap of different biogeographical zones and geological forms in coastal benthic zones, and environmental heterogeneity in physico-chemical properties such light, nutrients, temperature and salinity in the pelagic zone, contributes to high levels of biodiversity. The deep-sea bed is even more remarkable in this respect. Despite appearing as a featureless, homogeneous habitat, it is perhaps the highest-diversity environment in Europe, with large numbers of species coexisting in relatively small areas. Virtually all of this deep sea area claimed by the UK lies to the north and west of the Western Isles of Scotland; a suite of new initiatives, designed to protect and manage the biodiversity of marine habitats (Matthews *et al.*), provides a hopeful indication that their value is finally being recognised.

Diversity at the species level is reviewed in four chapters focusing on lower and vascular plants, invertebrates and vertebrates. Discrepancies in the level of our knowledge about these different groups are immediately apparent and the urgent need for information on distribution, status and habitat requirements of lower plants and invertebrates in Scotland is highlighted. The diverse micro- and macroscopic organisms that make up the fungi pose a particular problem: recognition, especially of non-lichenised fungi, is often extremely difficult (Watling). Although current knowledge makes status and trends of Scottish fungi almost impossible to assess accurately, high fungal diversity will only be maintained under conditions of high habitat diversity. The dependence of forests and crop plants on ecto- and endomycorrhizal fungi is a clear illustration of the economic as well as biological value of this understudied group. An estimated 14,000 species of insect are found in Scotland, making this the country's most species-rich group of terrestrial animals (Young and Rotheray). Although information on the distribution of these species is also relatively poor, it is still possible to identify priority habitats for insect conservation. Birch and aspen woodlands, montane communities and species-rich heathland are known to harbour many rare insects and deserve special attention in future conservation strategies.

The status and trends of vertebrate biodiversity (fish, amphibians, reptiles, birds and mammals) in Scotland are documented and reviewed in a chapter by Racey. The greatest threat to this group of organisms is the introduction of alien species. For example, few original pristine fish communities remain, having been disrupted by the introduction of species such as ruffe (*Gymnocephalus cernuus*) and chubb (*Leuciscus cephalus*), and the genetic integrity of native red deer (*Cervus elaphus*) is threatened by hybridisation with the introduced Asian sika deer (*C. nippon*). Dramatic declines in several common bird species such as skylarks (*Alauda arvensis*) and song thrushes (*Turdus philomelos*) are associated with changing land use and agricultural intensification. Conversely, large and growing populations of vertebrates such as geese and grey seals (*Halichoerus grypus*) may require management if their populations are not to outstrip resources to the detriment of overall biodiversity or sectoral economies.

Biodiversity at the genetic level is defined as all forms of genetic variation within a taxon that affect the ecological attributes of individuals. As an important determinant of the ecological amplitude and evolutionary potential of species, it underpins present and future biodiversity at the species and community levels. In a chapter by Ennos and Easton, the suitability of different methodologies for assessing genetic biodiversity are discussed, and the relatively few studies of genetic biodiversity of Scottish plants reviewed. The review points to the uniqueness of Scottish populations in terms of regional climatic adaptations and the presence of considerable diversity between and within local populations. In the species-poor higher-plant flora of Scotland (Sydes), perhaps as in many species-poor groups, genetic diversity is likely to comprise a significant proportion of total biodiversity.

By 2050 it is predicted that in the UK there will be a doubling of the atmospheric CO_2 concentration, a mean increase in temperature of 1.6°C, a 10% rise in annual precipitation and a 30% increase in the frequency of gales. Watt *et al.* draw on knowledge of past influences of climate change on plants and animals in reviewing the implications of such changes for biodiversity in Scotland. It is predicted that man-made reductions in abundance, range and genetic diversity of species will seriously restrict their response to future climate change and new strategies for species conservation must be developed if biodiversity in Scotland is to be maintained into the next millennium.

References

Jermy, A. C., Long. D., Sands, M. J. S., Stork, N. E. and Winser, S. (Eds.) 1995. *Biodiversity Assessment: a Guide to Good Practice.* London, Department of the Environment and HMSO.

Stork, N. E. and Samways, M. J. 1995. Inventorying and monitoring. In Heywood, V. H. (Ed.) *Global Biodiversity Assessment.* Cambridge, Cambridge University Press, 453–542.

3 HABITAT DIVERSITY IN SCOTLAND

J. Miles, G. Tudor, C. Easton and E. C. Mackey

Summary

1. Four current meanings of 'habitat' are discussed. Perhaps the commonest - and broadest - current use is as a synonym for ecosystem or biotope.

2. CORINE and the National Vegetation Classification are two plant-community classifications now used in Scotland. However, census information about habitat distribution and diversity only exists at a coarser scale in the Land Cover of Scotland 1988; some 56% of Scotland is under natural and semi-natural vegetation.

3. The National Countryside Monitoring Scheme has shown marked changes in land cover since the 1940s, notably in the conversion of moorland and blanket bog to conifer woodland, and in mire drainage, grassland improvement and urban expansion.

4. Five habitats for which Scotland is internationally important are discussed: blanket bog, heather moorland, machair, montane vegetation, and native woodland.

5. On land, there seem to be widespread albeit slow trends of declining habitat quality because of overgrazing, overburning, erosion and the deposition of atmospheric pollutants, particularly of fixed nitrogen.

6. Government policies currently being implemented to reverse some past trends should maintain and improve habitat diversity and quality.

3.1   Introduction

Whereas 'diversity', i.e. variety, has an unequivocal meaning, this is not true of 'habitat', which currently seems to be used in at least four ways! In origin it is a Latin verb meaning 'he, she or it inhabits'. In early scientific writings in Latin the word was used in just this sense, but by the late 18th century it was being used in English texts in its first scientific sense to mean the natural place of growth or occurrence of a species, a meaning that has persisted up to the present (Bates, 1961; Allaby, 1994).

Biologists increasingly described the habitats of plant species in terms of climatic and soil variables, and of the presence of other species (Sachs, 1887; Warming and Vahl, 1909), from which developed the second meaning of habitat as the environment of any organism (Lawrence, 1989). By both these definitions, the number of habitats in Scotland must equal the number of species found there. This, however, is not very helpful, because no one knows precisely how many species there are in Scotland (see Usher, this volume).

At least a century ago some biologists were beginning to think of habitat in a broader way. Marilaun (1895) wrote of 'the numerous habitats occupied by a species', noting that the nature of the habitat made different plant species 'join together in communities, each of which has a characteristic form, and constitutes a feature in the landscape of which it is a part', an early recognition of what we now think of as an ecosystem or biotope. In 1911, Tansley wrote 'a vegetation-unit is always developed in a *habitat* of definite characteristics', and in 1926, Tansley and Chipp wrote 'habitat means the sum of the effective environmental conditions under which the [vegetation type] exists'. The realisation that similar environmental conditions produced similar assemblages of plant and animal species became embedded in the thinking of ecologists in Europe and North America, and a third meaning of habitat was established: 'a *habitat* is an area of ground, small or large in its extent, over which the environment is essentially uniform' (Gleason and Cronquist, 1964).

However, many authors gradually began to use habitat in an even broader sense as a synonym for an ecosystem or community. This fourth meaning of the term, the vaguest yet, has now received the European Commission's imprimatur. The introduction to the *CORINE Biotopes Manual* (Commission of the European Communities, 1991) uses 'communities', 'biotopes' and 'habitats' indiscriminately as synonyms, although Council Directive 92/43/EEC *On the conservation of natural habitats and of wild fauna and flora* (the 'Habitats Directive') came down in favour of 'habitats'. This fourth meaning is also now used, albeit implicitly, by the UK Government (Anon, 1994a,b).

Taking the third definition, no one has yet tried to devise a system of classifying habitats in Scotland in the sense of sets of levels of environmental factors that sustain characteristic assemblages of organisms. The Institute of Terrestrial Ecology's Land Classification (Bunce *et al.*, 1983) used a wide range of environmental variables but stopped at a very broad-brush level with only 32 types. Certainly a very large number of types would be needed even as reference points to summarise adequately the intricately varying patterns of the soils, hydrology, topography and the latitudinal, longitudinal and altitudinal gradients of climate.

3.2 Habitat diversity

Taking the fourth definition of habitat, and considering Scotland's diversity of habitats in the sense of different types of ecosystem or biotope, also poses problems because there is no single system of classification accepted and understood throughout Scotland, Britain or Europe. The Commission of the European Communities (1991) has promoted a hierarchical classification of biotopes through its CORINE (Co-ORdination of INformation on the Environment) programme. A major drawback of this system, apart from lack of clear definitions, is that the biotopes vary from broad-brush, often geomorphological, environments (e.g. large shallow inlets and bays; caves not open to the public [sic]) to vegetation types characterised as Latinised phytosociological associations (e.g. *Astragalo-Plantaginetum subulatae phrygana*), with various hybrids in between.

However, the CORINE classification was used in the Habitats Directive to define in Annex I 'natural habitat types of community interest whose conservation requires the designation of special areas of conservation'. Counting the seven kinds of grey dune as one (it has recently been agreed that they should be re-aggregated), Annex I lists 169 habitat types that Member States have agreed shall be represented in Special Areas of Conservation (SACs). The text says they should be types that are (1) in danger of disappearance, or (2) now limited in extent either intrinsically or because of human activity, or (3) 'present outstanding examples of typical characteristics', although it is arguable whether these tests have always been met. Table 3.1 shows that Scotland has 38% of these habitat types, 54% of those believed to occur in the North Atlantic Biogeographic Region, and indeed 80% of those occurring in the UK. The Directive also introduces the term 'priority' natural habitat types for those in danger of disappearance and where the EU has a significant proportion of the total resource. Comparison of the relative number of priority habitat types is not useful because the definition was not applied consistently to the Annex I list; for example, actively growing blanket bogs have never been in danger of disappearance in the UK.

A finer grain for considering Scotland's habitat diversity is to use the vegetation types described in the National Vegetation Classification (NVC) (Rodwell, 1991–95). This has the advantage of setting Britain's vegetation in the same syntaxonomic context of that of Continental Europe, but because it kept the traditional bias of using samples selected subjectively from only relatively uniform patches of vegetation, much vegetational variability was not sampled. Table 3.2 shows that Scotland again does rather well: with only 35% of the area of Britain, it has 81% of the 184 NVC 'communities' published to date, including 23 communities not found south of the border; England and Wales, with 65% of Britain's area, muster about 88%.

Table 3.1 Numbers of habitat types listed in Annex I of the 1992 EC Habitats Directive 92/43/EEC that occur in Scotland and at wider geographic scales.

	Habitats[a]	*Priority habitats*
European Community (in 1992)	169	54
North Atlantic Biogeographic Region	120	31
United Kingdom	80	22
Scotland	65	16

[a]Excluding grey dune subtypes.

Table 3.2   Occurrence of National Vegetation Classification communities (Rodwell, 1991–95) in Scotland and in England and Wales.

	Scotland	England and Wales
Number of communities[a]	149	161
Number of communities exclusive to each 'country'	23	35

[a]Total of 184 communities described in the four volumes published (Vol. 5 has still to appear).

Another problem with both the CORINE and NVC classifications is that there are no good estimates of the extent of their different habitats/communities in Scotland. Cover estimates exist for Scotland from three projects; all have somewhat similar classifications but at a coarser grain than CORINE. The Countryside Survey 1990 (CS90) (Institutes of Terrestrial and Freshwater Ecology, 1993) gives estimates for Britain based on mapping a 0.2% sample of 1 km squares. The National Countryside Monitoring Scheme (NCMS) (Tudor *et al.*, 1994) is a 7.5% sample based on interpretation of air photographs. Thirdly, the Land Cover of Scotland (LCS) (Macaulay Land Use Research Institute, 1993) is a census map, available in digital form, derived from interpretation of 1987–89 air photographs.

CS90 is a unique data set, but estimates for Scotland must be used with caution because of the small sample size. LCS has the advantages of being a census, but has the limitations of air-photograph interpretation. For example, the cover of bracken (*Pteridium aquilinum*) has probably been underestimated, and blanket bog dominated by heather (*Calluna vulgaris*) can be particularly difficult to distinguish from heather moorland on air photographs. The NCMS has similar disadvantages to LCS, but is mapped at a higher resolution for accurate land-cover change estimation. Combining LCS and NCMS data (Table 3.3) shows that only some 24% of Scotland is under intensive agricultural use, 15% under woodland, 2% under freshwater bodies and some 56% under natural and semi-natural vegetation. There are, however, large variations in land-cover distribution across Scotland. If biogeographic zones are used to subdivide Scotland (Usher and Balharry, 1996), Table 3.4 shows that, whereas the proportion of cultivated land exceeds 50% in the lowlands of mainland Scotland, it is less than 10% of the mountainous areas of the Central and Western Highlands and in most of the Hebrides, and the

Table 3.3   Scotland's main vegetation and land cover types (% cover) in 1988. Data from Land Cover of Scotland and National Countryside Monitoring Scheme.

Heather moorland and blanket bog	38	Rivers and lochs	2
Tilled land	24	Montane	2
Woodland	15	Bracken	1.6
Unimproved grassland	14	Marshes	<0.5
Developed land	3	Dunes	<0.5

Table 3.4 Land cover in Scotland's 12 biogeographic zones (see Plate 2).

Number	Zone	Area (km²)	% cultivated land	% rough grassland	% heather moorland and blanket bog	% montane	% woodland	% other
1	Central and Southern Lowlands	16,500	61	11	4	<0.1	12	11
2	Grampian Fringe and Southern Uplands	15,000	24	25	20	<1	26	4
3	East Coast	4,320	72	4	2	<0.1	13	8
4	Cairngorm	3,650	2	10	66	9	9	5
5	Western Highlands	5,060	0.1	13	61	14	6	7
6	Western Highlands Fringe	10,900	2	23	47	3	18	7
7	Western Isles (North) and North Mainland	12,600	6	6	74	<1	9	4
8	Barra and Tiree	243	36	25	18	<1	3	17
9	Western Isles (South)	1,520	9	4	64	2	4	16
10	Northern Isles	990	47	16	24	<1	<1	11
11	Argyll and the Inner Hebrides	6,890	8	21	43	<1	21	6
12	Galloway Coast	617	70	8	2	0	10	7

proportion of heather moorland and blanket bog exceeds 40% in half of these zones and 60% in a third of them. Woodland cover also varies from less than 5% in the Northern Isles and most of the Western Isles to 26% in the Grampian Fringe and Southern Uplands zone.

3.3   Past changes

As trees recolonised Scotland after the last glaciation, perhaps a maximum of 70% of the land became covered in woodland and scrub, the remainder being too high, too rocky or too wet to support tree growth. However, Mesolithic people were burning woodland at least as early as 8,500 years BP (Robinson and Dickson, 1988; Edwards, 1990). Human disturbance gathered momentum as the Mesolithic hunter–gatherer lifestyle gave way to that of Neolithic farming, a gradual process that began at least 6,000 years ago in Scotland (Edwards, 1988), and deforestation was widespread by 4,000 BP (Huntley and Birks, 1983; Bennett, 1984; Bridge *et al.*, 1990; Birks, this volume). Deforestation was frequently followed by the onset of growth of blanket peat, especially in western and northern Scotland and in the Northern and Western Isles (Tallis, 1991). This coincided with a change to a cooler and wetter climate; there has been a long debate about the extent to which blanket bog growth or paludification was induced by the climate change, with the growth of *Sphagnum* mosses preventing tree regeneration, and was thus just facilitated by deforestation, or occurred only because of deforestation (Lowe, 1993). The accumulated evidence increasingly suggests a largely anthropogenic origin to Scotland's blanket bogs (Edwards, in press), and certainly both deciduous and coniferous woodland will regenerate today in the west of Scotland even with annual precipitation of 2,000–2,500 mm, and despite obvious tendencies for surface paludification. Blanket-bog growth probably reduced Scotland's potential woodland cover to a maximum of around 60%, a figure that steadily reduced during prehistoric and historic times as human activity spread and intensified.

There have also been appreciable changes in land cover in Scotland during the past 50 years (Mackey and Tudor, in press) (Table 3.5). The NCMS suggests that from the 1940s to the 1980s 2,423 km^2 of blanket bog were planted to coniferous woodland and a further 989 km^2 were converted to moorland through drainage;

Table 3.5   Changes in land cover (km^2) in Scotland, 1940s–1980s. Data from National Countryside Monitoring Scheme.

	1940s–1970s	*1970s–1980s*
Plantation woodland (mainly coniferous)	+ 5,830	+ 3,270
Heather moorland and blanket bog	− 4,130	− 3,980
Unimproved grassland	− 1,280	− 85
Broadleaved woodland	− 213	− 133

2,245 km^2 of heather moorland and 2,505 km^2 of unimproved grassland were also planted to conifers. The expansion of conifer plantation with mainly exotic species was accompanied by losses of the dwindling stock of native woodland, often through underplanting. There was a tendency towards grassland improvement, and from the 1970s to the 1980s the arable area expanded at the expense of improved grassland. The area of built land expanded by 46% and transport corridor (mainly roadway) expanded by 24%. However, trends in the reduction of semi-natural habitats have probably been greatly reduced in the past decade, and locally reversed, because of changes in fiscal support to agriculture and forestry.

3.4   Present trends and status

Here we focus for brevity on just five habitats for which Scotland is internationally important and for which the Habitats Directive requires the designation of examples as SACs: blanket bog, heather moorland, machair, montane vegetation, and native woodland.

3.4.1   Blanket bog

Despite some past afforestation and reclamation, blanket bog still covers some 15–20% of Scotland, depending on the minimum peat depth used to define it and hence distinguish it from heather moorland. Its development is a remarkable biogeomorphological phenomenon; it supports often rich assemblages of birds, especially waders and raptors (Stroud *et al.*, 1987) and is a vast archive of palaeo-ecological and palaeoenvironmental information. Most of the best examples are now protected as Sites of Special Scientific Interest, but two phenomena cause some concern. Firstly, a great deal of blanket peat is eroding; a recent survey (Grieve and Hipkin, 1996) showed peat erosion in 6% of the uplands sampled. Some erosion began at least 3,000 years ago (Bradshaw and McGee, 1988) but some only 300–400 years ago (Stevenson *et al.*, 1990), and some only 200 years ago (Jones *et al.*, 1987) or even more recently. The causes of blanket-bog erosion are unclear. Some may be a natural process, perhaps initiated in particularly wet weather (Carling, 1986), producing, for example, peat slides such as that on the Quirang on Skye in the mid-1980s, although, as in the Pennines, some may be an unanticipated result of burning (Tallis, 1981), perhaps exacerbated by trampling by red deer and sheep.

Secondly, although there is no evidence for catastrophic death of *Sphagnum* spp. and other mosses in blanket bog as a result of acidic depositions of the kind that occurred in the Southern Pennines following the industrial revolution (Tallis, 1964), there is evidence that many Scottish blanket peats have become more acid because of pollution (Skiba *et al.*, 1989). The effects of this, if any, are not known. Further, there are now concerns over increasing depositions of nitrogen as ammonia and nitrate. These currently contribute some half of the total acidic depositions in Britain and Europe, already exceed sulphur depositions over much of Scotland (S. Metcalfe, unpublished), and may be adversely affecting the performance of plants adapted to growing in nitrogen-poor soils, in particular the *Sphagnum* species (Press *et al.*, 1986) that are mainly responsible for continued bog growth. The critical load for nitrogen, i.e. the threshold above which damage from excess nitrogen may be caused, has been tentatively set for *Sphagnum* bogs as 5–10 kg N ha^{-1} yr^{-1} (Bobbink and Roelofs, 1995), a figure that is probably exceeded over more than half of Scotland, including most of the main areas of blanket bog (British Ecological Society, 1994).

3.4.2 Heather moorland

Although large areas of heather moorland have been converted into arable and livestock farms since the 18th century, and more has been lost to afforestation, heather moors still cover some 20% of Scotland (depending on the precise definition of blanket bog). Some heather moors and heaths may have natural origins (Edwards *et al.*, 1995) but most are ecologically degraded ecosystems, produced by past destruction of native woodland, which now have less fertile soils. However, heather moors are valued in economic terms for producing unnaturally high densities of red grouse (*Lagopus lagopus scoticus*) for sports shooting (Thompson *et al.*, this volume), for nature conservation as ecosystems that have mostly been lost in continental Europe, and as cultural landscapes often replete with archaeological remains.

Whereas about 23% of Scotland's heather moorland was converted to other cover types during 1947–88 (Tudor *et al.*, 1994; Mackey and Tudor, in press), the rate of change is now much reduced and up to a third of the remainder is under some form of protection or positive management (Ward *et al.*, 1995). Nevertheless, there is still concern about continued degradation of heather moorland from overburning and overgrazing (Thompson and Miles, 1995; Thompson *et al.*, 1995). Overburning causes changes in plant species composition and loss of surface organic matter, and leads in some instances to erosion of the surface mineral soil. Overgrazing leads to the loss of heather and other dwarf shrubs, and sometimes to erosion. Poor management in the past has been associated with declining numbers of game birds, particularly red grouse (Hudson, 1995) and black grouse (*Tetrao tetrix*) (Baines, 1996). Some areas of moorland have been degraded by stripping turf for use as fuel and bedding for livestock (Davidson and Simpson, 1984; Simpson, 1993, 1994).

There are also concerns that increased atmospheric depositions of fixed nitrogen may be changing the species composition of the vegetation (Lee *et al.*, 1992; UK Review Group on Impacts of Nitrogen Deposition on Terrestrial Ecosystems, 1994). Certainly, both nitrogen depositions and tissue-nitrogen contents of heather have increased significantly in Scotland since the 1960s (Pitcairn *et al.*, 1995). However, although eutrophication from atmospheric nitrogen deposition seems to have led to widespread replacement of heather and cross-leaved heath (*Erica tetralix*) by grasses in Dutch heathland (Heil and Diemont, 1983; Aerts *et al.*, 1990), and has been suggested as a cause of deterioration of heather stands in the Breckland (Pitcairn *et al.*, 1991), there is no evidence as yet of similar effects in

Scotland. Nitrogen deposition rates here, although enhanced, are still much lower than in The Netherlands and most of England.

3.4.3   *Machair*

Coastal sand dunes are quite common world-wide, but only in Scotland and Ireland have plains of shell-rich sand, termed machair, been developed. At least in the Western Isles, farming seems to have preceded formation of the machair plain, and periodic cultivation of the machair has continued up to the present day (Gilbertson *et al.*, 1995, 1996). Thus human use has not only contributed to the biodiversity seen today (Angus, 1994), but may indeed be the reason why the plain exists as such rather than as rolling dunes. Much of the machair is today under favourable management, e.g. within the Uists, Benbecula, Barra and Vatersay Environmentally Sensitive Area. However, there is concern that the almost complete replacement of cattle by sheep as grazers on the machair may be reducing the species-richness of the vegetation.

3.4.4   *Mountain vegetation*

Although the vegetation on Scotland's mountains is probably nearer a pristine state than in most other ecosystems, there are nevertheless specific concerns about the pervasive effects of overgrazing and atmospheric nitrogen deposition, and more locally about accelerated erosion of soil. Until recently, no one had studied the effects of grazing on montane vegetation. However, circumstantial evidence (Stevenson and Thompson, 1993; Thompson and Brown, 1992) suggests that stands of prostrate heather have been eliminated in southern Scotland by sheep grazing. Similarly, it has been suggested that the disappearance of ptarmigan (*Lagopus mutus*) from the Southern Uplands and Galloway in the early 1800s reflected the loss of their food plants because of sheep grazing (Galbraith *et al.*, 1988; Ratcliffe, 1990). There is also evidence for a decrease in extent of woolly fringe-moss (*Racomitrium lanuginosum*) heaths south of the Highlands because of sheep grazing (Ratcliffe, 1990; Thompson and Brown, 1992), changes that probably resulted in the associated loss of breeding dotterel (*Charadrius morinellus*).

Experiments have shown that high concentrations of nitrogen can inhibit the growth of certain bryophytes (Press *et al.*, 1986; Lee *et al.*, 1992); the decline of southern *Racomitrium lanuginosum* heaths may have been exacerbated by recent increases in atmospheric nitrogen deposition (Baddeley *et al.*, 1994). In addition, Woolgrove and Woodin (1996a,b) have shown that concentrations of nitrate (and sulphate) in the meltwater of Highland snowbeds are damaging to the characteristic snowbed moss *Kiaeria starkei*, and hence are likely also to be damaging other bryophyte species. Given the short growing season for snowbed plants, the slow recovery found after damage may threaten the survival of these assemblages.

Locally occurring sheet erosion of soils on mountains, particularly in the west of Scotland, is a common sight to hillwalkers. Although no estimates of the extent of this exist, a recent study of soil erosion in the wider uplands, using air-photograph interpretation, concluded that almost 12% of the surveyed areas was affected by erosion in one form or another (Grieve and Hipkin, 1996). Twice as much sheet erosion was identified above 550 m altitude as below. Little is known about the causes of this sheet erosion, or times of initiation, but there have been suggestions that introduction of sheep into the hills and mountains, and increases in numbers of red deer (*Cervus elaphus*), may have initiated or exacerbated this process.

3.4.5   *Native woodland*

Native woodlands recently covered only some 1.1% of Scotland (Mackenzie, 1987), i.e. some 98% of the former resource had been lost. During the past century, much native broadleaved woodland has been converted to conifer stands, a process that leads to the loss of most of the characteristic woodland herbs (Miles and Miles, 1997). Nevertheless, Scotland still has many internationally important stands of bryophyte- and lichen-rich oakwoods on the Atlantic seaboard and of Caledonian pinewood in the Highlands. The best examples of both will be given stronger protection as the Habitats Directive is implemented, while Government support for forestry now promotes the maintenance and expansion of all native woodland.

3.5   Conclusions

Given the millennia of increasingly intensive human use of Scotland's natural resources, and given that only a minority of ecosystems on land approach a natural condition, it is gratifying that Scotland still has an abundance of ecosystems of great nature conservation interest and value. This chapter has not mentioned the 2% of Scotland covered with freshwater bodies, mostly in a relatively pristine biological state (Maitland *et al.*, 1994), and where only a small proportion have experienced serious biological problems because of acidic depositions from the atmosphere (Harriman, 1988), although even here there has been some reversibility (Wright and Hauhs, 1991). Nor has it covered farming, so often criticised for producing species-poor habitats (see, for example, Arden-Clarke, 1988). Indeed, recent evidence suggests that arable farming may benefit invertebrates charac-teristic of disturbed ground – a naturally rare habitat – and that many ground beetles formerly thought of as rare in Scotland may not be rare at all (Abernethy *et al.*, 1996; Foster *et al.*, this volume).

This account may seem complacent, but it is not intended to be. For example, whereas atmospheric depositions of sulphate have been decreasing, those of fixed nitrogen have been increasing. Their effects are subtle and not readily detected, but are of concern because of the importance of many of Scotland's montane communities, blanket bogs and moorlands. Scotland's population alone cannot halt climate change, but overburning and overgrazing can be tackled locally. Yet the effects of these are generally slow and not readily detected except by experts, and although these effects seem to be widespread, no national survey of their extent has yet been done. Continued monitoring of Scotland's natural heritage is essential; Scottish Natural Heritage (Anon., 1995) has undertaken to do this.

Acknowledgements

We thank Stuart Gardner for the analysis that led to Table 3.4, and Hilary Anderson, Joyce Hunnam and Mike Shrewry for other help.

References

Abernethy, V. J., McCracken, D. I., Adam, A., Downie, I., Foster, J. M., Furness, R. W., Murphy, K. J., Ribera, I., Waterhouse, A. and Wilson, W. L. 1996. Functional analysis of plant-invertebrate-bird biodiversity on Scottish agricultural land. In Simpson, I. A. and Dennis, P. (Eds.) *The Spatial Dynamics of Biodiversity. Towards an Understanding of Spatial Patterns and Processes in the Landscape*. Aberdeen, International Association of Landscape Engineers (UK Region), 51–59.

Aerts, R., Berendse, F., De Caluwe, H. and Schmitz, M. 1990. Competition in heathland, along an experimental gradient of nutrient availability. *Oikos*, **57**, 310–318.

Allaby, M. 1994. *The Concise Oxford Dictionary of Ecology*. Oxford, Oxford University Press.

Angus, S. 1994. The conservation importance of machair systems of the Scottish islands, with particular reference to the Outer Hebrides. In Baxter, J. M. and Usher, M. B. (Eds.) *The Islands of Scotland. A Living Marine Heritage*. Edinburgh, HMSO, 95–120.

Anon. 1994a. *Sustainable Development. The UK Strategy*. London, HMSO.

Anon. 1994b. *Biodiversity: The UK Action Plan*. Cm2428. London, HMSO.

Anon. 1995. *The Natural Heritage of Scotland: An Overview*. Perth, Scottish Natural Heritage.

Arden-Clarke, C. 1988. *The Environmental Effect of Conventional and Organic/Biological Farming Systems*. IV. *Farming Systems Impacts on Wildlife and Habitat*. Oxford, Political Ecology Research Group.

Baddeley, J. A., Thompson, D. B. A. and Lee, J. A. 1994. Regional and historical variation in the nitrogen content of *Racomitrium lanuginosum* in Britain in relation to atmospheric nitrogen deposition. *Environmental Pollution*, **84**, 189–196.

Baines, D. 1996. The implication of grazing and predator management on the habitats and breeding success of black grouse *Tetrao terix. Journal of Applied Ecology*, **33**, 54–62.

Bates, M. 1961. *The Nature of Natural History*. Revised Edn. New York, Charles Scribner's Sons.

Bennett, K. D. 1984. The post-glacial history of *Pinus sylvestris* in the British Isles. *Quaternary Science Reviews*, **3**, 133–155.

Bobbink, R. and Roelofs, J. G. M. 1995. Nitrogen critical loads for natural and semi-natural ecosystems: the empirical approach. *Water, Air and Soil Pollution*, **85**, 2413–2418.

Bradshaw, R. and McGee, E. 1988. The extent and time-course of mountain blanket peat erosion in Ireland. *New Phytologist*, **108**, 219–224.

Bridge, M. C., Haggart, B. A. and Lowe, J. J. 1990. The history and palaeoclimatic significance of subfossil remains of *Pinus sylvestris* in blanket peats from Scotland. *Journal of Ecology*, **78**, 77–99.

British Ecological Society 1994. *The Ecological Effects of Increased Aerial Deposition of Nitrogen*. Shrewsbury, Field Studies Council.

Bunce, R. G. H., Barr, C. J. and Whittaker, H. A. 1983. A stratification system for ecological sampling. In Fuller, R. M. (Ed.) *Ecological Mapping from Ground, Air and Space*. Cambridge, Institute of Terrestrial Ecology, 39–46.

Carling, P. A. 1986. Peat slides in Teesdale and Weardale, Northern Pennines, July 1983: description and failure mechanisms. *Earth Surface Processes and Landforms*, **11**, 193–206.

Commission of the European Communities 1991. *CORINE Biotopes Manual. Habitats of the European Community*. Luxembourg, Office for Official Publications of the European Communities.

Davidson, D. A. and Simpson, I. A. 1984. Deep topsoil formation in Orkney. *Earth Surface Processes and Landforms*, **9**, 75–81.

Edwards, K. J. 1988. The hunter-gatherer/agricultural transition and the pollen record in the British Isles. In Birks, H. H., Birks, H. J. B., Kaland, P. U. and Moe, D. (Eds.) *The Cultural Landscape: Past, Present and Future*. Cambridge, Cambridge University Press, 255–266.

Edwards, K. J. 1990. Fire and the Scottish Mesolithic. In Harris, D. R. and Thomas, K. D. (Eds.) *Contributions to the Mesolithic in Europe*. London, International Academic Projects, 61–73.

Edwards, K. J. In press. Vegetation history of the Northern Isles and Outer Hebrides. *Botanical Journal of Scotland*.

Edwards, K. J., Whittington, G. and Hirons, K. R. 1995. The relationship between fire and long term wet heath development in South Uist, Outer Hebrides, Scotland. In Thompson, D. B. A, Hester, A. J. and Usher, M. B. (Eds.) *Heaths and Moorlands: Cultural Landscapes*. Edinburgh, HMSO, 240–248.

Galbraith, H., Kinnes, L., Watson, A. and Thompson, D. B. A. 1988. Pressures on ptarmigan populations. *Game Conservancy Annual Review*, **19**, 60–64.

Gilbertson, D., Grattan, J. and Schwenninger, J.-L. 1996. A stratigraphic survey of the Holocene coastal dune and machair sequences. In Gilbertson, D., Kent, M. and Grattan, J. (Eds.) *The Outer Hebrides. The Last 14,000 Years*. Sheffield, Sheffield University Press, 72–101.

Gilbertson, D., Kent, M., Schwenninger, J. K., Wathern, P., Weaver, R. and Brayshaw, B. 1995. The machair vegetation of South Uist and Barra in the Outer Hebrides of Scotland: its ecological, geomorphic and historical dimensions. In Roberts, N. and Butlin, R. (Eds.) *Human Impact and Adaptation: Ecological Relations in Historic Times*. London, Blackwell, 17–44.

Gleason, H. A. and Cronquist, A. 1964. *The Natural Geography of Plants*. New York, Columbia University Press.

Greenstreet, S. T. R. and Hall, S. J. 1996. Fishing and the ground-fish assemblage structure in the north-western North Sea: an analysis of long-term and spatial trends. *Journal of Animal Ecology*, **65**, 577–598.

Grieve, I. C. and Hipkin, J. A. 1996. Soil erosion and sustainability. In Taylor, A. G., Gordon, J. E. and Usher, M. B. (Eds.) *Soils, Sustainability and the Natural Heritage*. Edinburgh, HMSO, 236–248.

Harriman, R. 1988. Patterns of surface water acidification in Scotland. In *Acidification in Scotland 1988*. Edinburgh, Scottish Development Department, 71–79.

Heil, G. W. and Diemont, W. M. 1983. Raised nutrient levels change heathland into grassland. *Vegetation*, **53**, 113–120.

Hudson, P. J. 1995. Ecological trends and grouse management in upland Britain. In Thompson, D. B. A., Hester, A. J. and Usher, M. B. (Eds.) *Heaths and Moorland: Cultural Landscapes*. Edinburgh, HMSO, 282–293.

Huntley, B. and Birks, H. J. B. 1983. *An Atlas of Past and Present Pollen Maps for Europe 0–13,000 years ago*. Cambridge, Cambridge University Press.

Institute of Terrestrial Ecology and Institute of Freshwater Ecology. 1993. *Countryside Survey 1990. Main Report*. Eastcote, Department of the Environment.

Jones, V. J., Stevenson, A. C. and Battarbee, R. W. 1987. *A Palaeolimnological Evaluation of Peatland Erosion*. London, University College Department of Geography Research Paper 28.

Lawrence, E. 1989. *Henderson's Dictionary of Biological Terms*, 10th Edn. Harlow, Longman.

Lee, J. A., Caporn, S. J. M. and Read, D. J. 1992. Effects of increasing nitrogen deposition and acidification on heathlands. In Schneider, T. (Ed.) *Acidification, Research and Policy Implications*. Amsterdam, Elsevier, 97–106.

Lowe, J. J. 1993. Isolating the climate factors in early- and mid-Holocene paleaebotanical records from Scotland. In Chambers, F. N. (Ed.) *Climate Change and Human Impact on the Landscape*. London, Chapman and Hall, 69–82.

Macaulay Land Use Research Institute 1993. *The Land Cover of Scotland 1988. Final Report*. Aberdeen, MLURI.

Mackenzie, N. 1987. *The Native Woodlands of Scotland*. Edinburgh, Friends of the Earth.

Mackey, E. C. and Tudor, G. J. In press. Land cover changes in Scotland over the past 50 years. In Alexander, R. W. and Milington, A. C. (Eds.) *Vegetation Mapping from Patch to Planet*. London, Wiley.

Maitland, P. S., Boon, P. J. and McLusky, D. S. (Eds.) 1994. *The Fresh Waters of Scotland: A National Resource of International Significance*. Chichester, Wiley.

PLATE 5 FLUSHES

A high altitude flush, rich in bryophytes, in the Cairngorms (Photo: L. Gill).

Schleichers thread-moss (Bryum schleicheri var. latifolium), confined in Britain to a single flush on moorland near Stirling (Photo: L. Gill).

The endemic mountain scurvygrass (Cochlearia micacea) occurs in high altitude flushes and streamsides across the central Highlands (Photo: M.B. Usher).

PLATE 6 MOUNTAIN LEDGES

*Herb-rich ungrazed mountain ledges, Ben Lawers National Nature Reserve
(Photo: L. Gill).*

*Characteristic flowers of montane ledges and rocks
(from top): purple saxifrage (Saxifraga oppositifolia); tufted saxifrage (Saxifraga cespitosa); and
alpine forget-me-not (Myosotis alpestris)
(Photos: M.B. Usher).*

Marilaun, A. K. von 1895. *The Natural History of Plants. Their Forms, Growth, Reproduction and Distribution.* London, Blackie & Son.

Miles, J. and Miles, A. 1997. Plant indicators of long established woodland. In Smout, T. C. (Ed.) *Scottish Woodland History: Essays and Perspectives.* Edinburgh, Scottish Cultural Press, 37–43.

Pitcairn, C. E. R., Fowler, D. and Grace, J. 1991. *Changes in Species Composition of Semi-natural Vegetation Associated with the Increases in Atmospheric Inputs of Nitrogen. Final Report.* (C.S.D. Report No. 1246.) Peterborough, Nature Conservancy Council.

Pitcairn, C. E. R., Fowler, D. and Grace, J. 1995. Deposition of fixed atmospheric nitrogen and foliar nitrogen content of bryophytes and *Calluna vulgaris* (L.) Hull. *Environmental Pollution*, **88**, 193–205.

Press, M. C., Woodin, J. J. and Lee, J. A. 1986. The potential importance of an increased atmospheric nitrogen supply to the growth of ombrotrophic *Sphagnum* species. *New Phytologist*, **103**, 45–55.

Ratcliffe, D. A. 1990. *Bird Life of Mountain and Upland.* Cambridge, Cambridge University Press.

Robinson, D. and Dickson, J. H. 1988. Vegetational history and land use: a radio-carbon dated pollen diagram from Machrie Moor, Arran, Scotland. *New Phytologist*, **109**, 223–251.

Rodwell, J. S. (Ed.) 1991–95. *British Plant Communities.* Vols 1–4. Cambridge, Cambridge University Press.

Sachs, J. van 1887. *Lectures on the Physiology of Plants.* Oxford, The Clarendon Press.

Simpson, I. A. 1993. The chronology of anthropogenic soil formation in Orkney. *Scottish Geographical Magazine*, **109**, 4–11.

Simpson, I. A. 1994. Spatial constraints on anthropogenic soil formation in Orkney. *Scottish Geographical Magazine*, **110**, 100–104.

Skiba, U., Cresser, M. S., Derwent, R. G. and Futty, D. W. 1989. Peat acidification in Scotland. *Nature, London*, **337**, 68–69.

Stevenson, A. C. Jones, V. J. and Battarbee, R. W. 1990. The causes of peat erosion: a palaeolimnological approach. *New Phytologist*, **114**, 727–735.

Stevenson, A. C. and Thompson, D. B. A. 1993. Long-term changes in the extent of heather moorland in upland Britain and Ireland: evidence for the importance of grazing. *The Holocene*, **3**, 70–76.

Stroud, D. A., Reed, T. M., Pienkowski, M. W. and Lindsay, R. A. 1987. *Birds, Bogs and Forestry. The Peatlands of Caithness and Sutherland.* Peterborough, Nature Conservancy Council.

Tallis, J. H. 1964. Studies on southern Pennine peats. III. The behaviour of *Sphagnum. Journal of Ecology*, **52**, 345–353.

Tallis, J. H. 1981. Uncontrolled fires. In Phillips, J., Yalden, D. and Tallis, J. (Eds.) *Peak District Moorland Erosion Study, Phase I Report.* Bakewell, Peak Park Planning Board, 176–182.

Tallis, J. H. 1991. *Plant Community History.* London, Chapman and Hall.

Tansley, A. G. (Ed.) 1911. *Types of British Vegetation.* Cambridge, Cambridge University Press.

Tansley, A. G. and Chipp, T. F. (Eds.) 1926. *Aims and Methods in the Study of Vegetation.* London, The British Empire Vegetation Committee and The Crown Agents for the Colonies.

Thompson, D. B. A. and Brown, A. 1992. Biodiversity in montane Britain: habitat variation, vegetation diversity and some objectives for conservation. *Biodiversity Conservation*, **1**, 179–208.

Thompson, D. B. A., MacDonald, A. J., Marsden, J. H. and Galbraith, C. A. 1995. Upland heather moorland in Great Britain: a review of international importance, vegetation change and some objectives for nature conservation. *Biological Conservation*, **71**, 163–178.

Thompson, D. B. A. and Miles, J. 1995. Heaths and moorland: some conclusions and questions about environmental change. In Thompson, D. B. A., Hester, J. J. and Usher, M. B. (Eds.) *Heaths and Moorland: Cultural Landscapes.* Edinburgh, HMSO, 362–387.

Tudor, G. J., Mackey, E. C. and Underwood, F. M. 1994. *The National Countryside Monitoring Scheme: The Changing Face of Scotland, 1940s to 1970s. Main Report.* Perth, Scottish Natural Heritage.

UK Review Group on Impacts of Nitrogen Deposition on Terrestrial Ecosystems 1994. *Impacts of Nitrogen Deposition in Terrestrial Ecosystems.* London, Department of the Environment.

Usher, M.B. and Balharry, D. 1996. *Biogeographical Zonation of Scotland.* Battleby, Scottish Natural Heritage.

Ward, S. D., MacDonald, A. J. and Matthew, E. M. 1995. Scottish heaths and moorland: how should conservation be taken forward? In Thompson, D. B. A., Hester, A. J. and Usher, M. B. (Eds.) *Heaths and Moorland: Cultural Landscapes.* Edinburgh, HMSO, 319–333.

Warming, E. and Vahl, M. 1909. *Oecology of Plants. An Introduction to the Study of Plant Communities.* Oxford, The Clarendon Press.

Woolgrove, C. E. and Woodin, S. J. 1996a. Current and historical relationships between the tissue nitrogen content of a snowbed bryophyte and nitrogenous air pollution. *Environmental Pollution*, **91**, 283–288.

Woolgrove, C. E. and Woodin, S. J. 1996b. Effects of pollutants in snowmelt on *Kiaeria starkei*, a characteristic species of late snowbed bryophyte dominated vegetation. *New Phytologist*, **133**, 519–529.

Wright, R. F. and Hauhs, M. 1991. Reversibility of acidification: soils and surface waters. *Proceedings of the Royal Society of Edinburgh*, B**97**, 169–191.

4 THE NUMBER OF MARINE SPECIES THAT OCCUR IN SCOTTISH COASTAL WATERS

D. M. Davison and J. M. Baxter

Summary

1. An extensive literature review was carried out to obtain an estimate of the number of marine species occurring in Scottish coastal waters.

2. Provisionally it is estimated that there are of the order of 40,000 species occurring in these waters. This includes 21 non-native species.

4.1 Introduction

As a result of the 1992 Biodiversity Convention at the Earth Summit in Rio de Janeiro and the implementation of the European Union's Habitats Directive, the need to collate species information on a national basis has become increasingly important. Estimates of the numbers of marine species occurring in Scottish coastal waters have been collated and compared to estimates of the numbers of marine species occurring around the whole of the British Isles. Additionally, estimates of the numbers of endemic, vagrant and non-native marine species found in Scottish coastal waters have been collated and are detailed in Davison (1996).

Information on the number of marine species has been compiled from a variety of sources and experts. The primary source of species information has been Howson and Picton (1997), produced by the Ulster Museum and the Marine Conservation Society. For many of the phyla, there appear to be few differences between the number of marine species in the coastal waters around the British Isles and those around Scotland. However, for some phyla, which include many Lusitanian, Boreal, Arctic or deep-water species, the differences between the total numbers of species are more significant.

4.2 Overview of issues

During the course of compiling these data, four main issues were identified that need to be taken into consideration when referring to the estimates provided in Table 4.1.

First, there are taxonomic considerations. The traditional view of two kingdoms, Animalia and Plantae, has been regularly challenged. The taxonomic framework is a hierarchy of systematic categories based upon differences in species' levels of organisation and phylogenetic assumptions made about their evolutionary origin and derivation. Many of these assumptions were initially derived from studies of the fossil record and could not be proved. Developments in molecular genetics and ultrastructure continue to reveal that, while some of these assumptions are essentially correct, others are not. Recent research has challenged some phylogenetic assumptions, and the boundaries between the definitions of plants and animals have become increasingly blurred. As a result, a universally accepted taxonomic framework does not now exist. It is possible to divide living organisms into five kingdoms: Monera (Archaebacteria and Eubacteria), Protoctista (protists, eukaryotic algae and some fungi), Fungi (remainder of the fungi, i.e. Zygomycota, Ascomycota and Basidiomycota), Animalia (multicellular animals) and Plantae (mosses, liverworts and vascular plants) (Hoek *et al.*, 1995). Consequently, obtaining accurate totals for the number of species within each phylum, and its component classes, is difficult.

Second, there are geographic considerations. The geographical area of Scottish coastal waters reflects the sea area used by Howson and Picton (1997), generally defined by the 200 m isobath that surrounds the British Isles, within latitudes 48°N to 62°N and longitudes 13°W to 6°E. This sea area includes part of the coasts of France, Belgium and The Netherlands. The precise boundaries of sea areas vary for different phyla, and maps or information on the distribution of some phyla and classes are not readily available.

Third, there are biogeographical factors. Within this sea area there are no major biogeographic boundaries. However, features such as the Rockall Trough and the Wyville–Thomson Ridge correspond to abrupt changes in benthic fauna. The current geographical distribution of each species reflects its evolutionary history and its ecological and physiological tolerances. The British Isles' marine fauna and flora includes an interesting mixture of southern (Lusitanian) and northern (Boreal or Arctic) species that occur at or near the limits of their distribution. Distribution may vary with long-term climatic cycles where southern or northern boundaries of species may extend or recede, with distant, outlying populations persisting or disappearing accordingly. Many Lusitanian species reach their northern limits around the south and southwest of Britain but the situation is complicated by the influence of the Gulf Stream. Some Lusitanian species have spread along the southern English coast, around the Atlantic coasts of Ireland, across to west Scotland and occasionally around Scotland's north and east coast to reach their northern limit on the Northumberland coast. In comparison, many Boreal and Arctic species only extend throughout the northern part of the North Sea and in the British Isles are limited to the east coast of Scotland. Thus, in Scottish coastal waters, Lusitanian species tend to occur on the west coast, whereas Boreal and Arctic species tend to occur on the north and east coasts.

Few marine species are endemic to the British Isles. For some species, such as the dogwhelk (*Nucella lapillus*), the British Isles is the centre of their distributional range. No endemic marine species have been identified in Scottish waters. Many plankton and deep-water species are cosmopolitan, occurring all around the British Isles as the current systems and the continental shelf encourage dispersion. The current systems also contribute to the occurrence of seasonal vagrants: in summer from the south and in winter from the north. The most obvious vagrants are fish, turtle and cetacean species, but many plankton species and some mollusc species are observed seasonally in Scottish coastal waters.

Finally, there are aspects of sampling. The sampling effort for marine species is not uniform around the coast of the British Isles. Historically, sampling has been concentrated around certain favoured areas, often close to established centres of marine research. More recently, activities such as oil and gas exploration have resulted in an expansion of sampling effort. Sampling in the marine environment, especially sublittoral benthic sampling, is difficult, complicated and costly. Research has therefore tended to concentrate on marine macrofauna and macroalgae. Comparatively recently, new techniques have allowed quantitative sampling of microfauna and microalgae to be undertaken. It is, therefore, likely that a number of marine species that have been recorded elsewhere in British coastal waters may eventually be found in Scottish coastal waters.

4.3 The total number of marine species

Table 4.1 lists the best estimates for the numbers of marine species occurring in the coastal waters of the British Isles and Scotland (see Davison (1996) for greater detail). Some species are included that have not yet been recorded in the coastal waters of the British Isles but that have been recorded on the coastlines of neighbouring countries or, for other reasons, are thought to occur in this sea area. Other species have been included that occur in water deeper than the 200 m isobath and beyond the continental shelf, such as in the Rockall Trough. The majority of the deep-water species occur in four phyla: Pogonophora, Crustacea, Mollusca and Echinodermata.

The number of species that could potentially occur in the coastal waters of the British Isles have been added to the total number of species that are currently recorded. This produces a potential maximum number of species, excluding the discovery of new species, and is listed within parentheses in the British Isles column of Table 4.1. The best estimate for the numbers of marine species occurring in Scottish coastal waters is provided in the form of a range for those phyla where an exact figure does not exist. Four predominantly terrestrial phyla were excluded from this report: insects, birds, lichens and vascular plants.

4.4 Non-native marine species

Non-native species are defined by Eno *et al.* (1997) as species that have been introduced directly or indirectly by human agency (deliberately or otherwise) to an area where they have not occurred in historical times, and which is separate from and lies outside the area where natural range extension could be expected.

Table 4.1   The total number of marine species in 44 phyla in the coastal waters of Scotland and the whole of the British Isles (from Davison, 1996). See text for further explanation.

Phyla	Common name	Number of species	
		British Isles	Scotland
Protista	Protozoa	25,000–30,000	25,000–30,000
Mesozoa	(Microscopic parasites)	2 (10)	Unknown
Porifera	Sponges	353 (360)	250–300
Cnidaria	Jellyfish, hydroids, anemones and corals	375 (390)	250–350
Ctenophores	Comb jellies	3	3
Platyhelminthes	Flatworms and meiofaunal worms	355–375 (377)	300–350
Nemertea	Ribbon worms	67 (85)	60–70
Rotifera	'Wheel animalcules'	10–15	10–15
Gastrotricha	Meiofaunal roundworms	85 (140)	80–90
Kinorhyncha	(Microscopic worms)	15 (16)	6–10
Nematoda	Roundworms	408 (410)	350–400
Acanthocephala	Spiny-headed worms	10–20	10–20
Priapulida	(Worm-like animals)	1	1
Entoprocta	(Similar to sea mats)	35 (45)	35–40
Chaetognatha	Arrow worms	22	20
Pogonophora	'Beard' worms	2 (10)	2–10
Sipuncula	(Worm-like animals)	12 (21)	12–15
Echiura	'Spoon' worms	7	3
Annelida	Bristle worms, sludge worms and leeches	940 (995)	800–900
Chelicerata	Sea spiders and marine mites	91	60–70
Crustacea	Branchiopods, barnacles, copepods, ostracods, stomatopods, shrimps, crabs, lobsters, etc.	2,465 (2,665)	2,000–2,460
Uniramia	(Marine arthropods)	3–4	3–4
Tardigrada	'Water bears'	16	16
Mollusca	Chitons, limpets, sea snails, sea slugs, tusk shells, bivalves, octopus, etc.	1,395 (1,465)	650–700
Brachiopoda	Lamp shells	18	11
Bryozoa	Sea mats	270 (290)	120–150
Phoronida	Tube worms	3 (5)	2
Echinodermata	Feather stars, sea stars, brittle stars, sea urchins, sea cucumbers, etc.	145	100–130
Hemichordata	Acorn worms and planktonic larvae	>12	>11
Tunicata	Sea squirts and salps	120 (125)	100
Pisces	Fish	300 (332)	250
Reptilia	Sea turtles	5	4
Mammalia	Sea mammals	28 (35)	28–35
Archaebacteria and other Eubacteria	Bacteria	1,500–2,000	1,500–2,000

continued

Table 4.1 – *continued.*

Phyla	Common name	Number of species	
		British Isles	Scotland
Cyanobacteria (= Cyano-phyta)	Blue-green algae	41 (120)	41 (70)
Rhodophyta	Red algae	350 (450)	250
Heterokontophyta	Golden algae, diatoms, phyto-flagellates, etc.	>1,241 (1,341)	>1,201–1,321
Haptophyta (= Prym-nesiophyta)	Phytoflagellates	117	115
Cryptophyta	Phytoflagellates	78	78
Dinophyta	Dinoflagellates	440 (460)	250–450
Euglenophyta	Euglenoids	5–10	5–10
Chlorophyta	Green algae	160 (220)	100–150
Non-lichenised fungi	Non-lichenised fungi	120 (150)	120–150
Viruses	Viruses	2,000–2,500	2,000–2,500
Total		**38,625–44,666**	**36,207–43,605**

Non-native species have become established in the wild and have self-maintaining populations. The term also includes hybrid taxa derived from such introductions. 'Historical times' is taken as being since the beginning of the Neolithic age, circa 3500 BC. Table 4.2 lists the non-native marine species that have been recorded in Scottish coastal waters.

4.5 Conclusions

Marine taxonomy in the British Isles is subject to continual research, discussion and revision. Consequently, the figures for the total numbers of species should be regarded as best estimates. It should also be recognised that these figures will rapidly date. Whereas the marine macrofauna and macroflora of the British Isles is comparatively well studied, microscopic phyla and classes are more difficult to study and so continue to be poorly understood. Accurate and reliable distribution data are not comprehensive for British marine species and the estimates of the number of Scottish marine species are likely to be underestimates. Provisionally it can be said that there are in the order of 40,000 species occurring in Scottish coastal waters. However, this figure can only provide an indication of the marine biodiversity found in our coastal waters.

Acknowledgements

We are very grateful to the Ulster Museum and the Marine Conservation Society for permission to consult a draft of the second edition of *The Species Directory of the Marine Fauna and Flora of the British Isles and Surrounding Seas* and to all those who provided expert information and advice during the compilation of these data.

Table 4.2   Non-native marine species occurring in the coastal waters of the British Isles (from Eno *et al.,* 1997).

Phylum and number of species found in British coastal waters	Species found in Scottish coastal waters	Locations currently recorded
Cnidaria (3 species)	*Gonionemus vertens*	Most British coasts
	Haliplanella lineata	Locally restricted to estuaries, ports and harbours around Britain
Annelida (8 species)	*Marenzelleria viridis*	Humber Estuary, Firth of Forth and Firth of Tay
Crustacea (6 species)	*Elminius modestus*	Throughout
	Balanus amphitrite	South English coast and Shetland (S. Smith, pers. comm.)
	Acartia tonsa	Southampton, Tamar, Exe estuaries and Firth of Forth
Mollusca (9 species)	*Crassostrea gigas*	Throughout
	Aulacomya ater	Moray Firth (S. Smith, pers. comm.)
	Mya arenaria	Throughout
Tunicata (1 species)	*Styela clava*	South and west English coasts, Loch Ryan and other Scottish sites
Bacillariophyceae (= Diatomophyceae) (5 species)	*Thalassiosira teleata*	English Channel to Norway
	T. punctigera	English Channel and North Sea
	Odontella sinensis	Throughout
Rhodophyta (10 species)	*Antithamnionella ternifolia*	Throughout south and west
	A. spirographides	Throughout south and west
	Polysiphonia harveyi	Throughout south and west
	Bonnemaisonia hamifera	Throughout
	Asparagopsis armata	Throughout
Chlorophyta (2 subspecies)	*Codium fragile* subsp. *atlanticum*	Throughout
	C. fragile subsp. *tomentosoides*	Throughout
Phaeophyceae (3 species)	*Colpomenia peregrina*	Throughout

References

Davison, D. M. 1996. An estimation of the total number of marine species that occur in Scottish coastal waters. Scottish Natural Heritage Review No. 63.

Eno, N. C., Clark, R. A. and Sanderson, W. G. (Eds.) 1997. *Non-native Species in British Waters: A Review and Directory.* Peterborough, Joint Nature Conservation Committee.

Hoek, C. van den, Mann, D. G. and Jahns, H. M. 1995. *Algae: an Introduction to Phycology.* Cambridge, Cambridge University Press.

Howson, C. M. and Picton, B. E. (Eds.) 1997. *The Species Directory of the Marine Fauna and Flora of the British Isles and Surrounding Seas.* Belfast, The Ulster Museum and Marine Conservation Society

5 MARINE BIODIVERSITY

J. B. L. Matthews, J. D. Gage and D. G. Raffaelli

Summary

1. Life originated in an aquatic environment; relative to terrestrial biodiversity, marine diversity is greatest at higher taxonomic levels although it is still considerable at the species level. Extrapolation from deep-sea benthic samples suggests that the number of species present is potentially as high as that in any ecosystem in the world.

2. For practical reasons marine biodiversity is less well recorded and understood than biodiversity on land and in freshwater. There is an urgent requirement for much improved basic information on the taxonomic composition and geographic range of taxa, especially of the smaller macro- and meiofaunal groups in coastal and deep-sea communities and of soft-bodied pelagic organisms.

3. A marine strategy for conservation needs to be developed that takes into account the particular features of the aquatic environment, notably the fluid nature of the habitat and the flow across perceived geographical boundaries, the difficulties of access, and the international status of the high seas.

4. Predictions of the effects of human intervention, particularly in the deep sea, involve large uncertainties. This review points out where better understanding is necessary before specific impacts can be predicted with the necessary precision.

5.1 Introduction

Life began in an aquatic environment and for most of its history has been confined to the waters of the world. Although animals started to emerge onto dry land in the Cambrian, there are still relatively few groups that have done so successfully. On the other hand, the seas have provided a continuous environment for evolution to take place and at least a partial haven from the major catastrophes that seem to have afflicted the world's flora and fauna over geological time. Thus it is no surprise that the sea contains representatives of all the major animal taxa (Usher, this volume). In contrast, all but a very few marine plant species are found within the single taxon Algae; there are no marine bryophytes and only a few species of vascular plant that have become fully marine (Usher, this volume).

Cataloguing marine biodiversity began in earnest with the foundation of the great marine biological laboratories in Europe in the latter half of the 19th century and was mainly carried out by professional marine biologists. Despite this long history, there is a relative paucity of data on the biodiversity of many marine groups and their status, a feature explicitly recognised by the UK Biodiversity Steering Group (Anon., 1995). To a large extent this is due to the difficult access to all but the narrow strip of the intertidal zone so that sampling in most of the marine environment requires specialised and expensive technology. Although the shoreline is easily accessible, the biodiversity of intertidal areas is low relative to that of the sublittoral and offshore habitats because of the restricted suite of species that can cope with the harsh physical conditions of the shore. The stimulus for surveys offshore, at least on the continental shelf, has historically come not from collectors, but from a need to understand marine food chains in the context of fish stocks and environmental change and, more recently, from the need to monitor benthic assemblages around oil installations or dumping grounds. Cataloguing the biodiversity of deep-sea environments requires resources beyond the scope of all but a few research institutes and it is estimated that large numbers of species remain to be discovered in the deep sea (Gage and May, 1993).

Here we discuss the biodiversity of three major marine habitats: coastal benthic systems (including the intertidal zone), the pelagic zone and the deep sea. Where data are available we review the biodiversity status of these major environments and list those habitats and species for which action plans and statements are available. Finally, we discuss the special problems of managing marine biodiversity and highlight areas where further research is needed.

5.2   Coastal benthic systems

5.2.1   *Biogeographic considerations*

As used here, the term 'coastal' includes the seabed running from the maritime edge of the terrestrial environment to the edge of the continental shelf (approximately 200 m depth). This is broadly equivalent to the coastal area reviewed by Davison (1996) and extends much further than the *c.* 6 km coastal zone covered by the Marine Nature Conservation Review (Hiscock, 1996) of the Joint Nature Conservation Committee (JNCC), whose focus is mainly on benthic flora and fauna. Thus the coastal region described here includes the following habitats, as defined in the report of the UK Biodiversity Steering Group: estuaries, saline lagoons, inlets and enclosed bays, the open coast, shelf break, and offshore seabed (Anon., 1995). It does not include maritime cliff and slope, shingle above high tide mark, boulders and rock above high tide mark, coastal strandline, saltmarsh, machair and sand dunes.

There are two major biogeographic origins for marine species within the Scottish inshore region: southern (Lusitanian) and northern (Boreal and Arctic). Most of the Scottish coast lies largely within the Boreal–Lusitanian region, with Shetland in the Boreal or Arctic. The warming influence of the Gulf Stream extends along the Scottish west coast and much of the north coast, and its effects are felt far into the Moray Firth. This allows several species that would otherwise reach their northern limit at lower latitudes to extend their range much further north in Scotland. Many Boreal–Arctic species only extend into the northern North Sea and are limited to the north and east coasts of Scotland.

The high biodiversity afforded by the presence of these different bio-geographical zones is further enhanced by the variety and contrast of geological forms in Scotland which produce a rich diversity of habitat types. Scottish coastal habitats include the Shetland voes (flooded valleys with no sill at their entrance), the west coast fjords or sea-lochs (flooded, glaciated valleys with a shallow sill at the entrance that partly isolates the biota from the open sea), Hebridean lagoons (rock-bound ponds often several kilometres from the sea but connected to it by narrow channels), large estuaries and firths along the east (Cromarty, Moray, Tay, Forth), and the west (Clyde and Solway) coasts, thousands of kilometres of unspoilt rocky shores, sandy beaches and islands with various degrees of isolation. Offshore, the sediments range from coarse, gravelly sands on the outer shelf areas, such as the Stanton banks west of the Hebrides, to fine muds in deep depositional environments with water depths ranging to 200 m or more. Within each of these habitats local-scale processes operate to maintain biodiversity, as reviewed extensively by Raffaelli and Hawkins (1996).

5.2.2   How many species?

Species lists for the coastal area are based on intensive collections focused on Scotland's oldest marine stations on the Clyde at Millport (initially the Scottish Marine Biological Association, now the Universities Marine Biological Station, Millport) and at St Andrews (the Gatty Marine Laboratory), by the extensive and detailed checklists from Phycological Society field meetings, Royal Society of Edinburgh special volumes, Royal Museum of Scotland collections, atlases for particular taxa, extensive literature reviews and surveys of various sections of the coastline by the JNCC Marine Nature Conservation Review, taxonomic keys and guides, and original papers. Much of this information has been collated in a key report on the biodiversity of the Scottish coastal area by Davison (1996) and is summarised by Davison and Baxter (this volume). That report is based largely on Howson and Picton (1997) marine species directory for the British Isles, which in turn draws heavily on many of the above sources.

The numbers of species within each of the major phyla are shown in Table 4.1 of Davison and Baxter (this volume). Although this section emphasises benthic species, the data given by Davison (1996) are not habitat-based and include both benthic and pelagic coastal species. The total number of species of plant, invertebrate and fish in Scottish coastal waters is of the order of 8000. If proto-zoans, viruses and bacteria are included, the total, as estimated by Davison (1996), is about 40,000 species, of which 25–30,000 are protozoans, 2000–2500 are viruses and 1500–2000 are bacteria. The number of species within these groups, especially protozoans, are largely 'guestimates' and could be revised up or down (Davison, 1996). Only a few of the species recorded in Scottish inshore waters are not native (see Table 4.2 of Davison and Baxter, this volume) and all of them are found elsewhere in the UK, with the possible exception of the bivalve *Aulacomya ater*.

5.3 The pelagic zone

5.3.1 The paradox of the plankton

At first sight the water masses of the seas and oceans seem an unlikely habitat for a great diversity of species, plant or animal. The whole realm is a fluid continuum with seemingly invariate conditions: salinity varies within a few percentage points and the proportions of the various salts is constant throughout the world's oceans; seasonal changes of temperature, light and nutrients occur with considerable predictability, often within relatively narrow limits, and depth gradients, of pressure, light, etc., are generally smooth. In addition, it is a three-dimensional environment in which there are no obvious topographical or biological structuring features. If speciation requires the isolation of breeding populations, the pelagic realm would seem to provide fewer opportunities than almost anywhere else on earth (Angel, 1997). It was such thoughts, set against the observation that plankton can nevertheless contain a considerable diversity of species, that led G. E. Hutchinson to write his paper *The Paradox of the Plankton* (1961).

It is now clear that the biodiversity of the pelagic zone is associated with much greater environmental structure and heterogeneity than is apparent at first sight. Patchiness occurs in all physico-chemical properties, including light, nutrients, temperature and salinity, across temporal scales ranging from hours, to days, to seasons, to decades, and across spatial scales ranging from millimetres, to tens of metres, to hundreds of kilometres, with different-sized taxa responding appropriately to these different scales (Denman, 1994; Mann and Lazier, 1996). It is this patchiness that promotes high biodiversity at both local and regional scales. At the larger scales, physical and biological heterogeneity due to ocean currents and local fronts will contribute significantly to the biodiversity of Scottish waters in particular.

5.3.2 How many species?

The history of species descriptions of calanoid copepods and euphausiids (Figure 5.1), which have long been recognised as important and characteristic groups in the zooplankton, illustrates well the progress in cataloguing biodiversity in the pelagic zone. The bimodal pattern is repeated for many taxonomic groups and is a reflection of those periods when systematics was popular, at least with a number of enthusiasts who have had the freedom to indulge their enthusiasm. More recently, the search for new species has penetrated into less accessible habitats – caves and just above the deep sea-bed – which call for modern techniques of precision sampling. New species described in recent decades from the more

(a)

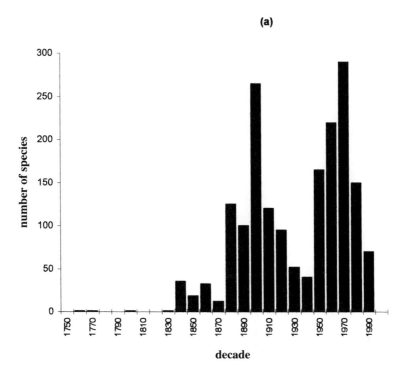

(b)

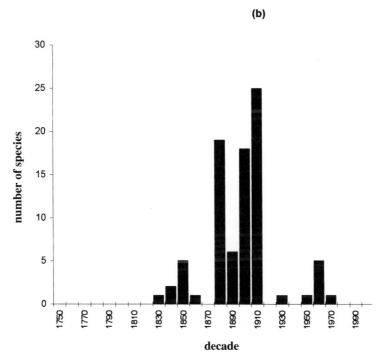

Figure 5.1   The number of new species of (a) calanoid copepod and (b) euphuasid described in each decade from 1750 to the present (Mauchline, pers. comm.).

accessible habitats are mostly due to the resolution of existing taxa into sibling species (Frost and Fleminger, 1968). With the development of molecular genetics the number of new species can be expected to continue to grow.

At higher taxonomic levels biodiversity is great, reflecting the aquatic origin of all animal phyla, with one possible exception, the Onychophora. Species diversity of such taxa as the Copepoda is considerable. More than 1,800 calanoid species have been described to date, the great majority being marine and a large proportion ubiquitous. The deep-water fauna is especially cosmopolitan and differences between closely related species indicate that speciation is a consequence of breeding separation rather than of ecological adaptation. The Calanidae and Metridiidae, families common in the upper layers of open waters where separation of breeding stocks can be supposed to be infrequent, contain a mere 35 and 36 species, respectively. By contrast, families well represented in deep water, such as the Aetideidae (with 232 species), Scolecithricidae (167 species) and Augaptilidae (131 species), have large numbers of species and the main differences between congenerics are in details of the genital apparatus, rather than in mouthparts or other features that might be related to a different mode of life (Knudsen, 1980). It is noteworthy that the exhaustive study of the copepods of the Izu region of Japan by Tanaka (1956–65) yielded a total of 309 species; of these 57 were new and over 200 had been recorded from distant parts of the world ocean, such as the Atlantic, the Antarctic and the Indian Oceans.

The zooplankton around the Scottish coasts contains representatives of at least 20 of the 31 phyla listed by Davison (1996) and Davison and Baxter (this volume, Table 4.1). Over 70 taxonomic groups, incorporating many more than that number of species and representing nine phyla, are taken regularly by the Continuous Plankton Recorder in the open North Sea alone (Williams *et al.*, 1993). Many of these species are meroplankton, larval forms of benthic and other species with a transient existence in the plankton. These contribute enormously to the abundance and diversity of plankton, particularly in coastal water and at certain times of year.

5.4 The deep-sea benthos

5.4.1 *Patterns and processes in deep-sea biodiversity*

Perhaps one of the most challenging intellectual questions facing marine biology lies in understanding the processes underlying the unexpected richness in species co-existing over relatively small areas of the deep-sea bed. This is the more surprising given their low density in what superficially seems to be a featureless, homogeneous habitat, lacking the structural complexity of the tropical rain forest or coral reefs. To understand this, we need to consider both regional- and local-scale patterns and processes.

5.4.2 *Regional scales*

Regional processes include those changes in species' distributions that occur over large geographic areas. These are essentially biogeographic: evolutionary consequences of the immense period of time over which speciation has occurred in the marine environment. In the North Atlantic, species diversity is relatively low on the continental shelf, increasing rapidly down the continental slope to a maximum at mid-slope depths before decreasing again on the abyssal plain (Rex, 1983; Paterson and Lambshead, 1995; but see also Rex *et al.,* 1997). The existence of a latitudinal gradient in benthic species richness in the deep sea (Rex *et al.,* 1993) seems less well-founded. For example, in the Atlantic the decreasing poleward trend in species richness that was demonstrated by Rex *et al.* (1993) for certain abyssal macrofauna in the North Atlantic is not apparent south of the equator. Here there is a pronounced variation from basin to basin as a result of very different tectonic and evolutionary histories (see, for example, Rex *et al.,* 1997).

Some explanations put forward for diversity patterns on land may be relevant to the deep sea. One attractive idea is related to the extent of geographic ranges (Stevens, 1989, 1992). Ranges are greatest at high latitudes (or altitude) where species need to adapt to seasonal extremes in conditions, and least in the tropics (or low altitude) where organisms are adapted to a much narrower range of conditions and thus species will be restricted to much smaller patches of suitable habitat (Rapoport's Rule). This results in many more species being able to co-exist in the tropics than in a similar area at high latitudes. Rapoport's Rule would predict that deep-sea benthic species would have narrower environmental tolerances than coastal species and hence show higher endemicity. Another prediction of Rapoport's Rule is that in areas with high endemicity or diversity there should be a large number of accidentals (individuals out of their preferred habitat). Although yet untested, the high incidence of rare species in deep-sea samples is consistent with the Rule.

Some support for Rapoport's Rule as an explanation for the parabolic depth-related pattern comes from the change in the vertical range of species along the bathymetric gradient on the continental slope. At least in the northwest Atlantic, the most diverse community occurs where bathymetric ranges are broadest (Pineda, 1993). Here environmental variability may be related to boundary currents along the slope and to tidal motion. However, there may also be much lateral variability related to submarine canyons and associated effects. Spatial variability related to hydrodynamics may be less evident at abyssal depths.

Resuspension of organic matter deposited seasonally on the deep-sea bed provides another source of small-scale heterogeneity (Rice and Lambshead, 1994). Such effects may persist for weeks or months and may be reflected in the geographical range of the organisms that exploit them (Gage and May, 1993).

5.4.3 *Local scales*

In contrast to the large-scale patterns, the processes determining species composition and richness at the local scale have been thought to involve the interactive processes of predation, competition and disturbance, similar to other benthic systems. Small-scale patchiness generated by a seasonal flux of organic phytodetritus to the deep-sea bed, and its subsequent patchy redistribution at the small scale (Billett *et al.*, 1983; Lampitt, 1985; Thiel *et al.*, 1989), as well as carcasses or wood sinking to the sea bed (Smith, 1986; Gooday and Turley, 1990), may also play a major role in structuring abyssal biodiversity, further enhanced by the ability of propagules from a potentially vast regional pool of species to colonise such an 'open' system (Grassle and Grassle, 1994; Snelgrove and Grassle, 1995). Another determinant of biodiversity may be increased productivity, as originally put forward by Connell and Orias (1964). However, there are many instances in aquatic systems where a negative, rather than a positive, relationship between species richness and productivity seems to occur, giving rise to the 'paradox of enrichment' (Rosenzweig, 1971). In the deep sea this is supported in the higher, rather than lower, diversity observed in the most oligotrophic abyssal areas of the Pacific (Hessler, 1974).

5.4.4   How many species?

The high species richness of deep-sea sediments has particular relevance to the United Kingdom and the rest of Europe as the deep-sea bed is the only high-diversity environment in Europe and lies within their direct management control. Virtually all the part of this area claimed by the UK lies to the north and west of the Western Isles (Figure 5.2). However, although it is clear that the deep-sea areas around Scotland harbour a high biodiversity, there are insufficient data as yet to provide a synthesis like that given by Davison and Baxter (this volume) for coastal waters. Estimates of even the order of magnitude of species richness in the deep sea remain highly speculative. Perhaps fewer than 200,000 species have so far been described world-wide from the marine biosphere, mostly from coastal and pelagic waters (Briggs, 1991; May, 1994). This compares with a rough estimate of about 1.8 million species on earth, of which a very large proportion are insects (Stork, 1988; May, 1994).

The relatively low estimates for the deep sea reflect the lack of resources for biodiversity research in this most challenging of marine environments and it is likely that great numbers of species remain to be described. For instance, Grassle and Maciolek (1992) found that 58% of deep-sea macrofaunal species encountered were undescribed. Using similar extrapolation methods to those employed for terrestrial biodiversity, they estimated that, globally, deep-sea sediments may support between 10^7 and 10^8 species, and the figure for smaller metazoans may be even higher (Lambshead, 1993). Preliminary results from the Southern Ocean and Pacific even suggest that Grassle and Maciolek's (1992) estimates based on macrofauna from the Atlantic may be too low (Poore and Wilson, 1993; Brey *et al.*, 1994).

5.5   Status of Scottish marine species and habitats

The UK Biodiversity Steering Group Report (Anon., 1995) lists key species and habitats for management and conservation action. For species to qualify for the list they must satisfy one or more of the following criteria:

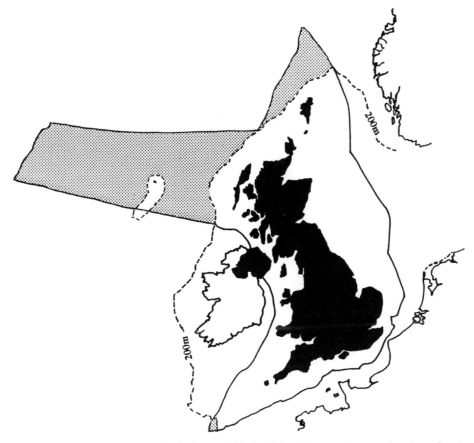

Figure 5.2   The approximate area of seabed over which the United Kingdom has agreed or claimed territorial jurisdiction (solid line) covers more than three times its land area; a third of this area lies at depths below 200 m (stippled). Virtually all of the deep-sea bed lies to the west and northwest of Scotland.

1. species that are threatened and endemic or globally threatened;
2. species where the UK has more than 25% of the population;
3. species whose numbers or range have declined by more than 25% in the past 25 years;
4. species found in fewer than fifteen 10 km squares in the UK;
5. species listed in the schedules and annexes of international agreements and conventions, or in the schedules of the Wildlife and Countryside Act (1981).

This process selected 1,250 species, of which only 68 are marine, and many of those do not occur in Scottish inshore waters. However, it is worth noting that the Steering Group emphasised the paucity of information on the status of many marine species. From this long list, an intermediate list was derived; this has recently been revised (M. B. Usher, pers. comm.) to 387 species and action plans are to be produced within the next three years for 295 of these. Finally, a short list has been drawn up of 116 species for which action plans have already been produced (Anon., 1995; Kerr and Bain, this volume).

Only two Scottish species, both inshore fish, appear on the short list. These are the two shad species, allis shad (*Alosa alosa*) and twaite shad (*Alosa fallax*), for which the Solway is the most likely habitat. Both these species are listed in the 'Bern' *Convention on the conservation of European wildlife and natural habitats* and the European Community (EC) *Directive on the conservation of natural habitats and of wild fauna and flora* (the 'Habitats Directive'), although only allis shad is protected under the Wildlife and Countryside Act (1981). SNH is currently funding basic research on these species in the Solway, and Action Plans exist for both species (Anon., 1995).

The intermediate list now contains three species that occur in Scottish inshore waters. These are the basking shark (*Cetorhinus maximus*), the northern hatchet shell (*Thyasira gouldi*) and the sturgeon (*Acipenser sturio*). All these species have declined and the last two are legally protected by inclusion on Schedule 5 of the Wildlife and Countryside Act (1981). The sturgeon is also listed in the annexes of the EC Habitats Directive. Many of the long-list species are not uncommon in Scottish waters (although they may be uncommon elsewhere in the UK) and it is not always clear which of the criteria listed above were met by these species.

Action plans for two marine habitats – saline lagoons and sea-grass beds – are included in the UK Biodiversity Steering Group Report (Anon., 1995). Habitat Statements also exist for a further six marine habitats: estuaries, inlets and enclosed bays, the open coast, the open-sea water column, the shelf break, and the offshore seabed. Future habitat plans are expected to include maerl beds and *Ascophyllum nodosum mackaii* floating communities (M. B. Usher, pers. comm.). It is noteworthy that the deep sea, probably our most biodiverse resource, is not specifically identified as a habitat in the Report. The offshore habitat (seabed more than 6 miles from the coast) does include communities at >200 m depth but the management of this habitat is clearly focused in the Steering Group Report on much shallower areas.

5.6 Managing marine biodiversity

From the point of view of management, there are some clear messages to be gained from our appreciation of marine biodiversity. First, the interdependence of species within a community indicates that effects cannot operate on one or a few species alone. Secondly, the fluidity of the boundaries, particularly in the pelagic environment, calls into question the whole idea of circumscribed areas for protection, with the corollary that 'anything goes' elsewhere. Both the pelagic and benthic environments are highly dynamic and open systems. Pelagic systems are best defined by reference to particular water masses rather than geographical coordinates; benthic systems are tightly coupled to the overlying pelagic system through the flux of larvae, nutrients and organic matter, and the immediate inshore environment is often affected by ecological processes operating in the terrestrial hinterland. These issues have been extensively discussed in recent years with particular regard to coastal waters and to fisheries (North Sea Task Force, 1994; Andersen *et al.,* 1996; Baxter and McIntyre, 1992).

5.6.1 *Legislation and the international context*

A major obstacle to the effective management of inshore, pelagic and deep-sea habitats is the huge variety of legislation, conventions and directives which cover these areas. Effective management of coastal biodiversity can only be achieved through a system of Integrated Coastal Zone Management, an issue currently under review for Scottish waters (Anon., 1996). Away from the coast, international legislation becomes more important, such as the United Nations Law of the Sea Convention (UNCLOS), Convention for the Prevention of Pollution from ships (MARPOL), European Union Common Fisheries Policy (CFP), and the Oslo and Paris Commission (OSPARCOM). The UK government assesses the potential impact of oil, gas and aggregate extraction as well as dumping of dredgings and other materials offshore. Management and protection of the Scottish deep-sea environment are hampered by a restricted knowledge base as well as limited national regulatory powers. For instance, it is unclear to what extent deep-sea communities will be resilient to perturbations associated with the increasing human impacts on this environment, whether through hydrocarbon prospecting or through deep-sea disposal of decommissioned oil and gas structures, and whether any such change will be long-lasting. Until there is much improved understanding of variability and associated processes controlling biodiversity in this environment it will be necessary in many specific cases to adopt the Precautionary Principle (Gray and Bewers, 1996).

5.6.2 *Marine biodiversity programmes*

Fortunately, research into the factors maintaining marine biodiversity now has a solid international scientific base on which to develop. Following an initial workshop sponsored by the International Union of Biological Sciences (IUBS), the Scientific Committee on Problems of the Environment (SCOPE) and the United Nations Educational, Scientific and Cultural Organisation (UNESCO) to discuss a research agenda for biodiversity (Solbrig, 1991), the international DIVERSITAS Programme has been developed and approved by the International Council of Scientific Unions (ICSU) on behalf of the full range of international science (di Castri and Younes, 1996). The Operational Plan has recently been published (Diversitas, 1996) and is seen to be a major international response to the Biodiversity Convention signed in Rio de Janeiro in 1992. This programme is relevant to the biodiversity on land as well as in the sea – to the whole of the subject as presented in this volume – but in the context of this chapter it is the Marine Biodiversity Programme Element to which we draw attention. Several nations, notably the United States (Butman and Carlton, 1995), have proposed comprehensive research programmes as their national contributions to DIVERSITAS. The international framework within which such programmes can operate has been reviewed by Waugh (1996) and in Europe the Marine Science and Technology

(MAST III) Programme includes marine biodiversity as a component element. One way in which the geographical scope of the problem can be tackled is through regional networks of marine laboratories (Lasserre *et al.*, 1994) who can undertake the necessary surveys, monitoring and data management. The 'MARS' Network of European Marine Stations has given this high priority ever since its inception (Matthews and Heip, 1994).

The UK Biodiversity Steering Group Report (Anon., 1995) provides a framework for the proposal and implementation of research and management of biodiversity in the UK, although it has to be said that the marine content is weak. The definitions of habitats exclude a major element of the Scottish natural heritage, namely the fjordic sea-lochs of the west coast, and taking the species as the basis for particular Action Plans is largely inappropriate in the marine environment for reasons given above. With these caveats, the habitat action plans that are included in the UK Biodiversity Steering Group Report are a positive beginning.

References

Andersen, J., Karup, H. and Nielsen, U. B. (Eds.) 1996. *Proceedings of the Scientific Symposium on the North Sea Quality Status Report.* Copenhagen, Danish Environmental Protection Agency.

Angel, M. V. (1997). Pelagic biodiversity. In Ormond, R. F. G., Gage, J. D. and Angel, M. A. (Eds.) *Marine Biodiversity: Causes and Consequences.* Cambridge University Press, Cambridge, 35–68.

Anon. 1995. *Biodiversity: The UK Steering Group Report.* London, HMSO.

Anon. 1996. *Scotland's Coasts. A Discussion Paper.* London, HMSO.

Baxter, J. M. and McIntyre, A. D. (Eds.) 1992. *Marine Conservation in Scotland, Proceedings of the Royal Society of Edinburgh,* B **100**, 1–204.

Billett, D. S. M., Lampitt, R. S., Rice, A. L. and Mantoura, R. F. C. 1983. Seasonal sedimentation of phytoplankton to the deep-sea benthos. *Nature,* **302**, 520–522.

Brey, T., Klages, M., Dahm, C., Gorny, M., Gutt, J., Hahn, S., Stiller, M., Artnz, W. E., Wägele, J.-W. and Zimmermann, A. 1994. Antarctic benthic diversity. *Nature,* **368**, 297.

Briggs, J. C. 1991. Global species diversity. *Journal of Natural History,* **25**, 1403–1406.

Butman, C. A. and Carlton, J. T. (Eds.) 1995. *Understanding Marine Biodiversity.* Washington, D.C., National Academy Press.

Castri, F. di and Younes, T. (Eds.) 1996. *Biodiversity, Science and Development. Towards a New Partnership.* Cambridge, International Union of Biological Sciences and CAB International, Cambridge University Press.

Connell, J. H. and Orias, E. 1964. The ecological regulation of species diversity. *American Naturalist,* **98**, 399–414.

Davison, D. M. 1996. An estimation of the total number of marine species that occur in Scottish coastal waters. Scottish Natural Heritage Review No. 63.

Denman, K. L. 1994. Scale-determining biological-physical interactions in oceanic food webs. In Giller, P. S., Hildrew, A. G. and Raffaelli, D. G. (Eds.) *Aquatic Ecology. Scale, Pattern and Process.* Oxford, Blackwell Scientific Publications, 377–402.

Diversitas 1996. *DIVERSITAS: An International Programme of Biodiversity Science. Operational Plan.* Paris, International Union of Biological Sciences.

Fischer, A. G. 1960. Latitudinal variations in organic diversity. *Evolution,* **14**, 64–81.

Frost, B. and Fleminger, A. 1968. A revision of the genus *Clausocalanus* (Copepoda: Calanoida) with remarks on distributional patterns in diagnostic characters. *Bulletin of the Scripps Institution of Oceanography,* **12** 1–235.

Gage, J. D. and May, R. M. 1993. A dip into the deep seas. *Nature,* **365**, 609–610.

Gooday, A. J. and Turley, C. M. 1990. Responses by benthic organisms to inputs of organic material to the ocean floor: a review. *Philosophical Transactions of the Royal Society,* B **331**, 119–138.

Grassle, J. F. and Grassle, J. P. 1994. Notes from the abyss: the effects of a patchy supply of organic material and larvae on soft-sediment benthic communities. In Giller, P. S., Hildrew, A. G. and Raffaelli, D. G. (Eds.) *Aquatic Ecology. Scale, Pattern and Processes*. Oxford, Blackwell Scientific Publications, 499–515.

Grassle, J. F. and Maciolek, N. J. 1992. Deep-sea species richness: regional and local diversity estimates from quantitative bottom samples. *American Naturalist*, **139**, 313–341.

Gray, J. S. and Bewers, J. M. 1996. Towards a scientific definition of the precautionary principle. *Marine Pollution Bulletin,* **32**, 768–771.

Hessler, R. R. 1974. The structure of deep benthic communities from central oceanic waters. In Miller, C. B. (Ed.) *The Biology of the Oceanic Pacific*. Corvallis, Oregon State University Press, 79–93.

Hiscock, K. 1996. *Marine Nature Conservation Review. Rationale and Methods*. Peterborough, Joint Nature Conservation Committee.

Howson, C. M. and Picton, B. E. (Eds.) 1997. *The Species Directory of the Marine Fauna and Flora of the British Isles and Surrounding Seas*. Belfast, The Ulster Museum and Marine Conservation Society.

Hutchinson, G. E. 1961. The paradox of the plankton. *American Naturalist*, **95**, 137–145.

Knudsen, T. 1980. Ecological differentiation in closely related bottom-living calanoid copepods (Aetideidae & Phaennidae). Cand. Real. thesis, University of Bergen, 114 pp. (In Norwegian.)

Lambshead, P. J. D. 1993. Recent developments in marine benthic biodiversity research. *Océanis*, **19**, 5–24.

Lampitt, R. S. 1985. Evidence for the seasonal deposition of detritus to the deep-sea floor and its subsequent resuspension. *Deep-Sea Research*, **32**, 885–897.

Lasserre, P., McIntyre, A. D., Ogden, J. C., Ray, G. C. and Grassle, J. F. 1994. Marine laboratory networks for the study of the biodiversity, function and management of marine ecosystems. *Biology International*, Special Issue No. 31, 33pp.

Mann, K. H. and Lazier, J. R. N., 1996. *Dynamics of Marine Ecosystems*. Oxford, Blackwell Science.

Matthews, J. B. L. and Heip, C. H. R. 1994. Inventorying and monitoring of coastal marine biodiversity. *Workshop Report, MARS Network of European Marine Stations*. Paris, UNESCO.

May, R. M. 1994. Biological diversity: differences between land and sea. *Philosophical Transactions of the Royal Society of London,* B **343**, 105–111.

North Sea Task Force 1994. *North Sea Quality Status Report 1994*. Oslo and Paris Commissions. Denmark, Olson and Olson.

Paterson, G. L. J. and Lambshead, P. J. D. 1995. Bathymetric patterns of polychaete diversity in the Rockall Trough, northeast Atlantic. *Deep-Sea Research*, **42**, 1199–1214.

Pineda, J. 1993. Boundary effects on the vertical ranges of deep-sea benthic species. *Deep-Sea Research,* **40**, 2179–2192.

Poore, G. C. B. and Wilson, G. D. F. 1993. Marine species diversity. *Nature*, **361**, 597–598.

Raffaelli, D. G. and Hawkins, S. H. 1996. *Intertidal Ecology*. London, Chapman and Hall.

Rex, M. A. 1983. Geographic patterns of species diversity in the deep-sea benthos. In Rowe, G. T. (Ed.) *The Sea*. Vol. 8. New York, John Wiley, 453–472.

Rex, M. A., Stuart, C. T. and Etter, R. J. 1997. Large-scale patterns of species diversity in the deep-sea benthos. In Ormond, R. F. G., Gage, J. D. and Angel, M. A. (Eds.) *Marine Biodiversity: Causes and Consequences*. Cambridge, Cambridge University Press, 94–121.

Rex, M. A., Stuart, C. T., Hessler, R. R., Allen, J. A., Sanders, H. L. and Wilson, G. D. F. 1993. Global-scale latitudinal patterns of species diversity in the deep-sea benthos. *Nature*, **365**, 636–639.

Rice, A. L. and Lambshead, P. J. D. 1994. Patch dynamics in the deep-sea benthos: the rôle of a heterogeneous supply of organic matter. In Giller, P. S., Hildrew, A. G. and Raffaelli, D. G. (Eds.) *Aquatic Ecology. Scale, Pattern and Processes*. Oxford, Blackwell Scientific Publications, 469–497.

Rosenzweig, M. L. 1971. Paradox of enrichment: destabilization of exploitation ecosystems in ecological time. *Science,* **171**, 385–387.

Smith, C. R. 1986. Nekton falls, low intensity disturbance and community structure of infaunal benthos in the deep sea. *Journal of Marine Research*, **44**, 567–600.

Snelgrove, P. V. R. and Grassle, J. F. 1995. The deep sea: desert and rainforest. *Oceanus*, **38**, 25–28.

Solbrig, O. T. (Ed.) 1991. *From Genes to Ecosystems: A Research Agenda for Biodiversity*. Cambridge, Massachusetts, International Union of Biological Sciences.

Stevens, G. C. 1989. The latitudinal gradient in geographical range: how so many species coexist in the tropics. *American Naturalist,* **133**, 240–256.

Stevens, G. C. 1992. The latitudinal gradient in altitudinal range: an extension of Rapoport's latitudinal rule to altitude. *American Naturalist,* **140**, 893–911.

Stork, N. E. 1988. Insect diversity: facts, fiction and speculation. *Biological Journal of the Linnean Society,* **35**, 321–337.

Tanaka, O. 1956–65. The pelagic copepods of the Izu Region, Middle Japan. Systematic account. I-XIII. *Publications of the Seto Marine Biological Laboratory,* **5** (251–272) - **12** (379–408).

Thiel, H., Pfannkuche, O., Schriever, G., Lochte, K., Gooday, A. J., Hemleben, Ch., Mantoura, R. F. G., Turley, C. M., Patching, J. W. and Rieman, F. 1989. Phytodetritus on the deep-sea floor in a central oceanic region of the Northeast Atlantic. *Biological Oceanography,* **6**, 203–239.

Waugh, J. 1996. The global policy outlook for marine biodiversity conservation. *Global Biodiversity,* **6**, 23–30.

Williams, R., Lindley, J. A., Hunt, H. G. and Collins, N. R. 1993. Plankton community structure and geographical distribution in the North sea. *Journal of Experimental Marine Biology and Ecology,* **172**, 143–156.

6 BIODIVERSITY OF LICHENISED AND NON-LICHENISED FUNGI IN SCOTLAND

R. Watling

Summary

1. Fungi are a diverse group of often unrelated microscopic and macroscopic organisms brought together by virtue of their heterotrophic lifestyle.

2. Lichens in the simplest form are an association of a fungus and an alga; their classification is based on the fungal component, although they are often treated as separate entities in ecological studies.

3. Fungi occur in terrestrial and aquatic habitats. Although little is known of the role of fungi in the aquatic environment, they play a significant role in terrestrial ecosystems as decomposers, mutualists and parasites. Some are intimately associated with non-vascular plants and with animals.

4. With the inability to identify the vegetative stage of many fungi, recognition relies on the presence of the fruiting body which, in all but the lichens, is usually not persistent.

5. The problems of identifying non-lichenised fungi lead to many inconsistences in recording; knowlege of their distribution is therefore often problematic. Although some groups have been very well documented, overall there has only been patchy coverage of Scottish fungi.

6. There has been no experimental work in the UK to assess species loss; one can only extrapolate to Scotland results from studies abroad. Habitat loss appears to be the main reason for species loss.

7. The total number of fungi in Scotland is unknown. An estimate of the number of fungal taxa including lichens in Scotland is of the order of 8,000.

6.1 Introduction

The diversity of Scottish cryptogams is immense. The compilation of the records of known fungi alone presents an enormous task. It is therefore only possible to indicate some trends and observations in this chapter.

The term fungus, although useful, covers a broad spectrum of often quite distantly related organisms brought together by virtue of their heterotrophic lifestyle (Table 6.1). For instance, in the traditional sense, fungi include (in addition to the familiar mushrooms and their allies) some protistids, such as slime moulds (Myxogastrales), and some algal relatives, such as water moulds (Saprolegniales). They may be invisible to the naked eye or form prominent macroscopic fruit-bodies, growing in some species to over a metre in size (Alexopoulos *et al.*, 1996; Hawksworth *et al.*, 1995). Lichens have traditionally been studied separately but they have now taken up their rightful place among the true fungi.

The lichens are a mutualistic group of ascomycetous and basidiomycetous fungi where usually a single fungus and an alga live intimately together. It is not generally well known that close relationships are also found between fungi and mosses and liverworts (bryophytes), and between fungi and brown algae. For instance, among the bryophytes a close association of fungi and the chlorophyll-less *Cryptothallus mirabilis* has been documented (Ligrone *et al.*, 1993), as have associations between various fungi in the Ascomycotina and a range of cushion mosses (Döbbeler, 1978, 1979). We are only now unravelling these complexities. The association of *Mycosphaerella* in the thallus of the brown alga *Pelvetia* is still obscure, nearly a century

Table 6.1   Diversity in the fungi: some of the groups of organism embraced by the term *fungi*.

Basidiomycotina	*basidiomycetes*
Mushrooms and toadstools (agarics)	
Bracket, hedgehog and club fungi (including lichenised forms)	
Stomach fungi (puffballs and allies)	
Jelly fungi and allied yeasts	
Rust and smut fungi	
Ascomycotina	*ascomycetes*
Loculate flask fungi and allies (including lichenised forms)	*pyrenomycetes* p.p.
Non-loculate cup and flask fungi (including lichenised forms)	*disco- and pyrenomycetes* p.p.
Powdery mildews	
Yeasts and cleistothecial fungi	*hemiasco- and plectomycetes*
Mastigomycotina	*oomycetes*
Water moulds (chytrids and *Saprolegniales*) and downy mildews	
Zygomycotina	*zygomycetes*
Mucoraceous moulds and arbuscular mycorrhizas (Endogonales)	
Deuteromycotina	*fungi imperfecti*
Moulds (hyphomycetes) and coelomycetes	
Myxomycotina	*myxomycetes* or *mycetozoa*
Cellular slime moulds	
Plasmodial slime moulds	

after its discovery (Sutherland, 1915). Vascular cryptogams also are involved in fungal associations, especially in their prothallus stage (Hepden, 1960).

Any survey of the diversity of fungi must therefore cover cryptogams in the same way as we accept that fungi are associated in mycorrhizal relationships with the majority of our trees, shrubs and herbaceous plants. Such mycorrhizas include arbuscular mycorrhizas, ectomycorrhizas, orchid and ericoid mycorrhizas (see Harley and Harley, 1987a,b, 1990). It must not be forgotten that mycorrhizal fungi are soil fungi, only producing prominent fruiting bodies at particular periods in their life-cycle. Mutualistic relationships are also commonplace between fungi and animals particularly in some groups of invertebrates.

These mutualistic associations, and some parasitic ones, are known as biotrophic: the fungus derives its organic nutrients from living host-cells. To these must be added (a) the fungi that obtain their nutrients by killing living cells of plants or animals, that is the necrotrophs; and (b) the saprotrophs, which release essential nutrients from organic material and often chelate inorganic substrates for recycling.

In parallel to the majority of necrotrophs, many of the saprotrophs are extremely specific in their requirements; for example, *Marasmius hudsonii* grows only on dead leaves of holly (*Ilex aquifolium*). Some ectomycorrhizal fungi also show a high degree of host specificity whereas others are associated with a whole range of vascular plants. It is because of this high specificity that the Scottish mycota numbers in the thousands and not the hundreds. This is despite observations in Scotland that have demonstrated that some fungi can switch hosts under different ecological conditions. For instance, *Boletus luridus*, normally associated with oak (*Quercus* spp.) occurs with rockrose (*Helianthemum nummularium*) in Perthshire (and Derbyshire) and with mountain avens (*Dryas octopetala*) in Strathnaver, Sutherland (Watling, 1988).

In general both micro- and macrofungi have been recorded by individual workers even though some bias towards a preferred area of discipline is seen. The tradition is changing with the increasing complexity of the systematics, not appreciated even twenty-five years ago. Watling (1986) has reviewed the excellent data natural historians have left Scotland, even though their information undoubtedly will need some adjustment in the future.

In the realm of microfungi Scotland is well known internationally for its extensive studies on fungi causing plant disease. The rust fungi (Uredinales) have been the subject of detailed study by Wilson and Henderson (1966) where enquiry has been made into physiological races and *formae speciales*. It is in the horticultural and agricultural areas, however, that strides have been made by a whole range of authors. Examples are studies of *Phytophthora infestans* (Boyd *et al.*,1968), *Plasmodiophora brassicae* (Hughes, 1964) and *Crumenulopsis sororia* (Ennos and Swales, 1991). Most physiological studies to date have been based on standard laboratory strains. Only now is it becoming fashionable to examine a wider variety of natural isolates for study by modern techniques.

The diversity of form that may occur in a single fungal species is well known.

In the microfungi the full expression of the asexual apparatus is affected by nutrient status of the substrate; thus the necessity to study such fungi in pure culture under controlled conditions is obvious. Although the basic mushroom shape is retained by the mushrooms and toadstools there may be considerable changes in size and colour and some modification in shape, even to the extent in some bracket fungi of incorporating plant material (Watling and Moore, 1994). It is the professional who judges whether differences in spore size and shape, or cystidial characters, are significant in taxonomic terms. The field mycologist must also use microcharacters for identification, and this adds to the difficulty of recording directly in the field. Fortunately, correlated field characters can sometimes be recognised.

R. Kemp has for sometime been examining the variation in populations of species of *Coprinus*; this study has led him to recognise homing devices (Kemp, 1977). These are mechanisms to discourage hybridisation but their existence allows previously undescribed taxa to be predicted. Chromatographic techniques have also shown variation within various populations of toxic fungi, for example in *Cortinarius* sp. (Tebbet *et al.*, 1983) and more recently in *Inocybe* spp. (N. Cumming, unpublished).

6.2   Fungi in Scotland

6.2.1   Numbers of fungi in Scotland

Despite the criticisms levelled by May (1992), the suggestion by Hawksworth (1990) that the number of species of fungi for the world will ultimately be in excess of 1,500,000 taxa is being supported, with the majority of fungi occurring in the tropics (Hawksworth *et al.*, 1996, in press). This calculation is made from the ratio (typically 6:1) of fungus to vascular plant species numbers in well-studied temperate countries. The resulting total excludes the numbers of fungi on insects, a group of which we have only a glimpse at present, although their numbers are undoubtedly significant (Hawksworth, 1990; Watling, 1997, in press). From Hawksworth's formula (Hawksworth, 1990) and Kent's estimate for the number of flowering plants in the British Isles (Kent, 1992), and excluding half the alien plant species, an estimate of 11,300 fungal taxa for Britain is achieved. However, when this formula is applied to Scotland, with 990 species of vascular plant, one can compute the possible number of fungi as 5,840 (i.e. 6 × 990). If aliens and/or ornamental plants (D. McKean, pers. comm.) are included in the calculation, the number is even greater, probably in excess of 8,000 species.

If we consider the lists of fungi for Yorkshire (Bramley, 1985), Warwickshire (Clark, 1980) and southeast England (Dennis, 1995), supplemented by unpublished data from selected areas in Scotland, a pattern emerges. It would appear that the ectomycorrhizal larger fungi make up 35–40% of the total Basidiomycota for any one list. This figure is supported by data from elsewhere in Europe. The more obvious Ascomycota appear to be about equal to the total Basidiomycota. Although these figures are upheld by data from other sites, this estimate is undoubtedly inaccurate, as the ascomycetes are the greater group taken world-wide. The lichenised ascomycetes, although there are a few lichenised basidiomycetous forms in Scotland, equal about half the non-lichenised ascomycetes. For British lists 10% is accepted as a reasonable figure to represent the remainder of the fungi, such as the downy mildews. Again this is an absurdly low figure because all those ascomycetes lacking sexual stages (the mitosporic fungi), and all those host-specific Trichomycetes associated with insects, along with other obscure groups, are excluded. Using the ratios outlined above a formula of $m \times$ 7.3 (6.87–7.86) where m = ectomycorrhizal fungal numbers gives a guesstimate of the total mycota depending on the limits of 35–40%. The value of m is dependent on full knowledge of the larger fungi of an area, and this requires in excess of five years' monitoring (Watling, 1995). Using the formula suggested above a series of estimates can be made for well-studied areas of Scotland (Table 6.2).

If these estimates are remotely correct, then on the specialist knowledge of one or two groups it is possible to make a guess at the expected totals of fungi for an area, or estimate how efficient the recording has been. Equally, one would be able to see where resources should be directed.

6.2.2 How good is the coverage for Scotland?

The map of Scotland shows the uneven spread of data for non-lichenised fungi (Figure 6.1). Some areas have records only from earlier generations (Stevenson, 1879) but there have been some significant recent contributions for discrete geographical areas, such as the Hebrides (Dennis, 1986, 1990) and Shetland (Watling, 1992), or specific groups, e.g. larger fungi of Upper Clydesdale (Silverside, 1991). A mycota of Orkney is currently being produced by R. Watling, T. Eggeling and E. Turnbull. The picture for lichens is very much better, as widespread collecting has been carried out since early in the last century.

Many of the data available are based, because of their prominence in the autumn, on the larger fungi; records stretch back for centuries, such as those of Robert Brown (Watling, 1986). This pattern of autumnal, but not necessarily

Table 6.2 Estimated number of species (see text for explanation) for selected areas of Scotland and actual recorded numbers.

Area	Estimated	Recorded	Source
Hebrides	2,578	2,905	Dennis, 1986
		1,219b	
Mull	2,200	2,007	Henderson and Watling, 1978;
		869b	Watling, 1985
Strathardle, Perthshire	2,025	645b	Watling, 1968, 1978, 1995
Cairngorms area	4,325	1,468b	Watling *et al.*, 1996

b = basidiomycetes only

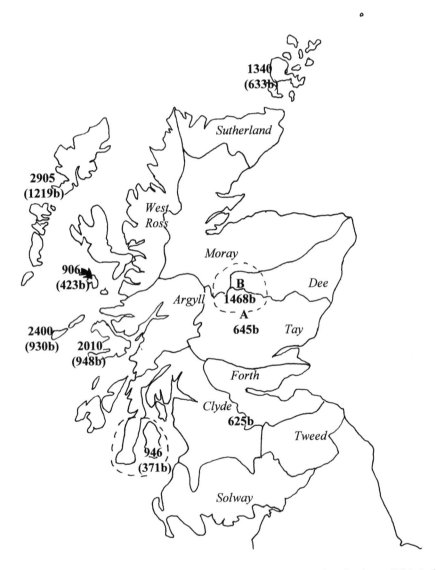

Figure 6.1   Numbers of fungi recorded for well defined areas superimposed on Buchanan White's division of Scotland based on watersheds. In Central Scotland **A** indicates the Strathardle valley and **B** the Greater Cairngorm area; b, number of basidiomycetes.

annual, fruiting emphasises the difficulty of recording non-lichenised fungi, and therefore the problems of estimating diversity. We are basically ignorant of the specific environmental conditions on which fruiting depends (Miller, 1995). A plea can in fact be made for the study of fungi to be a special case in any biodiversity programme (Watling, 1995). Lichenised fungi are generally easier to record and can be studied and monitored all the year around in much the same way as mosses.

At the other extreme we have only a small idea of the diversity of those mitosporic, aero-aquatic fungi found in running water (for example, see Ingold, 1973), and of marine moulds such as thraustochytrids (Gaertner, 1980), undoubt-edly because of a dearth of collecting and recording. Equally we have only scratched the surface in unravelling the range of free-living soil fungi to be found.

Watling (1986) examined the numbers of agaricoid fungi in Scotland and the percentage of taxa within those families outlined in the New British Check List (Dennis *et al.*, 1960). Buchanan White's division of Scotland into watersheds adopted by Stevenson (1879) in *Mycologia Scotica* is a useful basis on which to superimpose these data. In such a framework Sutherland, with a poor overall flora, has half those agaricoid fungi recorded for the Tay area, with its huge spectrum of natural communities and planted ornamentals. This low figure must in part be a reflection of the vast areas of peatlands in the former watershed.

Similarly, Moray, which includes the mycologically rich Speyside, has twice the number of fungi of Sutherland. However, Solway and Tweed, both of which have an excellent spectrum of flowering plants, are lower in fungal diversity than Sutherland! The reason is undoubtedly under-recording of these areas. At least these figures show where attention should be directed in the future. Even the setting up of a cryptogamic sanctuary and reserve at the Royal Botanic Garden of Edinburgh's garden at Dawyck in the Borders has increased the Tweed figures significantly with some very interesting finds.

The number of larger fungi in Scotland agrees favourably with that of the worked areas in England, e.g. the southeast (Dennis, 1995), Warwickshire (Clark, 1980) and Yorkshire (Bramley, 1985). Ireland, although slightly larger than Scot-land, is not as well documented. Many of the records for southeast England are of fungi that have never been seen in Scotland and indicate a southern element in the British mycota, species that might not be expected in Scotland except under unusual circumstances.

The figures for macromycetes for the various areas of Scotland probably also apply to the microfungi; we have some data but not nearly enough (Table 6.3). Our knowledge of the Scottish rust fungi (Wilson and Henderson, 1966), smut fungi (Mordue and Ainsworth, 1984) and downy mildews (Francis and Water-house, 1988) is very good but of more obscure groups we only have a few glimpses of the final picture. The British Lichen Society's Mapping Scheme was started in 1964 and, on the whole, Scotland has been very well documented for lichenised fungi. Even here, however, there are some unexplained anomalies: 440 lichens are known from Shetland but only 258 from Orkney (K. Dalby, personal com-

Table 6.3   Numbers of fungi recorded for well documented areas of Scotland compared with total UK numbers.

Group	UK	Scotland	Source
Stomach fungi (Basidiomycotina: gastero-mycetes)	104	52	Pegler *et al.*, 1995
Smut fungi (Basidiomycotina)	98	67	Mordue and Ainsworth, 1984
Lichenised fungi	1,667	1,486	O'Dare and Coppins, 1993; B. J. Coppins, pers. comm.
Laboulbenialean fungi (Ascomycotina)	46	24[a]	A. Weir, pers. comm.
Downy mildews (Mastigomycotina)	87	52	Francis and Waterhouse, 1988
Slime moulds (Myxomycotina)	336	275[b]	Ing, 1995

[a]Including a single endemic species.
[b]Including two endemic species.

munication). Ecological data and Scottish records have been incorporated into the recently published British Lichen Flora (Purvis *et al.*,1992).

Substrate studies are generally fairly complete and cover the whole spectrum of fungi; thus a good basic knowledge of Scottish dung fungi exists (Richardson and Watling, 1968, 1969). Even so, new species are being described (Richardson and Watling, 1997) or new Scottish records made, e.g. *Ascobolus hawaiiensis.*

6.2.3   *Which fungi are special to Scotland?*

Although the early mycologists undoubtedly recognised in the field that certain fungi appeared in certain ecological habitats or on certain substrates, or under certain trees, in early lists little ecological information is given. Indeed, the main difference between the modern mycotas, such as those for the Hebrides (Dennis, 1986, 1990) or Shetland (Watling, 1992), and their earlier counterparts is the emphasis placed on ecology. Some of the early compilations of fungal records are therefore of limited use, but they evolved before the teachings of Tansley (1934) and Godwin (1956).

As pointed out by Miles *et al.*(this volume), Scotland has a wide spectrum of habitats rarely equalled in variety in areas of similar size anywhere else in the world. The records of fungi for Scotland reflect northern, arctic–alpine, even oceanic or Lusitanian patterns against a backdrop of fungi that are naturally widely distributed, or that show the influence of human activities such as those associated with planted beech (*Fagus sylvatica*) (Watling, 1987). Even southern elements push their way up into the drier eastern side of Scotland.

Two main elements of fungal diversity can be found in Scotland, a general element (which may be found not only elsewhere in Britain but also in Europe), and an element dotted across Europe where exotic trees and ornamentals have been planted in historic times. Many of the introduced exotics were based not on cuttings or on seed but on bare-rooted stocks, which by accident were accompanied by associated fungi of the three trophic states. Thus larch (*Larix* spp.) has a very characteristic fungus flora associated with it, such as the bolete *Suillus grevillei* and the pathogenic ascomycete *Lachnellula wilkommii*.

Parallels are seen in more recent introductions such as *Eucalyptus* from Australia and its hypogeous associate *Hydnangium carneum*. Equally, common, widespread macrofungi may become associated with introduced trees, probably colonising from local populations on native hosts: examples include several species of *Cortinarius* subgenus *Dermocybe* with Sitka spruce (*Picea sitchensis*) (Alexander and Watling, 1987).

To these elements must be added a characteristically Scottish one. Scotland has some fine examples of habitats that support characteristics suites of fungi, but collecting in suitable habitats in other areas of the British Isles destroys the idea that certain species are exclusively Scottish. Thus collecting in northern Wales and the Lake District has yielded montane species such as *Amanita nivalis* not previously known outside Scotland (Watling, 1996).

Although Scotland has montane birchwoods and remnant oakwoods the species lists for macromycetes are little different from those of birch and oakwoods south of the border. Perhaps the only habitat supporting species rarely if ever seen in England and Wales is the Caledonian pine forest. Thus at Rothiemurchus, Glen Affric and Glen Tanar, *Hydnellum* spp., *Sarcodon* spp. and *Bankera fuligineoalbum* are widespread yet they are unknown or extremely rare elsewhere. Records of *Boletopsis leucomelaena* have only ever been made in Rothiemurchus Forest, where it was last found in 1963. Even after intensive collecting in the sand dunes of England and Wales, additional sites for *Laccaria maritima* have not been found; it is known in the UK from a single locality in Scotland, Culbin Forest.

Some plant parasites are only recorded for Scotland because of the distribution of their hosts, but they might be expected elsewhere in the British Isles; examples are *Arctiomyces warmingii* on purple saxifrage (*Saxifraga oppositifolia*) and *Bostrichonema alpestre* on alpine bistort (*Polygonum vivipara*). This is supported by the rust fungi, of which no taxon is unique to Scotland. In contrast eleven species of British smut fungus and one species of downy mildew, *Peronospora arthuri,* are known in Britain only from Scotland. In the little-known group of parasites of arthropods, the Laboulbeniales (Ascomycotina), one species is apparently endemic to Scotland but again it is considered that intense collecting will find it elsewhere in Britain (A. Weir, pers. comm.).

Among the British lichens there are many more species that are found only in Scotland than there are species unique to England; the final number of these Scottish specialities is being compiled at the moment (B. J. Coppins, pers. comm.).

6.3   Conclusions

When all the records of fungi for Scotland are added the picture, although needing some focusing, is reasonably good. As demonstrated above, we are a good way towards knowing the diversity of certain groups of fungi in Scotland but there are many obscure groups that are very poorly known. With a comparatively small financial input to support targeted fieldwork, general collecting by amateur mycologists (parataxonomists) and analysis of herbarium material and published records, it would be possible for Scotland to be the first country in the world to attempt a complete inventory of fungi.

What is desperately required is a check list of Scottish fungi. Ideally it would be best to consider this on a British Isles basis, but the completion of the first would make the second considerably easier. Even taking account of the problems outlined by Rossman (1994) and Watling (1995), it is within our grasp. A concerted effort is required to document this enormous component of our natural heritage.

Techniques are developing that address the problems inherent in recording fungi, and methodology has been proposed, for example by Hawksworth *et al.*, (1997), Harper and Hawksworth (1994), and Lodge *et al.* (1995). Fungi, as demonstrated above, are very dependent on the range of organisms available in the environment. High habitat diversity therefore maintains high fungal diversity; habitat loss means a decrease in fungal diversity (Watling, 1997, in press).

As fungi are particularly useful in wealth creation, any financial input would be money well spent (Hawksworth, 1992). After all, the grandeur of Scotland's forests is dependent on ectomycorrhizal fungi, as are its crop plants on endomycorrhizas. Plant diseases cause damage to crops and to the native flora and one must remember that penicillin is the best-selling antibiotic in the world, along with another fungal metabolite, cyclosporin. My case rests!

Acknowledgements

I thank B. Ing, J .Lennard, K. Dalby, M. J. Richardson, A. Weir and my colleagues Brian Coppins and S. Helfer, and Douglas McKean and especially Evelyn Turnbull for their help. I also thank Prof. D. Hawksworth and Dr J. Lodge for helpful discussions.

References

Alexander, I. J. and Watling, R. 1987. Macrofungi of Sitka spruce in Scotland. *Proceedings of the Royal Society of Edinburgh*, **93B**, 107–115.

Alexopoulos, C. J., Mims, C. W. and Blackwell, M. 1996. *Introductory Mycology,* 4th Edn. New York, John Wiley and Sons.

Boyd, A. E. W., Morton, A. P. and Lennard, J. H. 1968 . Effects of haulm destruction on crop yield and incidence of blight in seed tubers. *The Edinburgh School of Agriculture: Experimental Work (1968)*, 28.

Bramley, W. G. 1985. *A Fungus Flora of Yorkshire.* Leeds, Yorkshire Naturalists' Union.

Clark, M. C. 1980. *A Fungus Flora of Warwickshire.* London, British Mycological Society.

Dennis, R. W. G. 1986. *Fungi of the Hebrides.* Kew, Royal Botanic Gardens.

Dennis, R. W. G. 1990. Fungi of the Hebrides – Supplement. *Kew Bulletin*, **45**, 287–301.

Dennis, R. W. G. 1995. *The Fungi of South East England.* Kew, Royal Botanic Gardens.

Dennis, R. W. G., Orton, P. D. and Hora, F. B. 1960. A New Check List of British Agarics and Boleti. *Transactions of the British Mycological Society* (Suppl.), **43**, 1–225.

Döbbeler, P. 1978. Moosbewohnende Ascomyceten 1. Die Pyrenocarpen den Gametophyten besidesunder. *Mitteilungen Botanischen Staatssammlung München,* **14**, 1–36.

Döbbeler, P. 1979. Untersuchungen an moosparasitischen Pezizales aus der Verwandtschaft von *Octospora. Nova Hedwigia,* **31**, 817–864.

PLATE 7 MOUNTAINS/CLIFFS

Coire Fee, Caenlochan National Nature Reserve (Photo: M.B. Usher).

Raven (Corvus corax), a cliff-nesting scavenger of our uplands (Photo: L. Campbell).

The grey mountain carpet moth (Entephria caesiata) (Photo: M.B. Usher).

White stalk puffball (Tulostoma niveum), known in Britain only from moss-covered limestone boulders at Inchnadamph (Photo: L. Gill).

PLATE 8 MONTANE PLATEAU

Summit plateau, Beinn Udlamain (Photo: M.B. Usher).

Male dotterel (Charadrius morinellus) incubating
(Photo: D.B.A. Thompson).

Least willow (Salix herbacea) forms
carpets in summit plateau vegetation (Photo: M.B. Usher).

Ennos, R. A. and Swales, K. W. 1991. Genetic variability and population structure in the canker pathogen *Crumenulopsis sororia. Mycological Research,* **95**, 521–525.

Francis, S. and Waterhouse, G. M. 1988. List of Peronosporaceae reported from the British Isles. *Transactions of the British Mycological Society,* **91**, 1–62.

Gaertner, A. 1980. Quantitative studies on the marine phycomycetes, chytrids and higher mycelial fungi of the upper Tay estuary. *Proceedings of the Royal Society of Edinburgh,* **78B**, 57–78.

Godwin, H. 1956. *The History of the British Flora.* Cambridge, Cambridge University Press.

Harley, J. L. and Harley, E. L. 1987a. A checklist of mycorrhiza in the British flora. *New Phytologist,* (Suppl.) **105**, 1–102.

Harley, J. L. and Harley, E. L. 1987b. A checklist of mycorrhiza in the British flora – addenda, errata and index. *New Phytologist,* **107**, 741–749.

Harley, J. L. and Harley, E. L. 1990. A checklist of mycorrhiza in the British flora – second addenda, and errata. *New Phytologist,* **115**, 699–711.

Harper, J. L. and Hawksworth, D. L. 1994. Biodiversity: measurement and estimation. *Philosophical Transactions of the Royal Society of London,* B **345**, 5–12.

Hawksworth, D. L. 1990. Presidential address: The fungal dimension of biodiversity; magnitude, significance and conservation. *Mycological Research,* **95**, 641–655.

Hawksworth, D. L. 1992. Fungi: a neglected component of biodiversity crucial to ecosystem function and maintenance. *Canadian Biodiversity,* **1**, 4–7.

Hawksworth, D. L., Kirk, P. M., Sutton, B. C. and Pegler, D. N. 1995. *Ainsworth and Bisby's Dictionary of the Fungi,* 8th Edn. Wallingford, CAB International.

Hawksworth, D. L., Lodge, D. J. and Ritchie, B. J. 1996. Microbial diversity and tropical forest functioning. *Ecological Studies,* **122**, 69 –100.

Hawksworth, D. L., Minter, D. W., Kinsey, G. C. and Cannon, P. F. In press. Inventorying a tropical fungal biota: intensive and extensive approaches. In Janardhanan, K. K., Rajendran, C. and Natarajan, K. (Eds.) *Tropical Mycology.* New Delhi, Oxford, IBM.

Henderson, D. M. and Watling, R. 1978. 15 Fungi. In Jermy, A. C. and Crabbe, J. A. (Eds.). *The Island of Mull.* London, British Museum.

Hepden, P. M. 1960. Studies in vesicular-arbuscular endophytes 2. *Transactions of the British Mycological Society,* **43**, 559–570.

Hughes, W.A. 1964. Trials of organic compounds for clubroot control. *Edinburgh School of Agriculture Experimental Work (1964),* 27.

Ing, B. 1995. A review of the myxomycetes of Scotland with special reference to snowbed communities. Scottish Natural Heritage Review No. 38.

Ingold, C. T. 1973. Aquatic Hyphomycete spores from West Scotland. *Transactions of the British Mycological Society,* **61**, 251–255.

Kemp, R. F. O. 1977. Oidial homing and the taxonomy and speciation of basidiomycetes with special reference to the genus *Coprinus.* In Clémençon, H. (Ed.) The species concept in hymenomycetes. *Bibliotheca Mycologica,* **61**, 259–276.

Kent, D. H. 1992. *List of Vascular Plants of the British Isles.* London: Botanical Society of the British Isles.

Ligrone, R., Pocock, K. and Duckett, J. G. 1993. A comparative ultrastructural analysis of endophytic basidiomycetes in the parasitic achlorophyllous hepatic *Cryptothallus mirabilis* and the closely allied photosynthetic *Aneura pinguis* (Metzgeriales). *Canadian Journal of Botany,* **71**, 666–679.

Lodge, D. J., Chapela, I., Samuels, G., Uecker, F.A., Desjardin, D., Horak, E., Miller, O. K. Jr., Hennebert, G. L., Decock, C. A., Ammirati, J., Burdsall, H. H. Jr., Kirk, P. M., Minter, D. W., Halling, R., Laessoe, T., Mueller, G., Huhndorf, S., Oberwinkler, F., Pegler, D. N., Spooner, B., Petersen, R. H., Rogers, J. D., Ryvarden, L., Watling, R., Turnbull, E. and Whalley, A. J. S. 1995. A survey of patterns of diversity in non-lichenized fungi. *Mitteilungen Eidgenössische Forschungsanstalt für Wald, Schnee und Landschaft,* **70**, 157–173.

May, R. M. 1992. How many species inhabit the Earth? *Scientific American,* October, pp.18–24.

Miller, S. (Ed.) 1995. Assessment of fungal diversity. Symposium, IMC 10, Vancouver August 1994. *Canadian Journal of Botany* (Suppl.), **73**.

Mordue, J. E. M. and Ainsworth, G. C. 1984. Ustilaginales of the British Isles. *Mycological Paper,* **154**, 1–96.

O'Dare, A. M. and Coppins, B. J. 1993. *Biodiversity Inventory – Lichens.* Unpublished report to Scottish Natural Heritage.

Pegler, D. N., Laessoe, T. and Spooner, B. M. 1995. *British Puffballs, Earthstars and Stinkhorns.* Royal Botanic Gardens, Kew.

Purvis, O. W., Coppins, B. J., Hawksworth, D. L., James, P. W. and Moore, D. M. (Eds.) 1992. *The Lichen Flora of Great Britain and Ireland.* London, Natural History Museum Publication.

Richardson, M. J. and Watling, R. 1968. Keys to fungi on dung 1. *Bulletin of the British Mycological Society*, **2**, 18–43.

Richardson, M. J. and Watling, R. 1969. Keys to fungi on dung 2. *Bulletin of the British Mycological Society*, **3**, 86–88.

Richardson, M. J. and Watling, R. 1997. *Keys to Fungi on Dung,* new, revised and enlarged edn. London, British Mycological Society, 68pp.

Rossman, A. 1994. A strategy for an all-taxa inventory of fungal biodiversity. In Peng, C.-I. and Chou, C. H. (Eds.) Biodiversity and terrestrial ecosystems. *Institute Botany, Academia Sinica Monograph Series,* **14**, 169–194.

Silverside, A. J. 1991. *Fungus Flora of the Mid-Clyde Valley.* Private publication, 65pp.

Stevenson, J. 1879. *Mycologia Scotica. The fungi of Scotland.* Edinburgh, Cryptogamic Society of Scotland.

Sutherland, G. K. 1915. New marine fungi on *Pelvetia. New Phytologist,* **14**, 33–42.

Tansley, A. G. 1934. *The British Islands and their Vegetation.* Cambridge, Cambridge University Press.

Tebbett, I. R., Kidd, C. B. M., Caddy, B., Robertson, J. and Tilstone, W. 1983. Toxicity of *Cortinarius* species. *Transactions of the British Mycological Society*, **81**, 636–638.

Watling, R. 1968. Larger Fungi of Kindrogan. *Report of the Scottish Field Studies Association 1967*, 28–46.

Watling, R. 1978. More Fungi of Kindrogan. *Report of the Scottish Field Studies Association 1977*, 21–46.

Watling, R. 1985. *The Fungal Flora of Mull – Additions.* Edinburgh, Royal Botanic Garden.

Watling, R. 1986. 150 years of Paddockstools: A history of agaric ecology and floristics in Scotland. *Transactions of the Botanical Society of Edinburgh*, **45**, 1–42.

Watling, R. 1987. Pilzkartierung in England. *Beitrage zur Kenntnis der Pilze Mitteleuropas*, **3**, 31–39.

Watling, R. 1988. Presidential address. A mycological kaleidoscope. *Transactions of the British Mycological Society,* **90**, 1–28.

Watling, R. 1992. *The Fungus Flora of Shetland.* Edinburgh, Royal Botanic Garden.

Watling, R. 1995. Assessment of fungal diversity: macromycetes, the problems. *Canadian Journal of Botany* (Suppl.), **73**, 15–24.

Watling, R. 1996. Putative fungus-tree relationships: what do they tell us? *Documents Mycologiques,* **25**, 479–486.

Watling, R. 1997. Pulling the threads together: habitat diversity. In Hawksworth, D. L. and Watling, R. (Eds.) Fungal Biodiversity. *Biodiversity and Conservation,* **6**, 753–763.

Watling, R. In press. Myco-ecology in Scotland. In Proctor, J. and Dix N. (Eds.) Plant Ecology of Scotland. *Botanical Journal of Scotland* (Suppl.).

Watling, R., Coppins, B. J., Fleming, L. V. and Davy, L. M. 1996. *Cairngorm project: Inventory of restricted lower plants (fungi and lichenized fungi): Phase II.* Unpublished report to Scottish Natural Heritage.

Watling, R. and Moore, D. 1994. Moulding moulds into mushrooms: shape and form in higher fungi. In Ingram, D. I. and Hudson, A. (Eds.) *Shape and Form in Plants and Fungi.* London, Academic Press, 272–290.

Wilson, M. and Henderson, D. M. 1966. *British Rust Fungi.* Cambridge, Cambridge University Press.

7 Vascular Plant Biodiversity in Scotland

C. Sydes

Summary

1. With the exception of some apomictic groups, the Scottish flora is well described, although new species continue to be described by re-examination of existing taxa. Total diversity is low relative to the tropics but it is comparable for its size with other countries in NW Europe.

2. Within the British Isles, Scotland is particularly important for the range and abundance of its mountain species, its northern woodland species and aquatic plants.

3. Internationally, Scotland is significant for: its endemic species, although some are of fragile taxonomic status; populations of those species that are threatened throughout their world or European ranges; and the abundance of those species that are largely restricted to the Atlantic seaboard of western Europe. Our knowledge of the significance of these species is hampered by the difficulty of collating comparable data from other countries.

4. More than half the taxa in the flora of Scotland are now derived from introductions and as a result the total diversity of vascular plants has probably never been higher.

5. Very rare taxa are still liable to extinction of local populations despite our protection measures. However, the greatest loss of biodiversity has occurred, and appears to be still occurring, among more common but evidently vulnerable taxa.

7.1 The importance of Scotland's vascular plant biodiversity

7.1.1 The size of the flora

The taxa of Scotland are as well known as those of almost any comparable political unit in the world, although the total number depends on which taxa are included and the information source. One thousand, one hundred and seventeen native

Table 7.1 Total number of vascular plant species recorded
in Scotland and the UK (derived from Dring, 1994).

	Scotland	UK
Species found before 1930	1,186	1,522
Species found since 1930	1,117	1,476

plant species are recorded from Scotland (Table 7.1) in Dring (1994), which is mainly derived from the results of survey and collation carried out by members of the Botanical Society of the British Isles for the *Atlas of the British Flora* (Perring and Walters, 1962).

7.1.2 Hybrids, subspecies and microspecies

The total size of the Scottish flora swells to 1,559 with the addition of taxa from *Hieracium* and *Taraxacum* – hawkbits and dandelions – two of the better known apomictic genera (Anon., 1995). Some of these taxa, notably those most British botanists still call *Ranunculus auricomus*, are still undescribed in Britain (Stace, 1991).

The taxa of apomictic plants are often called microspecies, reflecting the fact that they do not have the range of variation associated with full sexual reproduction. Despite this, apomictic taxa have attracted conservation concern over the past 30 years. A National Nature Reserve (NNR) exists to increase protection for the Arran whitebeams, *Sorbus arranensis* and *S. pseudofennica*. Taxa of the better-known apomictic genera *Alchemilla* (lady's mantle) and *Taraxacum* were included in the British Red Data Book (Perring and Farrell, 1983). Three *Hieracium* species found only on Shetland are now listed on Schedule 8 of the Wildlife and Countryside Act. Inevitably, given the lack of knowledge about some apomictic groups, one suspects that our protection measures are not yet even-handed across these taxa.

Nevertheless this long-term concern for apomictic species now reflects the concerns addressed by the UNCED Convention on Biological Diversity. Microspecies are a visible indication of the range of biodiversity normally hidden within a species and that we should aim to protect. By analogy with other higher plant species, only a small proportion of the populations of microspecies are likely to require active protection. This protection may be easier to provide even within tiny relict populations, as loss of genetic variation by genetic drift cannot occur in apomicts.

The published distributions of the native Scottish species of an apomictic taxon like *Hieracium* indicate that these species are not simply examples of recent local variation. They show a range of distribution patterns from the expected endemic to Scotland, through endemic to Britain, to those also found in neighbouring countries such as the Faeroes or Scandinavia, and to some with widely disjunct

Table 7.2 Sample of the first 151 Scottish taxa from Stace (1991). See text for explanation.

Category	No.	Species	Taxa
Native species	76	58%	—
Introduced species	55	42%	—
All species	131	100%	87%
Native subspecies	8	—	5%
Native hybrids	12	—	8%
All taxa	151	—	100%

distributions, although one must always bear in mind that full knowledge of the distribution of such taxa may be lacking.

Subspecies also indicate recognisable noda in the variation within the range of a species, although the challenge of numbers appears to be less (Table 7.2). *Cerastium fontanum* subsp. *scoticum* is restricted to a few sites in Angus and might contain as much unique genetic variation as the endemic full species Shetland mouse-ear (*C. nigrescens*). Like many other Scottish endemic species, *C. nigrescens* was classified as a subspecies in the past.

Hybrids resulting from crosses between species are also examples of visible and novel variation, but the importance of hybrids is confused because human agency has increased the rate of hybridisation by bringing species together. Even without any impact from human management of the environment, infertile hybrids are probably doomed to extinction when their habitat faces natural changes. Fertile hybrids, by contrast, may have a far greater potential for survival and may be important for the maintenance of an element of biodiversity. For instance, at sites such as Loch Leven, the spearwort *Ranunculus* × *levenensis* could be maintaining genetic material from extinct British populations of its parent, creeping spearwort (*R. reptans*).

We cannot be sure of the unique contribution of any taxon to total biodiversity, because there is inevitably an element of judgement in the selection of taxonomic level. It is essential to bring an open and informed mind to the problem of assessing the significance of any threatened population of plants to the maintenance of biodiversity. In particular, while we need to determine priorities for protection, we should not automatically restrict protection to the species level and we should regularly reassess the vulnerability and value of all Scottish plant taxa.

7.1.3   International importance

Scotland's total diversity of higher plants is low, particularly when compared with that of tropical areas (Groombridge, 1992). The La Selva Forest Reserve in Costa Rica alone is known to have about 1,500 plant species (Myers, 1988), about as many as the Scottish flora. However, Scotland's diversity is comparable to that of countries of similar size, especially those in northern Europe. Areas closer to the equator and centres of endemism show much greater diversity (Groombridge, 1992), but Scotland is apparently more diverse for its area than some more southerly countries, such as those in north-central Africa.

Regardless of how British diversity compares, like all signatories to the Bio-diversity Convention, Britain now has international obligations to conserve its biodiversity and, when exploiting it, to do so in a sustainable manner. However, biodiversity is not a matter of maximising, or even maintaining, the number of species. The UK Biodiversity Action Plan (Anon., 1994) indicates objectives for Britain:

> To conserve and where practical enhance the overall populations and natural ranges of native species ... international and threatened species ... species characteristic of local areas ... declining species ... the biodiversity of natural and semi natural habitats where this has diminished. To contribute to the conservation of biodiversity on a European scale.

Internationally, Scotland is particularly significant for: (a) its endemic species, which by definition occur nowhere else; (b) populations of those species that are threatened elsewhere in their range; and (c) the abundance of those species that are largely restricted to the Atlantic seaboard of western Europe.

7.1.4   Endemics

Endemic species, that is those restricted to a limited area or 'geo-political unit', may be of great antiquity and some are the last relics of once widespread groups. Northwest Europe does not have a high level of endemism but Scotland has 9 endemic, sexually reproducing plant species that are found nowhere else and also three endemic plants that are found elsewhere in the UK. Microspecies show an even higher degree of endemism, as expected with their apomictic breeding system. For instance, there are at least 160 species of *Hieracium* in Scotland of which more than 100 are unique to that country.

The best-known Scottish endemic, the Scottish primrose (*Primula scotica*), has declined on Orkney (Berry, 1985) where agricultural improvement appears to have been more intense than elsewhere in its range. However, it is known from thirty-one 10 km × 10 km squares (Bullard, 1994) and on the mainland appears to be little threatened and may be increasing (Cowie *et al.*, 1996), perhaps owing to recent mild winters (Bullard *et al.*, 1987). Its long-term future in the face of climate change could be less certain (Sydes, 1993). Seeds of *Primula scotica* dating from the early Quaternary have been found in England (Dovaston, 1954). The history of most British endemic species is not fully known but they are not likely to be of great antiquity.

A sample survey of mountain scurvygrass (*Cochlearia micacea*), funded by Scottish Natural Heritage (SNH), demonstrated that this species is relatively widespread in its mountain habitat, where it appears to be thriving under present land use (Rich and Dalby, 1996). The lack of previous information on *C. micacea* reflects the fact that British endemic species have not attracted great interest in the past. Even now, most are only confidently named by a few specialist botanists. *C. micacea* is only reliably identified when the seed capsule is present. Similarly, Scottish small-reed (*Calamagrostis scotica*) is distinguishable from its circumpolar congener, narrow small-reed (*C. stricta*), only by small differences in its flower. This has led in the past to understandable confusion over the distribution of this species in Scotland. It is at present only known from one site despite a resurvey of its previously recorded locations sponsored by SNH.

Such difficulties of identification lead to practical difficulties in the protection of endemic plants. At present, most Scottish endemic species need to be referred to one authority for final identification. Perhaps this is also the cause of the apparent taxonomic fragility of some endemic species. Many have changed their taxonomic status between authors; they should be protected with the proviso that, although these species are unique to Britain, other taxa might contribute more unique variation to British biological diversity.

Welsh groundsel (*Senecio cambrensis*) and another British endemic, Young's helleborine (*Epipactis youngiana*), are thought to have evolved recently. It is an exciting development that we have apparently recognised speciation occurring within the British flora. There is evidence that the Scottish *S. cambrensis,* only discovered in 1982, was the result of an origin separate from that of the plants found in Wales. Although rare, *E. youngiana* has been found on two new sites in Scotland in the past few years. However, this species has revealed little evidence of a unique genetic contribution to biodiversity.

The presence of these two species in abandoned anthropogenic habitats reflects their origin, which results from the creation of new habitats bringing together species that formerly would not have met. Redevelopment threatens the sites of both; *S. cambrensis* has not been relocated recently despite a new resurvey of its urban habitat in Leith, sponsored by SNH. Protecting these species of 'derelict' ground may be difficult, but neglected artificial habitats such as these are an important element of the local biodiversity of many people in urban Scotland.

Recovery effort may be required for other endemic species. Surveys of an endemic eyebright (*Euphrasia rotundifolia*) largely failed to find the species in 1995 or 1996. Past records were relatively widespread along the north coast. Increased interest in British endemics resulting from the UK Biodiversity Action Plan may lead to its rediscovery, perhaps at new sites.

7.1.5   *Species threatened with extinction throughout their world range*

The plant conservation strategy of the Joint Nature Conservation Committee (Palmer, 1994) recognises two non-endemic species, Norwegian mugwort (*Artemisia norvegica*) and Killarney fern (*Trichomanes speciosum*), which occur in Scotland and are threatened with extinction throughout their world range. *T. speciosum* is found on Arran, where its sporophyte populations are very small and show little evidence of regeneration. One of Britain's largest populations occurs in Argyll. It was collected to extinction at one other site in Scotland in the 19th century. *A.*

norvegica is found in northwest Scotland in only three locations in two mountain areas. Only one of its populations is large and evidently thriving. It is not threatened with habitat loss in its remote locations but it is possible that it is affected by habitat changes resulting from increased densities of red deer or from increasing atmospheric pollution (see below). A third species, Highland cudweed (*Gnaphalium norvegicum*), has recently been suggested to be threatened world-wide. The international status of *A. norvegica* and *G. norvegicum* is not clear; they may not be threatened with extinction in Scandinavia (H. J. B. Birks, pers. comm.).

The cause of such confusion is the difficulty of collating disparate information from a large number of countries. Even the UK may be inadvertently guilty of presenting inadequate information. Pillwort (*Pilularia globulifera*), endemic to western Europe, is regarded as threatened or extinct in most countries except Britain and France (Smith, 1988). However, in Britain the species occurs in more than fifteen 10 km × 10 km grid squares and therefore has attracted no detailed re-survey before scarce plants (those recorded recently in 16–100 British 10 km squares) were reviewed (Stewart *et al.*, 1994). Although it has clearly invaded new sites in the past 100 years, the species has been lost from many more and appears to be rapidly declining. It is likely that collation of comparable data may remove two of these scarce species from the internationally threatened category but is likely to replace them with many more.

Two Scottish species, yellow marsh saxifrage (*Saxifraga hirculus*) and slender naiad (*Najas flexilis*), are protected by European legislation because they are declining throughout the European Community. Both are present in other parts of the world, where they are not necessarily under any threat. The majority of species found in Scotland are found widely across Eurasia and many also occur in North America. A substantial minority are confined to Europe and it is among these that we are most likely to discover more threatened species such as *Pilularia globulifera*.

Species that are restricted to the Atlantic fringe of Europe are well represented in Scotland. The vascular plants in this category do not appear to be a prime conservation concern because they are relatively widespread in Scotland, but their Scottish populations may be very significant internationally precisely because of their abundance. Hay-scented fern (*Dryopteris aemula*), for instance, may have more populations in Britain than elsewhere (A. Jones, pers. comm.). However, such species have attracted no detailed work here because of their relative abundance and we can only speculate about the influences on, and trends within, their populations.

One of the important features of British vegetation is that a few species largely restricted to the west of Europe, such as bluebells (*Hyacinthoides non-scripta*), heath rush (*Juncus squarrosus*) and western gorse (*Ulex gallii*), achieve unusual dominance in British vegetation. Heather (*Calluna vulgaris*) also dominates heathland at low altitudes only in the western fringe of northwest Europe. The known decline in the area dominated by this species indicates the changes that can occur in such species under pressure of human management.

7.1.6 *Importance of Scottish vascular plant biodiversity within the UK*

Within the British Isles, Scotland is particularly important for the range and abundance of its montane species (Ratcliffe, 1977). Scotland's mountain habitats are extensive. Although affected by forestry, acidification, and increasingly intense grazing, they provide a contrast with lowland areas where many taxa are restricted to pockets of suitable semi-natural habitat separated by large areas of more heavily modified vegetation. Even among those montane species that also occur in England or Wales, the bulk of the UK population is usually in Scotland. Alpine catchfly (*Lychnis alpina*), for instance, is found in only one population in Scotland but 98% of the 10,000 plants estimated to occur in Britain are in Scotland.

There are 76 species of vascular plant that are found only in Scotland within the UK; the mountain habitat grouping is the largest category within these species (Table 7.3). The importance of the montane habitat for Scotland's most restricted plants means that any analysis of the distribution of populations of the rarest plants in Scotland picks out the mountains of the southern Highlands as biodiversity 'hotspots'. This particularly applies to hills with the greatest area of base-releasing rocks, such as the Breadalbane hills and Caenlochan. However, the Cairngorms, although predominantly acidic, also emerge as an area with many populations of restricted plants. The diversity of the entire vascular plant flora does not necessarily follow this same trend (Usher, this volume).

The woodland habitat is notable, as rare woodland species in Scotland are associated particularly with the relict pine woods. This habitat has been greatly reduced and these species have large recorded declines, indicating a small group of species for which we must have concern and for which Scotland has particular responsibility within the UK.

Table 7.3 Habitat associations of plant species restricted to Scotland within the UK (derived from Dring, 1994).

Habitat	No. of species
Mountains	45
Coast	9
Marsh	5
Rocks	4
Woods	4
Bogs and flushes	4
Aquatic	3
Open	2
Total	**76**

7.2 Trends in Scotland's vascular plant biodiversity

7.2.1 Introduced species

An estimate obtained from a sample of about 10% of Scottish species suggests that about 40% of Scottish plant species are now introduced (Table 7.2). As a result, the total diversity of vascular plant species in Scotland can never have been higher! Introduced species can be a threat to native biodiversity. Some may be existing in vacant niches in habitats that have been created by human modification of native habitats. However, others threaten the native diversity of Scottish habitats to a remarkable degree and with little hope that we can control them.

Rhododendron (*Rhododendron ponticum*) shades out most other plants where it dominates, particularly in woods in western Scotland. Local control can be effective but is expensive. Giant hogweed (*Heracleum mantegazzianum*), which is still spreading rapidly, must have an impact on riverside habitats, where it forms a closed canopy. There is no evidence yet that native animals are adapting to these well-established species and thus limiting their impact. The negative effect of other species is less certain. New Zealand willow-herb (*Epilobium brunnescens*), introduced from New Zealand and first seen in Britain in 1908, is now widespread in Scotland and has reached the most remote mountain locations. It rarely dominates its open stony habitats to a degree where competition with native species is evident. Few-flowered leek (*Allium paradoxum*) is spreading rapidly in shaded habitats in Southern Scotland. It forms a dense carpet very early in the season; although the species is small, it is conceivable that it may shade out seedlings of native vernal plants. Its impact on the vernal ground flora could become significant, especially in the Borders, where native woodland is very restricted.

Introduced species are an element of change to Scottish biodiversity that is almost certainly largely beyond our ability to remedy. Although intensive chemical control of the most aggressive invasive species is practised locally on some important sites, e.g. National Nature Reserves, only biological control is economically feasible nationwide: introducing the invasive plant's natural predators or pathogens. Large sums have been spent by local authorities on attempts at chemical control of some species with little evidence of a significant reduction in infestation. Biological control would have to be restricted to species that were clearly having a deleterious impact since attempting to control all introduced plants in this way could well lead to as many problems as it cured. In addition we may not be able to attempt to control some species. The native hogweed (*H. sphondylium*) could be vulnerable to predators of *H. mantegazzianum,* for instance, and extensive use of rhododendrons in gardens is likely to prevent the release of control agents for *R. ponticum.*

7.2.2 Extinctions

Two bogland species found in only single sites, alpine butterwort (*Pinguicula alpina*) and cotton deergrass (*Trichophorum alpinum*), were lost from Scotland in the 19th century and so became extinct in the UK, either because of habitat destruction or

Table 7.4   Extinctions of plant species recorded in Scotland prior to 1930 (derived from Dring, 1994).

Category	No. of species
Extinct in UK before 1930	2
Extinct in UK 1930 –1960	3
Extinct in UK since 1960	1
Extinct in Scotland but <100 UK 10 km squares since 1930 or 'scarce'	18
Extinct in Scotland but >100 UK 10 km squares since 1930	14

through excessive collecting by early botanists. At least one other species, Arctic bramble (*Rubus arcticus*), has become extinct this century, although it has been suggested that records might have been the result of temporary introductions of seed by migrating birds (Raven and Walters, 1956). These examples emphasise the vulnerability of species with few populations. Some rare taxa of very restricted distribution are probably still liable to extinction despite our mechanisms for protection. Far more are likely to continue to lose populations.

The large number of extinctions within Scotland of plants that still occur in other parts of the UK (Table 7.4) at first sight appear of little interest. A few of these are errors. For example, rigid buckler-fern (*Dryopteris submontana*) probably never occurred in Scotland despite earlier records (Jermy, 1994). Others are likely to be rediscovered. Brown bog-rush (*Schoenus ferrugineus*) was thought to be extinct in 1960 after the destruction of its known population, but was later found in a number of unexpected locations (Smith, 1980). However, the substantial remainder are significant in indicating the scale of local attrition of populations, made evident in these examples by causing their loss from Scotland entirely. At least one of the species, lamb's succory (*Arnoseris minima*), is now believed to be extinct throughout the UK, indicating that such trends, if unchecked, can lead ultimately to national extinction.

Although not all 'scarce' species are threatened with decline (Stewart *et al*, 1994) a disproportionate number of them are represented in this list of species extinct in Scotland. Local extinction is evidently more likely among scarce plants.

In Britain at least, perhaps because of its protection measures, *extinction* of vascular plant species is not the primary cause for concern in the maintenance of biodiversity but, as the United Kingdom Biodiversity Action Plan emphasises, the central issue is to 'conserve and where practical enhance the overall populations and natural ranges of native species' (Anon., 1994).

7.2.3   Losses among rare species

Scotland has approximately 1,500 populations of rare plants (endemic species and those listed in the Red Data Book). As part of the necessary background work for the five-yearly review of the Schedule 8 of the Wildlife and Countryside Act (Fleming, this volume), SNH organised visits to all known populations of these plants to confirm their existence, estimate population sizes and assess potential recruitment. Surveyors not only revisited the recorded locations of populations but were also asked to search the vicinity for new populations.

Table 7.5 reveals that about 20% of previously recorded populations could not be relocated. How significant are these apparent losses? There has not previously been such a comprehensive check of Scotland's rare plant populations and so the period over which the losses occurred is not clearly defined. Some populations may yet be refound, although repeat searches were carried out by independent surveyors as part of the process. However the crucial factor may lie in natural dynamics. Ecological theory predicts that populations are liable to become extinct because of normal changes to their habitat and also by chance effects alone. If we take into account the new populations discovered, the overall net loss of 2% of populations is scarcely significant. We need only be concerned for Scotland's rare species as a whole if this rate of loss is repeated in subsequent remonitoring. However, a few species lost far more populations than they gained and it is these species that need to be re-examined more urgently (Table 7.6).

It may be tempting to assume that very rare species are moribund relics at the end of their evolutionary life. However, there is little evidence of this. Alpine catchfly (*Lychnis alpina*) and string sedge (*Carex chordorrhiza*), for instance, are both found in only two locations in the UK but have large and dynamic populations in Scotland. Blue heath (*Phyllodoce caerulea*) and oblong woodsia (*Woodsia ilvensis*) have few and apparently declining UK populations but are clearly thriving in Scandinavia. Our evidence that there is a relatively high turnover of populations of rare plants within Scotland implies that these species are capable of establishing new populations to replace those that become extinct, as long as suitable habitat is maintained and the rate of population loss does not exceed establishment of new populations.

Even in seeking to maintain biodiversity we should not be concerned with those populations that are declining because of natural forces but only with those

Table 7.5   Preliminary results of SNH's monitoring of rare and endemic plant species.

Populations	Number	Percentage
Found before 1990	1,479	100
Confirmed 1990–1996	1,157	78.2
Not relocated 1990–1996; presumed extinct	217	32.4
Found for first time 1990–1996	281	19.0
New total	1,456	98.4
Net loss	23	1.6

Table 7.6   Plant species found from monitoring to have suffered the greatest losses of populations.

Species	Populations found before 1990	Not relocated 1990 –1996	Not known before 1990
Irish lady's tresses *Spiranthes romanzoffiana*	69	58 (84%)	6
Iceland purslane *Koenigia islandica*	30	24 (80%)	18
Wavy meadow-grass *Poa flexuosa*	15	10 (66%)	1
Tufted saxifrage *Saxifraga cespitosa*	16	6 (37%)	0
Highland saxifrage *Saxifraga rivularis*	50	18 (36%)	6

that are being adversely affected by human activity. Threats to populations from our actions can be considered in medical terminology as acute or chronic. Acute changes are relatively sudden impacts, which lead to the destruction of the habitat and include forestry, agricultural improvement, quarrying and building. These have been the main conservation concern in Britain in the past. Chronic changes allow the habitat to persist but lead to its slow modification, owing to changing management. Chronic change may include alterations to grazing intensity or season in mountain pastures and lowland grasslands (see, for example, Sydes and Miller, 1998), changing intensity of burning in heathland (Hestor and Sydes, 1992), use of herbicides and seed cleaning in arable fields (Wilson, 1992), and increased mechanisation in forestry. It is possible that chronic trends are now the greatest threat to wild plant populations, particularly for rare species, which mainly occur on Sites of Special Scientific Interest (75% of populations) where acute change is likely to be resisted.

Populations need to be monitored more closely to detect the effects of chronic impacts. Monitoring the number of plants alone may, given sufficient time, demonstrate current trends but further research then has to be instigated to discover the reason for any decline. Assessing mortality and regeneration on a range of populations of a species and relating that to management factors may indicate not only the effects of changing management but also safe limits. Application of this technique to *Schoenus ferrugineus* demonstrated that two populations (on one farm) had higher mortality and lower regeneration than the other British populations (Cowie and Sydes, 1995). This impact coincided with much higher grazing intensity on these two populations and indicated an upper limit of grazing above which the survival of this species is compromised.

7.2.4   *Losses among more common species*

The data collated by the Biological Records Centre (Dring, 1994) suggest that the greatest loss of biodiversity over the past century has occurred among more common taxa. Although rare species as a whole show a larger mean percentage loss than more common species, the total loss of range, measured as 10 km squares, is greatest among species that occur in more than 100 such squares. These are therefore considered too common to receive protection by specific legal mechanisms or even to be classified as threatened or scarce, so the greatest losses may be occurring in species in which hitherto we have taken little direct interest.

The greater loss of range in these species is to some extent a product of their greater initial frequency. However, the commonest species in Britain do not show these large declines, so it is not simply that species with greatest range have more to lose. Stemming these losses will require a new approach that is likely to depend on partnership with land users and organisations outside traditional conservation circles.

Plants of aquatic habitats have been lost from the greatest number of 10 km squares (Table 7.7). Aquatic plants are notoriously difficult to survey (Preston, 1995) but it is hard to see that our abilities are likely to have decreased with time. Conversely, the impacts of human management (including eutrophication, from agricultural run-off and sewage, and acidification, from combustion and forestry) are known to have increased.

It is possible that some mountain species have also been affected by acid precipitation although increasing numbers of sheep and red deer may have had a greater impact. Acute change to forestry and by agricultural improvement has been extensive. Any form of land management benefits some species (which become 'weeds') while others must weather the change. 'Blanket' afforestation with non-native conifers is one of the more radical changes and for many years was applied with few concessions to native biodiversity. Despite that most plants survive the change within gaps in the forest (Hill 1978), although clearly many have greatly reduced populations and are more vulnerable to local extinction in the longer-term. Designing land use to ensure that plant populations left within

Table 7.7 Habitats of species showing loss of more than thirty-five 10 km squares in Scotland.

Habitat	No. of species
Aquatic	24
Rough grazing	6
Coast	5
Mountains	4
Woods	4
Bogs	3
Marsh	2
Flushes	2
Arable	1
Total	51

its interstices have long-term viability would provide a considerable contribution to biodiversity because of the overwhelming and increasing predominance of areas managed by intensive methods.

Juniper (*Juniperus communis*) has, apparently, been lost from more grid squares in Scotland than has any other plant (Dring, 1994). Originally a woodland element, particularly characteristic of the Scottish pine woods, it survived the conversion to heathland or rough grassland on grazed areas. Nevertheless, the losses imply that 20th century management has been inimical to it. Finding ways to maintain species like this in modern land use is the challenge set by the Biodiversity Convention.

7.3 Conclusion

Scottish vascular plant biodiversity is impoverished in numbers of taxa compared with that of equatorial regions of the world, but it is comparable to that of other countries in northwest Europe. Nevertheless it has elements that are internationally important. There are a number of unresolved issues relating to our endemic plants: for instance, the taxonomic status of some, and the relative importance we should attach to neo-endemics and to apomictic plant taxa. However, the most significant issue is that we are probably underestimating the international significance of Scottish populations of a range of species, most notably those restricted to Europe. This is because data to compare Scottish populations with those in the rest of the world are lacking, either from Scotland or from the other countries. Very few, if any, individual botanists have the breadth of experience to be able to add an international dimension to our understanding when, even within Britain, we do not have an adequate grasp of species that are declining rather than simply rare.

Numbers of taxa alone do not add greatly to our understanding of biodiversity when it comes to attempting to maintain and enhance it. Introduced species are one of the most striking examples of this: they add greatly to total biodiversity but can adversely affect the diversity of native species. Maintaining native biodiversity, whether in Scotland, a part of Scotland or an individual site, depends on looking at the dynamics of individual populations, rather than overall biodiversity, in order to amend the pressures from modern human impacts. At first sight this appears to make an incomprehensibly large task impossible, but that is not so. The evidence we are collecting suggests that the majority of plants are able to tolerate a range of management conditions and are probably establishing new populations roughly as fast as they lose existing ones. Only a few rare species appear to be losing populations at a rate that threatens their survival. Conversely, some more common species have shown considerable declines. We have to identify the declining species of all types, identify the factors responsible, and find ways to enable these plants to survive with land management that is likely to continue to change, perhaps at an increasing rate.

The greatest problem in the conservation of many groups of organisms is to obtain information on a wide enough range of species to be sure that we can make an objective assessment of just which species and populations are at risk. For the rarer plants we have made great progress with our recent review of all Scottish locations. However, the historic data from the Biological Records Centre (BRC) reveal that there are, or were, many much more common species that were showing rapid rates of decline and which have had no recent attention. The new Atlas of vascular plants, to be produced by the Botanical Society of the British Isles with the support of the Department of the Environment, will provide the basic information that we need to bring up to date our understanding of what is happening to all other plants in the UK. We will then have to seek the partnerships in land use necessary to halt and reverse these declines.

References

Anon. 1994. *Biodiversity: the UK Action Plan*. Cm2428. London, HMSO.

Anon. 1995. *The Natural Heritage of Scotland: an Overview*. Perth, Scottish Natural Heritage.

Berry, R. J. 1985. *The Natural History of Orkney*. London, Collins.

Bullard, E. R. 1994. *Primula scotica*. In Stewart, A., Preston, C. D. and Pearman, D. A. (Eds.) *Scarce Plants in Britain*. Peterborough, Joint Nature Conservation Committee, 339.

Bullard, E. R., Shearer, H. D. H., Day, J. D. and Crawford, R. M. M. 1987. Survival and flowering of *Primula scotica* Hook. *Journal of Ecology*, **75**, 589–602.

Cowie, N. C., Harvey, M. L. and Legg, C. J. 1996. A re-survey of *Primula scotica* in Caithness and Sutherland in 1995. Unpublished report to Scottish Natural Heritage.

Cowie, N. R. and Sydes, C. 1995. Status, distribution, ecology and management of brown bog-rush *Schoenus ferrugineus*. Scottish Natural Heritage Review No. 43.

Dring, J. C. M. 1994. Support for the Biological Records Centre 1993/4: 1st annual report. JNCC Report No. 187. Peterborough, Joint Nature Conservation Committee.

Dovaston, H. F. 1954. *Primula scotica* Hook, a relict species in Scotland. *Notes from the Royal Botanic Garden Edinburgh*, **21**, 289–291.

Groombridge, B. (Ed.) 1992. *Global Biodiversity: Status of the Earth's Living Resources*. Cambridge, World Conservation Monitoring Centre.

Hestor, A. J. and Sydes, C. 1992. Changes in burning of Scottish heather moorland since the 1940s from aerial photographs. *Biological Conservation*, **60**, 25–30.

Hill, M. O. 1978. Vegetation changes resulting from afforestation of British uplands and bogs. Unpublished report. Bangor, Institute of Terrestrial Ecology.

Jermy, C. 1994. *Dryopteris submontana*. In Stewart, A., Preston, C. D. and Pearman, D. A. (Eds.) *Scarce Plants in Britain*. Peterborough, Joint Nature Conservation Committee, 141.

Myers, N. 1988. Threatened biotas: 'hot spots' in tropical forests. *The Environmentalist*, **10**, 243–256.

Palmer, M. 1994. *A UK Plant Conservation Strategy*. Peterborough, Joint Nature Conservation Committee.

Perring, F. H. and Farrell, L. 1983. *British Red Data Books: Vascular Plants*, 2nd Edn. Lincoln, Royal Society for Nature Conservation.

Perring, F. H. and Walters, S. M. 1962. *Atlas of the British Flora*. London, Thomas Nelson.

Preston, C. 1995. *Pondweeds*. London, Botanical Society of the British Isles.

Ratcliffe, D. A. (Ed.) 1977. *A Nature Conservation Review*. Cambridge, Cambridge University Press.

Raven, J. and Walters, S. M. 1956. *Mountain Flowers*. London, Collins.

Rich, T. C. G. and Dalby, D. H. 1994. The status and distribution of mountain scurvygrass (*Cochlearia micacea* Marshall) in Scotland. *Botanical Journal of Scotland*, **48**, 187–198.

Smith, A. 1988. European status of rare British vascular plants. CSD Contract Report No. 33. Peterborough, Nature Conservancy Council.

Smith, R. A. H. 1980. *Schoenus ferrugineus* L. – two native localities in Perthshire. *Watsonia*, **13**, 128–129.

Stace, C. 1991. *New Flora of the British Isles*. Cambridge, Cambridge University Press.

Stewart, A., Preston, C. D. and Pearman, D. A. (Eds.) 1994. *Scarce Plants in Britain.* Peterborough, Joint Nature Conservation Committee.

Sydes, C. (1993). Highland wildlife and landscape change in response to global warming. In *The Effects of Global Warming on the Highlands of Scotland.* Grantown-on-Spey, Highland Green Party.

Sydes, C. and Miller, G. R. (1988). Range management and nature conservation in the British Uplands. In Usher, M. B. and Thompson, D. B. A. (Eds) *Ecological Change in the Uplands.* London, Blackwell, 332–337.

Wilson, P. J. 1992. Britain's arable weeds. *British Wildlife*, **3**, 149–161.

8   INSECT BIODIVERSITY IN SCOTLAND

M. R. Young and G. E. Rotheray

Summary

1. Of an estimated 14,000 Scottish insect species, there are only three endemics, but about 1,300 are restricted to Scotland in Britain. Some groups, such as flies (Diptera), are well represented, whereas others, such as grasshoppers (Orthoptera), are scarce.

2. Poor information exists on the distribution, status and habitat preferences of most Scottish insects, so that priority must be given to gaining this information.

3. Nevertheless, some habitats, for example birch, aspen and pine woodlands, montane communities and botanically rich heath, are known to harbour many rare insects that are only found in Scotland and so deserve particular attention.

4. So little information exists on the past status of Scottish insects that it is difficult to substantiate the declines that we believe to have taken place. Consequently, regular monitoring must be started to allow trends to be detected quickly.

5. The habitat requirements of Scottish insects, which need to be known if realistic action plans are to be produced, are little studied and/or differ from the requirements known from southern populations.

6. The conservation of insects has been rather neglected; this imbalance should be redressed if their biodiversity is to be conserved in Scotland.

8.1   Setting the scene: counting insects in Scotland

No-one knows exactly how many insect species are found in Scotland, but Rotheray (1996) estimated 14,000, making them the most species-rich group of terrestrial animals here. He further estimated that 1,300 of these may be restricted to Scotland in Britain. However, few are endemic, although many species have distinct forms that only occur in Scotland.

Some insect groups have a high proportion of species occuring in Scotland, whereas others are poorly represented (Table 8.1). In very general terms species with aquatic larvae, such as Plecoptera (stoneflies) or Trichoptera (caddisflies), tend to be well represented whereas Odonata (dragonflies) are an exception. The Orthoptera (grasshoppers and crickets) is an example of an Order with more species in warm climates and with few Scottish species (Table 8.1).

The distribution of resident species in Scotland generally falls into one of a small number of patterns, as illustrated by Rotheray (1996). Most are at the northern edge of their range and are more common and widely distributed further south (for Lepidoptera this is 91% of the Scottish fauna). Many fewer are widespread in Scotland but are either confined here, or reach a southern limit in England (7% of Lepidoptera); amongst these are some that are restricted to high altitudes on the hills. Fewer still have a disjunct distribution and occur both in localised places in Scotland and elsewhere in the British Isles (2% of Lepidoptera).

Species reaching a southern limit in Scotland tend to have a northerly distribution elsewhere, often ranging into the Arctic and occasionally including Nearctic regions, for example the whirligig beetle (*Gyrinus opacus*). Some are also found in mountains further south and are truly arctic–alpine, such as the black mountain moth (*Psodos coracina*). The species reaching a northerly limit in Scotland are frequently found widely on the Continent, ranging down to the Mediterranean, for example the common field grasshopper (*Chorthippus brunneus*). However, many of these are western European and the predominantly Atlantic climate of Scotland

Table 8.1   The numbers of species of selected 'groups' of insects found in Scotland, expressed as a percentage of the total British list for each group.

Insect group	Number of species recorded in Scotland	Percentage of total British list for each group
Plecoptera (Stoneflies)	29	85
Odonata (Dragonflies)	21	55
Orthoptera (Grasshoppers)	7	24
Neuroptera (Lacewings) and allied orders	45	63
Trichoptera (Caddis-flies)	151	82
Lepidoptera (Moths)		
Nepticulidae	54	55
Zygaenidae	7	70
Coleophoridae	32	31
Pyralidae	64	42
Geometridae	189	63
Hymenoptera		
Formicidae (Ants)	17	37
Coleoptera (Beetles)		
Dytiscidae	79	71
Elateridae	35	48
Cryptophagidae	33	69
Diptera (Flies)		
Sepsidae	23	82
Overall total number	*c.* **14,000**	**60**

Table 8.2 UK endemic[a] insect species that are confined to Scotland.

Species	Distribution
Apion ryei	A weevil found in the Hebrides, Orkney and Shetland
Anapsis septentrionalis	A beetle found at Aviemore in 1876
Ceratophyllus fionnus	A flea from Manx shearwaters on Rum

[a]Three other UK endemic coleopterans and a plecopteran are found in Scotland and in other parts of theUK.

excludes most eastern species. There is a small number of Lusitanian species, which are restricted to the extreme western fringe of Scotland, Ireland and Iberia, such as the anthomyiid fly *Delia caledonica*, found on the Scottish hills, Puffin Island, Ireland and the Sierra Nevada (Rotheray and Horsfield, 1995).

It is likely that Scotland was too inhospitable for insect survival during the last full ice age (Young, 1997) and so the arrival of current residents must post-date 10,000–12,000 years before present. This is probably too brief a time to allow full speciation, even for insects with annual life-cycles, so accounting for the almost complete lack of endemic species (Table 8.2).

8.2 Which habitats host insect biodiversity in Scotland?

Action to conserve insect diversity in Scotland will have most success if it is focused on species for which Scotland is a key location. It is therefore important to identify these species or groups of species, as well as their locations and habitats. Ravenscroft (1995) compiled a provisional list of the habitats of Red Data Book (RDB) and Notable Scottish insects (as identified in Shirt, 1987) (Table 8.3).

Certain habitat types are predominantly Scottish (in a British context) but not all also host many rare insects. Blanket bog, exemplified by the Flow Country, is a priority habitat and includes rare plant and bird assemblages, but so far it has not proved particularly important for insects, except for one water beetle, *Oreodytes*

Table 8.3 Examples of habitat associations of Red Data Book and Notable species confined to Scotland.

Habitat type	Number of species using habitat
Woodland	
Broad-leaved	39
Native pine	47
Montane habitats	34
Riverine habitats	
Shingle and sand	10
Combined riverine habitats	23
Other habitats	33

alpinus, found only in sandy lochs on the edge of the peatlands (Foster and Spirit, 1986). Further survey is needed in such habitats.

Other habitats, such as Caledonian pine forests and aspen and birch woods, host a substantial fraction of scarce Scottish insects. The general location, nature and conservation status of the pine forests are described by Aldous (1995). They are very much reduced but at Rannoch and in Strathspey and Deeside, the extent of forest, together with the range of tree ages and stand structures, has allowed the typical insect fauna to survive. Many of these species use dead wood; others use fungi or other components of the forest habitat, whereas relatively few feed directly on the foliage of the healthy trees (Table 8.4). It is the structure of the forests and the full range of 'microresources' that are important. Ants provide an excellent example of this. They are the most obvious and characteristic insects of pine forests, and include the nationally rare *Formica exsecta*, as well as the commoner species *F. lugubris* and *F. aquilonia*, both of which are considered to be near threatened in global terms. They forage on pine trees and may use fallen pine needles as nest material, but it is clear that it is the correct forest structure and continuity of sheltered but sunny microhabitats that really governs their presence.

All but one of the RDB and Notable insect species breeding in association with aspen in Scotland depend on sap runs and fallen timber. Aspen woods are scattered remnants of their former size, sometimes surviving in ungrazed gorges or on steep hillsides and vulnerable to change and loss.

Change and loss may apply less to highland birch woods, which still cover a larger area. In fact birch is the tree species with the largest number of associated insect herbivores in Scotland, some of which are RDB species (Young, 1997). Some are exceedingly rare, such as the moth *Eutromula diana* (Choreutidae), which is restricted to Glen Affric in Britain, and many others are rare or localised. Some species depend upon very old trees. The Rannoch sprawler moth (*Brachionyx nubeculosa*) is found only among very old birches in Strathspey and Rannoch (Bretherton *et al.*, 1983). Others require regenerating trees. The Kentish glory

Table 8.4   The microhabitats of Red Data Book woodland insects in Scotland and of Lepidoptera from a woodland in Aberdeenshire.

Microhabitats	*Number of species*
RDB woodland insects	
Polyphagous on trees	5
Feeding on dead wood	28
Feeding on fungi	11
Feeding on leaf-litter	4
Woodland Lepidoptera	
Feeding on trees	100
Feeding on shrubs	44
Feeding at ground layer under canopy	42
Feeding at ground layer on woodland edge	146

moth (*Endromis versicolora*), which now occurs only in Scotland, will lay its eggs only on birches below about 3 m in height and so is restricted to areas of extensive regeneration, even if these are along ride-sides in older woodland (Barbour and Young, 1993).

The montane zone is particularly important for insects, with many species localised there. Despite the continuing lack of recording effort on the mountains there have been many RDB species found there, mainly beetles and flies. Young and Watt (in press) list 57 montane RDB species from Cairngorm alone, including such species as the beetles *Eudectus whitei* (Staphylinidae)and *Amara alpinus* (Carabidae), the flies *Alliopsis albipennis* and *Calliphora alpina* (Calliphoridae) and the wasp *Chrysura hirsuta* (Chrysididae).

Scotland contains a high proportion of the montane habitats in Britain, but many European countries have much more extensive Arctic or alpine areas. However, ericaceous heath is at least as well represented in Scotland as anywhere else in the world and this applies particularly to maritime heath. Even so there does not seem to be a characteristic insect assemblage restricted to maritime heath and those insects found there are usually also found inland. One heathland community that does have an interesting insect fauna is *Arctostaphylos* species-rich heath. This is restricted to slightly base-rich soils in relatively dry areas and its fauna includes *Coleophora arctostaphyli* (Coleophoridae), a moth with a case-bearing larva that has recently been found only in two or three sites in Strathspey and Deeside.

A number of species have a disjunct occurrence in the British Isles, with a southern range and a distinct northwestern range. An extreme example is the small moth *Periclepsis cinctana* (Tortricidae), which was known only on the South Downs of England, but in 1984 was discovered on areas of dark hornblende rock on Tiree (Harper and Young, 1986).

A final example of a Scottish habitat that is important for insects is river shingle. The fast-flowing upland rivers have the energy to transport stones and gravel extensively and so to produce semi-stable shingle-beds. Those of the River Tay, the River Spey and particularly the River Feshie are renowned for their extensive insect fauna. Flies and beetles predominate on shingle; at the Feshie Fan a total of 39 RDB and Notable species are found, including the cranefly *Nephrotoma aculeata* and the rove beetle *Stenus incanus*.

8.3   Scotland's insects: rare species and monitoring schemes

Table 8.5 lists those RDB insect species found in, and restricted to, Scotland, based on Ravenscroft (1995) and recent reviews. It is striking that Diptera are in the majority, with substantial numbers of Coleoptera and some Lepidoptera but very few of other orders. In general it seems that virtually unworked groups, such as Neuroptera, have very few listed RDB species; moderately worked orders, with recent reviews, such as Diptera, have many RDB listings; whereas relatively well worked groups, such as Lepidoptera, have an intermediate number.

The Lepidoptera have around 1,500 Scottish species, some of which are strays

Table 8.5   UK Red Data Book (RDB) insect species of six orders found in Scotland. RDB1, Endangered; RDB2, Vulnerable; RDB3, Rare.

Insect order	Numbers of RDB1 species	Total numbers of RDB1–3 species	Percentage of total number of Scottish RDB insect species
Diptera	28	207	59
Coleoptera	13	72	21
Lepidoptera	20	51	15
Hymenoptera	1	9	3
Odonata	0	2	<1
Hemiptera	0	0	0

or regular migrants. Only one species (a butterfly) has been listed as globally threatened but 53 are listed as RDB, or are proposed for listing. This RDB list is a mixture of very well-known species, such as the chequered skipper butterfly (*Carterocephalus palaemon*), which is now RDB4 (out of danger) in the UK and was formerly considered to be globally threatened, and species such as *Acleris abietana* (Tortricidae), which is an introduced species that feeds on exotic conifers and has been found in many areas. Regrettably, the allocation of insect species to RDB categories has failed to produce a consistent categorisation of species on which conservation action can always be reliably based.

It is necessary to consider the British context of any species when assessing its conservation status, and not to overvalue species that are large and attractive, like butterflies. Bland and Young (1996) recategorised scarce Scottish Lepidoptera (Table 8.6).

Only eight species fall into their Category X. These should be considered in any surveys but there are no rational conservation plans that can be prepared for

Table 8.6 Criteria for assigning priorities to Lepidoptera needing conservation action in Scotland, with the number of species in each category.

Category	Definition	No. of species
X	Apparently extinct or close to it	8
1	Mostly restricted to Scotland in UK	
1A	Very restricted distributions and so potentially in urgent need of protection	18
1B	In urgent need of research into biology or distribution; potentially in need of protection	34
1C	Apparently reasonably widespread but need for better distribution data	28
2	Present in only a few restricted areas in Britain, including Scotland	
2A	Very restricted British distribution, with colonies in Scotland and in urgent need of protection	8
2B	In urgent need of research into biology or distribution; potentially in need of protection	27
3	Scarce and edge of range in Scotland but more widespread elsewhere	22

them until they have been rediscovered. Since 1979 a further seven species have been moved from Category X to other categories, a cheering statistic.

Eighteen species are in Category 1A, if Scottish subspecies are included, and these should be considered, with the eight species in Category 2A, for immediate conservation plans. More worryingly, there are 61 species in Categories 1B and 2B in urgent need of research and for only two of these has any such research been undertaken so far (although some survey work has been carried out for a further two).

Bland and Young's criteria do not explicitly include 'decline'. This is not because it is not important and valid as an indicator of conservation need, but because there are very few species for which monitoring data exist to allow an estimate of whether a decline has occurred in Scotland. However, several species have been rediscovered (Table 8.7) or found to be more secure than thought, often following discovery of the best way to survey them. Bland (1993) found the larval habitat of the moth *Callisto coffeella* (Gracillariidae), previously known from only one location, and subsequently found it quite widely on the southern fringe of the Cairngorms.

Some have certainly declined; for example, the marsh fritillary butterfly (*Euphydryas aurinia*) is not now known in Aberdeenshire and Perthshire, nor the slender Scotch burnet (*Zygaena loti*) from the mainland on Morvern.

If monitoring can only be undertaken for a very limited number of species, then these must be chosen so as to produce the maximum return for the effort. For example, the large and distinctive fly *Hammerschmidtia ferruginea* (Syrphidae) is the easiest to find and monitor of the RDB species restricted to old aspen and its conservation will also benefit other species.

Table 8.7 Lepidoptera extinct in Scotland (Category X) or recently rediscovered and so removed from this category.

Species	Last seen	Rediscovered
Psychoides verhuella	1878	1987
Paraleucoptera sinuella	1954	No
Callisto coffeella	1984	1992
Plutella haasi	1954	No
Kessleria fasciapennella	1851	No
Roeslerstammia pronubella	1854	No
Ethmia pyrausta	1853	1996
Gnorimoschema streliciella	1920s	No
Eana argentana	1920	1985
Apotomis infida	1919	1979
Gypsonoma nitidulana	1908	No
Catoptria speculalis	1890	No
Stenoptilia islandicus	1954	1993
Pselnophorus heterodactylus	1920	1990
Isturge limbaria	1901	No

There is a case to be made for including critically rare species but not as part of a general recording scheme. The New Forest burnet moth (*Zygaena viciae*) is found at one site and its population numbers 15–20 adults in any one year. It is vital to continue working to conserve this species, but recording its numbers will contribute little to general information on the status of Scotland's Lepidoptera.

It is more useful to include species that will provide information on a range of different features and factors. For example, great concern is currently expressed about the likely effects of climatic warming. This may well cause a loss of species of the montane zone, or of the extreme north. It is difficult to carry out effective monitoring for montane species and Lepidoptera may be less suited for monitoring than species such as ground-dwelling Coleoptera and some Diptera, for which pitfall traps have proved effective (Rotheray and Horsfield, 1995). Nevertheless there are some candidate Lepidoptera, such as the mountain burnet moth (*Zygaena exulans*) and the black mountain moth (*Psodos coracina*), which are amenable to transect recording, and there is need for a pilot study to discover what can be achieved.

It is important to include insects from key habitats in monitoring schemes. For *Arctostaphylos* heath a suitable species is the netted mountain moth (*Semiothisa carbonaria*), found only on long-established and extensive heaths and easily disturbed by day. The status of regenerating birch in the highlands could be assessed by consideration of the Kentish glory moth (*Endromis versicolora*) and of juniper by the chestnut-coloured carpet moth (*Thera cognata*).

Habitats that are currently subject to changing management should also be included, so that there is early warning of changes in their fauna that may be caused by different management. Such changes could include reduced burning of heather moorland, changing grazing intensity on grasslands, extension of commercial forestry on moorland, or reduced red deer grazing in pine forest remnants. The Malloch Society (devoted to the study of Diptera in Scotland) are developing monitoring schemes for dead-wood insects in relation to changing woodland management.

There are two monitoring schemes that provide generalised information, because they cover common insect species, rather than rare ones. These deserve encouragement and extension.

First, the Rothamsted Insect Survey runs many light traps throughout Scotland, of standard design and on every night of the year; the larger Lepidoptera are then counted, providing an unparallelled data set (Woiwod and Harrington, 1994) that permits observation of long-term changes in the status and distribution of a wide range of species and so an assessment of general changes in the countryside. Scotland has proportionately fewer traps than lowland England and extra traps should be established, with a conscious effort to include wider geographic cover and representation of all general land-use types. Furthermore, the taxonomic scope of the identification should be widened. Only orders that are attracted by light can be included but these include microlepidoptera, some families of flies and caddis-flies (Trichoptera).

Second, the Butterfly Monitoring Scheme runs a series of transects, which are walked each week throughout the butterfly season (Pollard, 1977). The data thus obtained allow clear identification of year-to-year changes in status, as well as changes in range, which have been related to climate and land use.

8.4   Changes in Scotland's insect fauna

Despite the problems associated with lack of standardised recording, there have been many observable changes in Scotland's insect fauna. For Lepidoptera, losses and declines have been balanced by rediscoveries and by range extensions (except in the heavily populated lowlands).

An apparent decline has affected the rare hoverfly *Blera fallax*. From about 1900 to 1940 it was recorded regularly in Strathspey but, despite focused searches, there are only two post–1970 records and no definite signs of breeding, except a possible emerged puparium in the wet decaying roots of a dead pine, its specialised breeding microhabitat. Another RDB species, *Callicera rufa* (Syrphidae), had been found in only three localities between 1905 and 1987 but in this case a focused search showed that it is in fact quite widespread in northern Scotland and that it was its secretive breeding habits that prevented it from being recorded more easily (Rotheray and MacGowan, 1990).

Habitat change was implicated in the decline of the marsh fritillary butterfly in eastern Scotland and in that of two closely related burnet moths. The New Forest burnet moth (*Zygaena viciae*) has always been restricted to one site in Scotland, where it was discovered in 1963 (Best, 1963). After some years of apparent stability it declined dramatically in the 1980s and became limited to one or more cliff ledges. It seems quite certain that this was caused by an increase in the sheep stocking rate on the site, which became closely grazed, so that the moth could only survive on the ungrazed ledges. The slender Scotch burnet moth (*Z. loti*) now occurs commonly only in one part of the island of Mull but it is thought that its decline was caused by a reduction in grazing, for this species is found only where there is short turf, with bare ground and unstable slopes. Exactly opposite habitat changes have been responsible for the decline of two very similar species!

Some species are well documented colonists. The moth *Caloptilia rufipennella* (Gracillariidae) has larvae that make very conspicuous cones on sycamore leaves; these would not have been overlooked. It was first recorded in Britain in 1970 and has since spread throughout mainland Scotland and onto the nearer, larger islands (Agassiz, 1996). Two other examples are the congeneric *Blastobasis decolorella* and *B. lignea* (Blastobasidae), both of which feed on decaying leaves and other debris and have colonised Britain this century. They are both now found to the northern shores of the mainland (Agassiz, 1996). The speckled wood butterfly (*Pararge aegeria*) has recently spread eastwards along the Moray coast (Barbour, 1986).

8.5   Priorities for conservation of insects in Scotland

In our view the following are the priorities for assessing insect biodiversity and conservation in Scotland. Their interaction is set out in Figure 8.1.

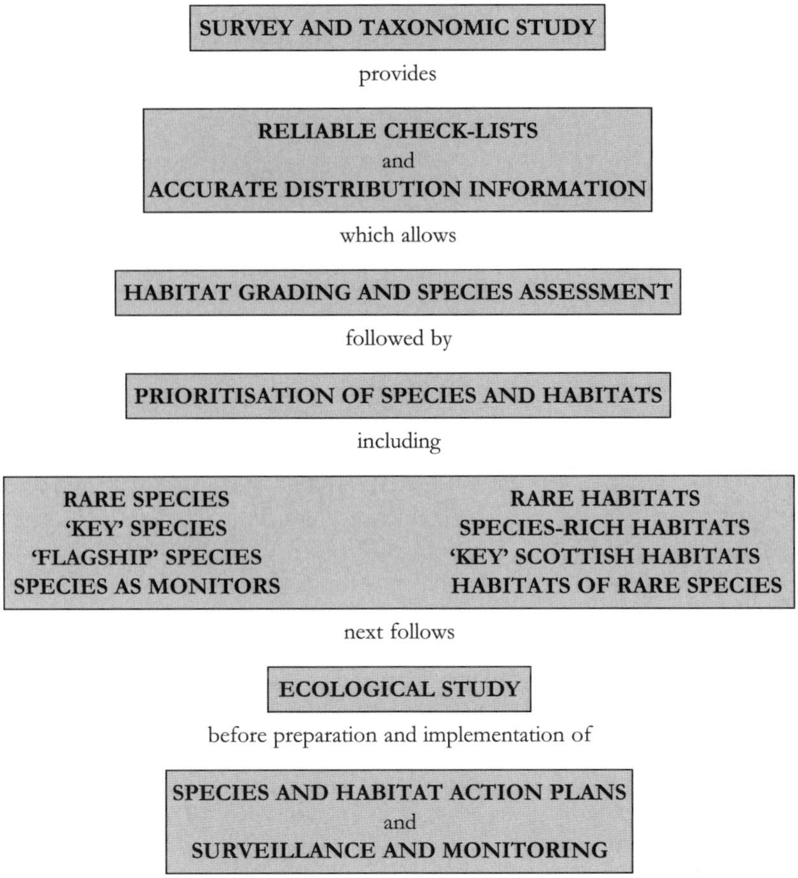

```
┌──────────────────────────────────────┐
│    SURVEY AND TAXONOMIC STUDY         │
└──────────────────────────────────────┘
                 provides
┌──────────────────────────────────────┐
│         RELIABLE CHECK-LISTS          │
│                 and                   │
│   ACCURATE DISTRIBUTION INFORMATION   │
└──────────────────────────────────────┘
               which allows
┌──────────────────────────────────────┐
│ HABITAT GRADING AND SPECIES ASSESSMENT│
└──────────────────────────────────────┘
                followed by
┌──────────────────────────────────────┐
│  PRIORITISATION OF SPECIES AND HABITATS│
└──────────────────────────────────────┘
                including
┌────────────────────────┬─────────────────────────┐
│   RARE SPECIES         │    RARE HABITATS        │
│   'KEY' SPECIES        │ SPECIES-RICH HABITATS   │
│  'FLAGSHIP' SPECIES    │ 'KEY' SCOTTISH HABITATS │
│ SPECIES AS MONITORS    │ HABITATS OF RARE SPECIES│
└────────────────────────┴─────────────────────────┘
               next follows
┌──────────────────────────────────────┐
│          ECOLOGICAL STUDY             │
└──────────────────────────────────────┘
     before preparation and implementation of
┌──────────────────────────────────────┐
│  SPECIES AND HABITAT ACTION PLANS     │
│                 and                   │
│      SURVEILLANCE AND MONITORING      │
└──────────────────────────────────────┘
```

Figure 8.1   A scheme for assessing insect biodiversity and conservation needs in Scotland.

8.5.1   Survey

More so than elsewhere in Britain, a major effort to survey the many under-recorded species, areas and habitats is needed in Scotland, including the key species and habitats identified as priorities above, and extending the taxonomic coverage. Shaw (1996) makes an eloquent plea for special attention for parasitic species and there are many other priority groups. This knowledge is essential as the foundation on which to build initiatives. Much more reliable 'Red Data' lists are needed. In relation to southern Britain, Scotland is very poorly known and so survey assumes a much greater priority here.

8.5.2   Assessing habitats for insects

Foster and Eyre (1992) have demonstrated that it is possible to develop realistic assessment techniques for sites, based on insect records, when the sites are clearly delimited and the status of the species is well known. They have assessed wetland sites by using water-beetle occurrence. Once habitats are graded then it is possible to make decisions on where the greatest biodiversity lies and which need immediate conservation action. What has not yet been tested is whether Foster's key sites for beetles are also key sites for other insect groups. Recently, Prendergast *et al.* (1993) have shown that in a British context the best 10 km squares for one group of organisms are not necessarily the best for other groups. This greatly complicates general area-based conservation schemes. Priority for such assessment should be the important Scottish habitats noted in Table 8.3.

8.5.3   *Prioritising species and habitats*
On the basis of the reliable check-lists of species and the assessment of habitats, priorities for immediate action can be identified. For species the choice will include rare species, declining species, 'key' species (those that have more influence on ecological functioning than expected from their abundance), 'flagship' species and those that act as effective monitors. For habitats the choice will include habitats of priority species, species-rich habitats, characteristically Scottish habitats and declining habitats. These are indicated in Figure 8.1.

8.5.4   *Studying ecological requirements*
Once a species or habitat has been identified as in need of biodiversity assessment and conservation, then action can follow. However, for all but a handful of species we know nothing of their ecological requirements. For many species we do not even know the larval food plants, let alone their more detailed needs. Consequently, basic ecological studies are urgently needed. Comparatively little attention has been paid to the ecological requirements of Scottish species but there is already evidence that there may be differences between these in Scotland and in England, so that southern research may not be applicable. The pearl-bordered fritillary (*Boloria euphrosyne*) has suffered a serious decline in England, consequent on loss of its coppice woodland habitat, but it is stable in Scotland, where it lives in open, sheltered grassland on woodland edges (Thomas and Lewington, 1991).

8.5.5   *Surveillance and monitoring schemes*
As well as survey work there is a clear necessity to instigate more effective surveillance and monitoring schemes to allow early warning of changes and of threats to either species or habitats. This is certainly possible for readily identifiable species or groups, especially where there is a good pool of volunteers to undertake recording. However, it should also be extended to less well-known groups as soon as possible. This is a special Scottish priority because of the serious lack of existing baseline information.

8.5.6   *Insects are important*
Finally, we urge that insects, the most species-rich group of terrestrial animals in Scotland, be considered and treated as major players in biodiversity and conservation assessment. Some species require conservation now; others, although

relatively safe from extinction, are excellent indicators of change, because they respond more quickly to subtle environmental influences than do plants or vertebrates.

Acknowledgements

We are most grateful to many fellow entomologists for helpful discussion, especially Keith Bland, David Barbour, the members of the Malloch Society, and to the steering group for the Initiative for Scottish Insects. This last group aims to raise the profile of insects in Scotland and to encourage their conservation. We are also grateful to our referees, Professor Michael Usher and Dr Ian McLean, for their helpful comments.

References

Agassiz, D. J. L. 1996. Invasions of Lepidoptera into the British Isles. In Emmet, A. M. (Ed) *The Moths and Butterflies of the British Isles*. Vol. 3. Colchester, Harley Books, 9–36.

Aldous, J. (Ed.) 1995. *Our Pinewood Heritage*. Edinburgh, Royal Society for the Protection of Birds, Forestry Commission and Scottish Natural Heritage.

Barbour, D. A. 1986. Expansion of the range of the Speckled Wood butterfly *Pararge aegeria* L. in north-east Scotland. *Entomologist's Record and Journal of Variation*, **98**, 98–105.

Barbour, D. A. and Young, M. R. 1993. Ecology and conservation of the Kentish Glory moth (*Endromis versicolora* L.) in eastern Scotland. *Entomologist*, **112**, 25–33.

Best, F. C. 1963. *Zygaena viciae* Schiff. (*meliloti* Esp.)(Lep.) in Scotland. *Entomologist's Gazette*, **14**, 149.

Bland, K. P. 1993. The status of *Callisto coffeella* (Zetterstedt, [1839])(Lepidoptera: Gracillariidae) in Scotland. *Entomologist's Gazette*, **44**, 15–18.

Bland, K. P. and Young, M. R. 1996. Priorities for conserving Scottish moths. In Rotheray, G. E. and MacGowan, I. (Eds.) *Conserving Scottish Insects*. Edinburgh, Edinburgh Entomological Club, 27–36.

Bretherton, R. F., Goater, B. and Lorimer, R. I. 1983. Noctuidae: Cuculliinae to Hypeninae. In Heath, J. and Emmet, A. M. (Eds.) *The Moths and Butterflies of Great Britain and Ireland*. Vol. 9. London, Curwen Press, 36–413.

Foster, G. N. and Eyre, M. D. 1992. *Classification and ranking of water beetle communities*. UK Nature Conservation No. 1.

Foster, G. N. and Spirit, M. 1986. *Oreodytes alpinus* new to Britain. *Balfour-Browne Club Newsletter*, **36**, 1–2.

Harper, M. W. and Young, M. R. 1986. *Periclepsis cinctana* [D. and S.] and other Lepidoptera on Tiree in 1984. *Entomologist's Gazette*, **37**, 199–205.

Pollard, E. 1977. A method for assessing changes in the abundance of butterflies. *Biological Conservation*, **12**, 115–134.

Prendergast, J. R., Quinn, R. M., Lawton, J. H., Eversham, B. C. and Gibbons, D. W. 1993. Rare species, the coincidence of diversity hotspots and conservation strategies. *Nature*, **365**, 335–337.

Ravenscroft, N. O. M. 1995. *Priorities for Invertebrate Conservation in Scotland: a Review*. Scottish Natural Heritage Review No. 39.

Rotheray, G. E. 1996. Why conserve Scottish insects. In Rotheray, G. E. and MacGowan, I. (Eds.) *Conserving Scottish Insects*. Edinburgh, Edinburgh Entomological Club, 11–16.

Rotheray, G. E. and Horsfield, D. 1995. Insects of Scottish mountains. *British Wildlife*, **6**, 160–167.

Rotheray, G. E. and MacGowan, I. 1990. Re-evaluation of the status of *Callicera rufa* Schummel (Diptera: Syrphidae) in the British Isles. *The Entomologist*, **109**, 35–42.

Shaw, M. R. 1996. Hymenoptera in relation to insect conservation in Scotland. In Rotheray, G. E. and MacGowan, I. (Eds.) *Conserving Scottish Insects*. Edinburgh, Edinburgh Entomological Club, 55–64.

Shirt, D. B. (Ed.) 1987. *British Red Data Books*. Vol. 2, *Insects*. Peterborough, Nature Conservancy Council.

Thomas, J. A. and Lewington, R. 1991. *The Butterflies of Britain and Ireland*. London, Dorling Kindersley.

Woiwod, I. P. and Harrington, R. 1994. Flying in the face of change: the Rothamsted Insect Survey. In Leigh, R. A. and Johnston, A. E. (Eds.) *Long-term Experiments in Agricultural and Ecological Sciences.* London, CAB International, 212–223.

Young, M. R. 1997. *The Natural History of Moths.* London, Poyser.

Young, M. R. and Watt, K. In press. Insects of the Cairngorms. In Gimingham, C. H. (Ed.) *The Cairngorms: Ecology, Land Use and Conservation.* Chichester, Packard Publishing.

9 CURRENT STATUS OF VERTEBRATES AND TRENDS IN VERTEBRATE BIODIVERSITY IN SCOTLAND

P. A. Racey

Summary

1. Introductions of alien species are threatening the integrity of Scotland's vertebrate biodiversity. The negative effect of such introductions on communities of fish, on red deer, and on water voles should prompt affirmative action against further introductions, such as muntjac deer.

2. Scotland has some of the rarest and some of the commonest species of amphibians and reptiles in Britain. Amphibians are indicator species of the health of standing freshwater habitats and are likely to benefit from the creation of more ponds and improvement in water quality. The commoner reptiles are often found in unimproved and non-intensively managed grassland and heathland; Scotland is characterised by a paucity of such grassland.

3. Intensification of farming has resulted in dramatic declines in bird populations and has also resulted in a decline in some reptile species.

4. Large and growing populations of large-bodied vertebrates such as red deer, grey seals and pink-footed geese continue to raise questions of sustainability. Such populations may require active management for different reasons, from the restoration of woodland biodiversity in the case of deer, to sectoral economic interests in the case of geese.

5. Populations of some hitherto threatened vertebrate species such as pine martens and otters are increasing and expanding their range.

6. Species action plans have done much to raise awareness of species not previously included in Scotland's conservation agenda, such as marine turtles.

9.1 Introduction

The aim of this contribution is to review the current status of Scottish vertebrates and to draw attention to population decreases that may require active intervention to prevent further species extinctions. Conversely, populations of some species may be outstripping their resources to the detriment of overall diversity or sectoral economies. Particular attention will be paid to introductions of alien species, which are potentially the greatest threat to the integrity of Scotland's vertebrate biodiversity, because once established, they are generally irreversible. Constraints of space require that such a review is selective, and it will focus on those species for which change in numbers or status has recently been documented.

9.2 Fish

There have been more introductions among fish than among any other vertebrate group (Table 9.1). The most striking example of this is the ruffe (*Gymnocephalus cernuus*), a small percid fish introduced from England into Loch Lomond, west Scotland, as live bait and discovered there in 1982 (Maitland *et al.*, 1983). Since then numbers have increased dramatically (Figure 9.1). At one time, Loch Lomond contained an almost full representation of the original stenohaline fish community with 15 species, but in addition to ruffe this is now mixed with introductions of gudgeon (*Gobio gobio*) dace (*Leuciscus leuciscus*) and chubb (*L. cephalus*) with crucian carp (*Carassius carassius*) in one of the feeder rivers, the Endrick. The threat posed by these introductions is that they may outcompete the native species. Original pristine fish communities are under substantial threat and few remain, mainly in the oligotrophic rivers and lakes of northern and western Scotland, such as Lochs Eck, Hope, and Meallt, Langavat in Lewis, and the river Skealter (Maitland and Lyle, 1991). The status of several of the individual species comprising these communities is of concern. Apart from competition with introduced species, the threats to most of these species are the same: eutrophication, acidification and other forms of pollution, overfishing and the installation of barriers such as hydroelectric dams. However, unlike introductions, such threats are often reversible and as rivers become cleaner, native species return.

The powan (*Coregonus lavaretus*) has declined in several parts of Europe and is now known only from two sites in west central Scotland (Loch Lomond and Loch Eck), although translocations have been carried out to two new Scottish sites. The allis shad (*Alosa alosa*) has declined throughout its range, and there are now no breeding sites in the British Isles. Ripe fish have been reported in the Cree estuary in southwest Scotland but there is no proof of spawning at this site (Maitland and Lyle, 1992). The twaite shad (*A. fallax*) is now much less common throughout its range although there are still a few spawning sites in England and Wales. Recent work on the Cree has established the existence of possibly the only spawning population in Scotland. The vendace (*Coregonus albula*) is now extinct from the two Scottish sites from which it was previously known (see Maitland, this volume); the sturgeon (*Acipenser sturio*) is also facing extinction throughout its range, although

Table 9.1 Scottish freshwater fish and their occurrence since the last ice age. Source: P. S. Maitland, pers. comm.

Original colonisers: eurybaline species	Introductions			
	by 1790	by 1880	by 1970	by 1985
Sea lamprey *Petromyzon marinus*	Pike *Esox lucius*	Brook charr *Salvelinus fontinalis*	Rainbow trout *Oncorhynchus mykiss*	Ruffe *Gymnocephalus cernua*
River lamprey *Lampetra fluviatilis*	Roach *Rutilus rutilus*	Grayling *Thymallus thymallus*	Pink salmon *Oncorhynchus gorbuscha*	
Brook lamprey *Lampetra planeri*	Stone loach *Neomacheilus barbatulus*	Tench *Tinca tinca*	Common carp *Cyprinus carpio*	
Atlantic salmon *Salmo salar*	Perch *Perca fluviatilis*	Common bream *Abramis brama*	Goldfish *Carassius auratus*	
Trout *Salmo trutta*	Minnow *Phoxinus phoxinus*	Chub *Leuciscus cephalus*	Gudgeon *Gobio gobio*	
Arctic charr *Salvelinus alpinus*		Crucian carp *Carassius carassius*	Rudd *Scardinius erythrophthalmus*	
Powan *Coregonus lavaretus*			Orfe *Leuciscus idus*	
Vendace *Coregonus albula*			Dace *Leuciscus leuciscus*	
Eel *Anguilla anguilla*			Bullhead *Cottus gobio*	
3-Spined stickleback *Gasterosteus aculeatus*				
9-Spined stickleback *Pungitius pungitius*				
Sea bass *Dicentrarchus labrax*				
Common goby *Pomatoschistus microps*				
Thick-lipped mullet *Chelon labrosus*				
Thin-lipped mullet *Liza ramada*				
Golden mullet *Liza aurata*				
Flounder *Platichthys flesus*				

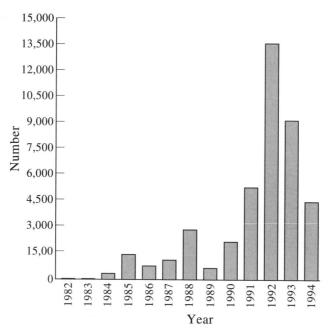

Figure 9.1   Numbers of ruffe collected from screens at Ross Priory, Loch Lomond, West Scotland 1982–
94. (P. S. Maitland, pers. comm.)

it was probably a vagrant in Scotland since there is no evidence that it ever spawned here. In contrast, the sparling or smelt (*Osmerus eperlanus*) was once a common species in the British Isles and occurred in most larger rivers from the Clyde and Tay in central Scotland southwards. Up to six tonnes were taken from the spawning run in the Cree some years ago (Hutchinson and Mills, 1987). However, over the last century it has disappeared from many rivers so that, of at least 16 known spawning sites in Scotland, only 3 remain. The river lamprey (*Lampetra fluviatilis*) in Loch Lomond is the only stock with a purely freshwater life cycle, which spawns in only one short stretch of the River Endrick.

Although there are about 200 individual populations of arctic charr (*Salvelinus alpinus*) in Scotland, this species has declined, particularly in southern Scotland where only one of at least six former stocks remain. It has, however, been successfully translocated to two sites in southern Scotland. Charr farming and the movement of alien stocks, some from Canada, threaten the genetic integrity of indigenous charr stocks, which, unlike those of most other species, have remained pristine since the last Ice Age and are a valuable scientific and economic resource in Scotland. Similar problems may result from the release or escape of farmed salmon (*Salmo salar*) (Ferguson *et al.*, 1985) or from inappropriate restocking of rivers (Nickson, 1997).

Elasmobranchs, because of their low reproductive rate, are potentially at greater risk than teleosts. The common skate (*Raja batus*) has almost disappeared from the Irish Sea. Concerns about the status of basking sharks (*Cetorhinus maximus*) have led to recent proposals for their inclusion in Schedule 5 of the Wildlife and Countryside Act and the only British-based fishery for this species has now ceased.

9.3 Amphibians

The six amphibians native to Scotland depend on natural and artificial ponds, lochs and lochans. Although the non-governmental organisation (NGO) Pond Action suggests that 70% of the UK's ponds have been lost, many new ones have been dug in Scotland in the past twenty years. The amphibian that is most naturally rare and is also threatened in Scotland is the natterjack toad (*Bufo calamita*) whose last four colonies on the Solway coast account for 500–1,000 of the estimated 20,000 natterjacks in the UK (Langton and Beckett, 1995). Only 78 spawn strings were counted in these colonies in 1996, a reduction on previous annual counts (T. E. S. Langton, pers. comm.). A major cause for concern so far as the commoner anurans is concerned is an apparently new frog disease recorded in garden ponds, the primary cause of which is probably a virus, giving rise to secondary infections of the bacterium *Aeromonas hydrophila* (T. E. S. Langton, pers. comm.).

The great crested newt (*Triturus cristatus*) is thought to have an estimated 1,000 adults remaining in the 50 or so Scottish populations. In the past 10 years, successive counts of great crested newts in ponds in southern Scotland suggest that a critical situation exists for this species. In contrast, palmate newts (*Triturus helveticus*) occur in many small bog pools in the Highlands.

9.4 Reptiles

The native turtle species include the leatherback turtle (*Dermochelys coriacea*) and the loggerhead turtle (*Caretta caretta*); the former moves regularly into Scottish waters to feed in autumn months when the water is warmest. Scottish waters are part of their northeast Atlantic range, although they must return to subtropical and tropical beaches to lay their eggs. The many occurrences of the leatherback turtle in Scottish waters over the past 20 years have been documented by Langton *et al.* (1996) and indicate that the species is much commoner in Scotland than had formerly been appreciated. However, throughout its range, it appears to be declining dramatically: it is now absent from the Mediterranean and may have been reduced to 100,000 breeding females. Many crashes in numbers have also been documented at breeding beaches such as those on the Pacific Coast of Mexico (Eckert and Eckert, 1996). The causes for decline include net and line drownings, and blockage of the alimentary canal with marine debris such as plastic bags.

The inclusion, in 1991, of the adder (*Vipera berus*) in Schedule 5 of the Wildlife and Countryside Act (1981) prompted an investigation into its distribution and past and present status in Scotland. It is widespread, mainly in areas with varied land use, although it is absent from much of the Central Valley, from the Outer Hebrides and the Northern Isles and from much of the mountainous region between Inverness and Glasgow. Results from farm questionnaire surveys provide strong evidence of a perceived decline in adder abundance during the past 10 years, and less strong evidence of some contraction in its range, both probably linked to changing land use and intensification of farming (Reading *et al.*, 1995).

The sand lizard (*Lacerta agilis*) is considered to be threatened in Britain and mainland Europe. Although not native to Scotland, it has persisted on the Island of Coll since it was experimentally introduced in 1970, and eight individuals were seen there in 1994 (Langton and Beckett, 1995).

9.5 Birds

NGOs have led the way in setting conservation priorities for birds and in 1996 produced a review of Birds of Conservation Concern, which sets out red and amber lists of priorities (Royal Society for the Protection of Birds, 1996). The Red list contains those species that are globally threatened and/or rapidly declining in numbers, or species that have declined greatly since 1800. The Joint Nature Conservation Committee (JNCC) has carried out a similar exercise but has sub-divided further to produce four lists (Table 9.2). There is almost complete coincidence between the species on first three JNCC lists and the NGOs' Red List (Table 9.3). The NGOs have omitted the chough (*Pyrrhocorax pyrrhocorax*) but have included the common scoter (*Melanitta nigra*) and twite (*Carduelis flavirostris*), resulting in a list that is larger by one species.

These lists contain many bird species that were historically persecuted by people and still are illegally persecuted in Scotland, the hen harrier (*Circus cyaneus*) being an example.

9.5.1   *Red List birds occurring only in Scotland*

Successful reintroduction has been achieved for the white-tailed or sea eagle (*Haliaeetus albicilla*). The programme is now in its 21st year, and by 1995 ten territorial pairs had become established and a total of 45 chicks had been fledged. The osprey (*Pandion haliaetus*) returned to Scotland of its own accord and there were 86 breeding pairs in 1995. The capercaillie (*Tetrao urogallus*) had previously become extinct in Scotland; after a successful reintroduction in the 19th century, the population rose to an estimated 20,000 birds in the early 1970s. It then declined dramatically and currently numbers about 2,000 birds as a result of poor breeding success and collisions with deer fences, which are a major cause of mortality

Table 9.2   British breeding birds of conservation concern. Source: Joint Nature Conservation Committee (1996).

Group	Status	n
1.	Globally threatened	3
2.	Uncommon and rapidly or historically declining	24
3.	Rapidly declining but common	8
4.	Moderately declining, historically declining but common, internationally important, localised, threatened in Europe	117

Table 9.3   Red List species that are globally threatened and/or rapidly declining and/or have undergone massive historical decline in the UK. Source: M. I. Avery, pers. comm.

Only Scottish	*Mostly Scottish*	*Found in Scotland*	*Not found in Scotland*
White-tailed eagle	Common scoter	Bittern	Stone curlew
Haliaeetus albicilla	*Melanitta nigra*	*Botaurus stellaris*	*Burhinus oedicnemus*
Osprey	Hen harrier	Red kite	Aquatic warbler
Pandion haliaetus	*Circus cyaneus*	*Milvus milvus*	*Acrocephalus paludicola*
Capercaillie	Merlin	Marsh harrier	Marsh warbler
Tetrao urogallus	*Falco columbarius*	*Circus aeruginosus*	*Acrocephalus palustris*
Red-necked phalarope	Black grouse	Grey partridge	Dartford warbler
Phalaropus lobatus	*Tetrao tetrix*	*Perdix perdix*	*Sylvia undata*
Scottish crossbill	Corncrake	Quail	Cirl bunting
Loxia scotica	*Crex crex*	*Coturnix coturnix*	*Emberiza cirlus*
	Wryneck	Black-tailed godwit	
	Jynx torquilla	*Limosa limosa*	
	Twite	Roseate tern	
	Carduelis flavirostris	*Sterna dougalli*	
		Turtle dove	
		Streptopelia turtur	
		Nightjar	
		Caprimulgus europaeus	
		Woodlark	
		Lullula arborea	
		Skylark	
		Alauda arvensis	
		Song thrush	
		Turdus philomelos	
		Spotted flycatcher	
		Muscicapa striata	
		Red-backed shrike	
		Lanius collurio	
		Tree sparrow	
		Passer montanus	
		Linnet	
		Carduelis cannabina	
		Bullfinch	
		Pyrrhula pyrrhula	
		Reed bunting	
		Emberiza schoeniclus	
		Corn bunting	
		Miliaria calandra	

(Ramsay, 1997). Although red-necked phalaropes (*Phalaropus lobatus*) are on the edge of their range in Scotland, they formerly occurred in the Inner and Outer Hebrides and Orkney but now the 40 British pairs breed mainly on Fetlar in Shetland. They require good-quality open water with emergent vegetation.

The Scottish crossbill (*Loxia scotica*) may be Britain's only endemic bird. The common (*L. curvirostra*) and Scottish crossbills have been joined by the parrot crossbill (*L. pytyopsittacus*) which bred here in 1984 and 1985 as a result of failure of the pine crop in Scandinavia (Batten *et al.,* 1990) and has frequently been recorded since. It is apparently difficult to distinguish between Scottish and parrot crossbills by their songs. Mitochondrial DNA analysis by Groth (1995) showed relatively low levels of overall sequence divergence between the three species; the nuclear genome is now being investigated by examining microsatellite and single-copy number polymorphisms (S. B. Piertney, pers. comm.). Until all these data are available to complement morphological and vocal information, questions about the systematic status of the Scottish crossbill remain speculative.

9.5.2 Birds of farmland

Many of the species on the Red List are associated with farmland habitats and are vulnerable either to agricultural intensification, such as the song thrush (*Turdus philomelos*) and the skylark (*Alauda arvensis*), or to the abandonment of traditional land management practices, such as the corncrake (*Crex crex*). Ten Scottish species are included among 18 nationally declining agricultural bird species. (Table 9.4). Although some of these declines may reflect edge-of-range effects, many were both predictable and inevitable; agricultural set-aside was and remains a lost opportunity to counteract them.

Table 9.4   Declines in agricultural bird species in the past 25 years.

Species	Estimated 1991 population (pairs)	Percentage decline in range	Percentage decline in numbers
Red-backed shrike *Lanius collurio*	2	−86.5	?
Cirl bunting *Emberiza cirlus*	350	−83.2	?
Corncrake *Crex crex*	500	−75.6	?
Stone curlew *Burhinus oedicnemus*	160	−41.9	?
Corn bunting *Miliaria calandra*	30,000	−32.1	>−80
Turtle dove *Streptopelia turtur*	75,000	−24.9	−77
Tree sparrow *Passer montanus*	110,000	−19.6	−89
Grey partridge *Perdix perdix*	150,000	−18.7	−82
Reed bunting *Emberiza schoeniclus*	220,000	−11.7	−61
Lapwing *Vanellus vanellus*	210,000	−9.0	c−50
Yellowhammer *Emberiza citrinella*	1,200,000	−8.6	−17
Bullfinch *Pyrrhula pyrrhula*	190,000	−6.5	−76
Linnet *Cardulus cannabina*	520,000	−4.6	−52
Spotted flycatcher *Muscicapa striata*	120,000	−2.3	−73
Song thrush *Turdus philomelos*	990,000	−2.1	−73
Blackbird *Turdus merula*	4,400,000	−1.9	−42
Skylark *Alauda arvensis*	1,500,000	−1.6	−58
Swallow *Hirundo rustica*	570,000	+1.1	−43

9.5.3 Seabirds and wildfowl

In contrast to such declines, Scotland is important in providing breeding sites for a considerable proportion of temperate north Atlantic seabirds: the majority of the world population of gannets (*Sula bassana*), manx shearwaters (*Puffinus puffinus*) and great skuas (*Stercorarius skua*). Scotland also provides winter quarters for a significant proportion of some of the world's wildfowl species such as the pink-footed goose (*Anser brachyrhynchus*) (*c.* 85%) and barnacle goose (*Branta leucopsis*) (*c.* 19%). The numbers of overwintering Iceland/Greenland pink-footed geese (Figure 9.2) have climbed steadily to a UK peak of *c.* 260,000 in October 1994 (five-year running mean = *c.* 225,000 (Mitchell 1996)). Over 80% of these geese either winter in Scotland or pass through on migration to England. Ninety percent of the *c.* 82,000 Iceland greylag geese (*Anser anser*) overwintering in the UK do so in Scotland; after a thirty-year increase, this population appears to be slowly declining (Figure 9.2). The reasons for this decline are not clear, but the fact that an estimated 35,000 were shot in Iceland in 1995 may be a contributing factor (Sigfússon 1996). The question remains, however, whether further increases in numbers of pink-footed geese are sustainable (The Scottish Office, 1996).

9.6 Mammals

9.6.1 Deer

The Scottish Natural Heritage (SNH) policy paper *Red Deer and the Natural Heritage* (1994) records a doubling of numbers of red deer from 150,000 in the 1950s to more than 300,000 today. Because six different sources of information are quoted in this data set, and counting methods have improved, it is more instructive to consider two areas of Scotland, the East and the West Grampians, where deer were originally counted by Fraser Darling (1953–55), and then by the Red Deer

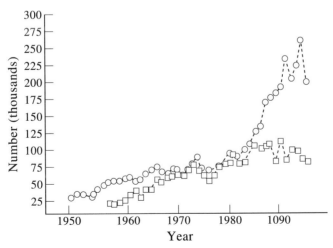

Figure 9.2 Numbers of pink-footed (circles) and greylag geese (squares) wintering in the UK in October–November (Mitchell, 1996).

Commission (RDC) (1966–94) (Figure 9.3). In both areas a doubling of numbers is apparent in the last thirty years of RDC counts. Present numbers prevent regeneration of native woodlands (Sottish Natural Heritage, 1994). When red deer numbers are reduced on sheep-free heather moor, as at Creag Meagaidh National Nature Reserve (NNR), birch (*Betula* spp.) regenerates (Figure 9.4). The results are striking and make Creag Meagaidh an important demonstration project (Ramsay, 1997). Similar effects have been observed for Caledonian pine (*Pinus sylvestris*) at Inshriach and Abernethy NNRs.

An additional problem is the spread of the Asian sika deer (*Cervus nippon*), which has been introduced into several locations in Scotland and has subsequently spread (Abernethy, 1994a). There are an estimated 7,000–10,000 sika in Scotland, and

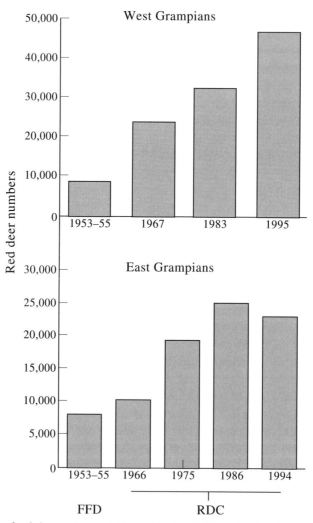

Figure 9.3   Counts of red deer in Scotland 1953–95 by Frank Fraser Darling (FFD) and The Red Deer Commission (RDC) (B. Staines, pers. comm.).

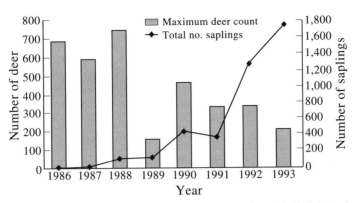

Figure 9.4   Maximum number of red deer counted on Creag Meagaidh NNR 1986–93 and total number of saplings recorded on five transects (Stewart, 1996).

they are found over 35% of the possible deer range, mainly in Sutherland, Argyll, Inverness-shire and Peeblesshire. Concern about this spread is heightened by the fact that sika interbreed with red deer and hybrids are fertile. Introgression occurs in both directions; the genetic integrity of one of Scotland's flagship species is clearly at risk (Abernethy, 1994b) and may be preserved in future only on islands off the west coast.

We are, however, slow to learn about the problems caused by alien species. Chinese muntjac deer (*Muntiacus reevesi*) were introduced to Woburn Park, England early this century, and have spread at the rate of about 1 km per year, aided by translocations. For example, in 1994, 15 animals were transported from Hertfordshire to Cumbria (N. Chapman, pers. comm.). There have been subsequent reports of muntjac shot near Forfar, after a release in Montreamont Forest, and a buck shot near Portpatrick, Wigtownshire. Opinions vary as to whether this species could become established in Scotland. It is intolerant of cold winters, when many animals starve, but it might survive the milder climate of western Scotland. Although the population of deer in Monks Wood NNR in Huntingdonshire is high, at one deer per hectare, because the deer are not culled, it provides a good example of their negative effect on ground flora such as bluebells (*Hyacinthoides non-scripta*), primroses (*Primula vulgaris*) and common spotted orchids (*Dactylorhiza fuchsii*) and the damage they do to regrowing coppice (Cooke, 1994; Cooke and Farrell, 1995). As a result of these effects and because range expansion of muntjac has resulted in the loss of roe deer, this species has now been included on Schedule 9 of the Wildlife and Countryside Act 1981, prohibiting unlicensed releases.

9.6.2   *Marine mammals*

9.6.2.1   Seals

Both grey seals (*Halichoerus grypus*) and harbour seals (*Phoca vitulina*) are native to Scotland and are included in Annex 2a of the Habitats and Species Directive. Grey seal population size is derived from counts of pups at breeding sites and at the start of the 1994 pupping season was estimated to be 108,500, of which 99,300 are associated with breeding sites in Scotland (Hiby *et al.*, 1996). This represents about half the world population of this species. If the present rate of increase of 6–7% per annum continues, the numbers of grey seals in Scotland will double in less than 12 years (Figure 9.5). Sandeels (*Ammodytidae*), cod (*Gadus morhua*), ling (*Molva molva*) and whiting (*Merlangius merlangus*) make up over 80% of the diet of grey seals in the North Sea; it is estimated that the seals ate approximately 76,000 tonnes of fish in 1992 (Hiby *et al.*, 1996). In 1985 the estimated consumption of cod by grey seals was only around 3% of the commercial catch, but as a result of increasing seal numbers and declining cod stocks (Cook *et al.*, 1997), the equivalent figure for 1992 is 11% (Hiby *et al.*, 1996). Although these figures must be treated with caution because seal diet varies significantly between years (Tollitt and Thompson, 1996) the increase in numbers of grey seals continues to raise questions of sustainability.

A contraceptive vaccine for grey seals has been developed in Canada, where more than half of the total seal population breeds on one island, and where females can be approached closely without causing desertion of their pups (Brown *et al.*, 1996). In Britain, where seals are distributed across more than 30 major colonies and animals are susceptible to disturbance, the administration of such a vaccine is likely to be more difficult. In addition, new ways of monitoring population size will have to be developed if contraceptive methods are used, because contraceptives alter the simple relation between the number of pups born each year and the population size, which forms the basis of current monitoring methods. Nevertheless, there have been repeated calls for further consideration of the use of non-lethal methods of controlling seal populations (Nickson, 1997).

The minimum size of the British harbour seal population is about 29,000, of which 90% is found in Scotland (Hiby *et al.*, 1996). However, British harbour seals account for only 6% of the world's population of this species.

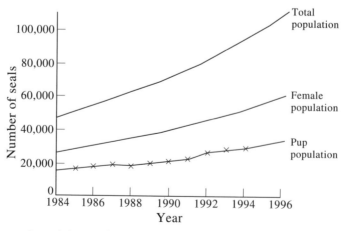

Figure 9.5   Grey seal population trends in the UK 1984–94, extrapolated to 1996 (Gardiner *et al.*, 1994).

9.6.2.2   Cetaceans

The species richness of cetaceans in waters around Scotland, particularly the northwest, has been documented by Evans (1987). Thirteen species are observed regularly (Thompson, 1992), nine of which are sometimes accompanied by newborn calves and may give birth in Scottish waters. Establishing the abundance of such species is difficult. However, recent intensive shipboard and aerial sightings of small cetaceans in the North Sea and adjacent waters have resulted in estimates of 352,523 harbour porpoises (*Phocaena phocaena*), 7,856 white-beaked dolphins (*Lagenorhyncus albirostris*) and 8,445 minke whales (*Balaenoptera acutorostrata*) (Hammond *et al.*, 1995). Of the two resident populations of bottlenose dolphins (*Tursiops truncatus*) in UK waters – in Cardigan Bay and the Moray Firth – the latter has been particularly well studied (Wilson, 1995) and there may be a much smaller resident population in the Sound of Barra (Grellier, in press). The Moray Firth population is the most northerly breeding population of this widespread species and is estimated to consist of 130 individuals. With only one such estimate, the status of the population cannot be determined (Wilson, 1995). Furthermore, power analyses showed that if annual estimates of population size were conducted, a considerable period of time would be required to detect trends in the size of the population: 11 years if the population changed at $\pm$ 5% p.a. Moreover if the population were declining, it would be substantially smaller by the time a trend could be detected. Ninety-five percent of the Moray Firth individuals had epidermal lesions, which may be associated with disease, and 6% had deformities (Wilson, 1995).

9.6.3   Rodents

The water vole (*Arvicola terrestris*) has suffered the greatest decline of any British mammal this century; The Wildlife Trust in England have predicted a 94% reduction in numbers. As a result, the water vole has recently been proposed for inclusion in Schedule 5 of the Wildlife and Countryside Act (1981). In Scotland, water voles are absent from 97 of 98 previously occupied sites in the catchment of the river Don in Aberdeenshire. In the adjacent Ythan catchment, 39% of sites surveyed (*n* = 44) still have water voles, compared with 90% seven years ago. Surviving populations are confined to the upper reaches of the Ythan catchment and their distribution pattern cannot be explained by the distribution of suitable habitat (Lambin *et al.*, 1996). Studies in lowland England have revealed that water voles are important items in the diet of mink (*Mustela vison*) newly arrived in an area but with the consequent decline in vole numbers they cease to be a food resource (Strachan *et al.*, in press). Nevertheless drastic local reductions of water voles occur within a year of the arrival of mink and they are effectively extinct within two years (Strachan *et al.*, in press). The results in the Ythan suggest that the intensity of mink predation may decline with distance from the main stem of the river (Telfer, 1996).

The red squirrel (*Sciurus vulgaris*) is the only rodent currently protected by the Wildlife and Countryside Act 1981. In England and Wales red squirrels have declined dramatically in recent decades but in Scotland the population has remained relatively stable. The grey squirrel (*Sciurus carolinensis*), which was introduced to southern England about 200 years ago, appears to have outcompeted the red throughout much of its former range, largely confining it to coniferous woodland. Consequently Scotland now holds three-quarters of the British population and there are concerns that as the grey squirrel population in Scotland spreads, red squirrel range will be further reduced.

9.6.4 Carnivores

At the beginning of the 19th century, the range of the pine marten (*Martes martes*) extended throughout Britain. It is now the rarest terrestrial mammal, mainly as a result of human persecution. However, results of a recent field survey suggest that martens have now recolonised some of the northern and western parts of Grampian, Tayside, Central and Strathclyde regions and are no longer confined to the Highlands. Comparisons of the area of woodland contained within the 1982 marten population distribution (Velander, 1983) and that of the 1994 distribution suggest that the marten population has approximately doubled in 12 years, from 1,200 to 2,600 adults (Balharry *et al.,* 1996).

The indigenous wildcat (*Felis sylvestris*) and the introduced domestic cat (*F. catus*) have been sympatric in Britain for more than 2,000 years, and a range of morphological criteria and genetic techniques have been used in an attempt to identify distinct groups. However, as a result of interbreeding, there appears to be considerable overlap between the wild type and the domestic form and genetic analysis provides little evidence for considering wildcats and domestic cats as separate species (Balharry *et al.,* 1997). Apparent morphological differences may be the result of selection pressures exerted by environmental factors rather than being genetically or phylogenetically determined (Daniels *et al.,* in press).

The otters (*Lutra lutra*) of Scotland's sea lochs and coasts are among the most important populations in Europe and have not undergone the dramatic decline, due mainly to pesticides, which affected otters in England from the 1950s to the 1970s (Kruuk, 1995). The decline now appears to have halted throughout the UK, and significant increases in distribution have been reported in Eastern Scotland and in the Clyde Valley (Green and Green, 1997).

9.6.5 Bats

In terms of numbers of species, bats are the most important contributors to Britain's mammalian biodiversity, although this declines with latitude so that of the 15 species found in England only eight occur in Southern Scotland and only two north of Inverness. Knowledge of the occurrence and distribution of bats has improved in recent years as a result of the use of ultrasonic receivers. Jones and van Parijs (1993) thus described two phonic types of the commonest British bat, the pipistrelle, *Pipistrellus pipistrellus*, which have subsequently been found segregated into separate colonies of different size, with significantly different diets (K. E. Barlow, pers. comm.). In addition to mating assortatively (Park *et al.,* 1996), they show an 11% DNA sequence divergence and are clearly separate species, both of which occur in Scotland (Barratt *et al.,* 1997).

Other bat species may be extending their range into Scotland. In recent years, increasing numbers of Nathusius' pipistrelle (*P. nathusii*) which is common in Northern Europe, have been found in Scotland (Speakman *et al,* 1991, 1993). Records are mainly of single individuals with numbers reaching a peak in September and May, suggesting that the bats may overwinter in Scotland (Speakman *et al,*, 1991; Hutson, 1993), although Rydell and Swift (1995) recorded it with bat detectors during summer. Recent evidence that *P. nathusii* forms mating groups in England (Barlow and Jones, 1996) raises the possibility that this species may breed in the UK.

9.7   The future

Many vertebrate species are in decline and many are now the subject of concerted action. The UK government has signed the Biodiversity Convention and produced a Biodiversity Action Plan (Anon., 1994) and a report from the Biodiversity Steering Group (Anon., 1995). The Steering Group Report built on the two Biodiversity Challenge reports (Wynne *et al.,* 1993, 1995) produced by NGOs, and has now been endorsed by government. It includes specific action plans for individual species, many of which occur in Scotland (Table 9.5). These species,

Table 9.5   Vertebrates for which species action plans have been produced by the Biodiversity Steering Group.

Endemic to Scotland	*Scotland very important*	*Occur in Scotland*	*Do not occur in Scotland*
Scottish crossbill *Loxia scotica*	Otter *Lutra lutra*	Water vole *Arvicola terrestris*	Dormouse *Muscardinus avellanarius*
	Harbour porpoise *Phocaena phocaena*	Brown hare *Lepus europaeus*	Mouse-eared bat *Myotis myotis*
	Red squirrel *Sciurus vulgaris*	Pipistrelle *Pipistrellus pipistrellus*	Aquatic warbler *Acrocephalus paludicola*
	Corncrake *Crex crex*	Skylark *Alauda arvensis*	Stone curlew *Burhinus oedicnemus*
	Capercaillie *Tetrao urogallus*	Bittern *Botaurus stellaris*	Greater horseshoe bat *Rhinolophus ferrumequinum*
	Allis shad *Alosa alosa*	Grey partridge *Perdix perdix*	
	Twaite shad *Alosa fallax* \	Song thrush *Turdus philomelos*	
		Sand lizard *Lacerta agilis*	
		Great crested newt *Triturus cristatus*	
		Natterjack toad *Bufo calamita*	

either listed by IUCN as globally threatened or thought to be rapidly declining in the UK, were chosen not only because they were each of conservation importance in their own right but also because they cover a wide range of habitats: marine, freshwater and terrestrial, forest and open country. Many of these species have been selected by the Country Agencies for specific action.

Scotland is rich in biodiversity and this biodiversity is sometimes seen to be rather less threatened than that of the more heavily populated areas of southern England. However, that merely allows more time in Scotland for species recovery. Intensification of agriculture, fisheries and forestry, together with tourism, industrial development, acid rain and climate change, are all live issues in Scotland which threaten biodiversity. Biodiversity really is part of Scotland's Natural Heritage, and that isn't just a good name for a conservation organisation but something that many of those who live in Scotland wish to help to conserve.

Acknowledgements

This paper could not have been written without the generous assistance of Mark Avery, David Balharry, Callan Duck, Vin Fleming, Steven Harris, Sandy Kerr, Tom Langton, Xavier Lambin, Peter Maitland, Carl Mitchell, Alan Morton, Stuart Piertney, Chris Smout, Brian Staines, Fiona Stewart, Ian Patterson, Sandra Telfer and the Biological Records Centre, Institute of Terrestrial Ecology, Monks Wood. To them all I am very grateful. Errors of fact or interpretation remain my own.

References

Abernethy, K. 1994a. *Sika deer in Scotland*. Report to the Red Deer Commission Sika Working Group.

Abernethy, K. 1994b. The establishment of a hybrid zone between red and sika deer (genus *Cervus*). *Molecular Ecology*, **3**, 551–562.

Anon. 1994. *Biodiversity: the UK Action Plan*. Cm2428. London, HMSO.

Anon. 1995. *Biodiversity: the UK Steering Group Report*. London, HMSO.

Balharry, E. A., McGowan, G. M., Kruuk, H. and Halliwell, E. 1996. Distribution of pine martens in Scotland as determined by field survey and questionnaire. Scottish Natural Heritage Research, Survey and Monitoring Report No. 48.

Balharry, D., Daniels, M. and Barratt, E. M. 1997. Wildcats: can genetics help their conservation? In. Tew, T. E., Crawford, T. J., Spencer, J. W., Stevens, D. P., Usher, M. B. and Warren, J. (Eds.) *The Role of Genetics in Conserving Small Populations*. Peterborough, Joint Nature Conservation Committee, 102–111.

Barlow, K. E. and Jones, G. 1996. *Pipistrellus nathusii* (Chiroptera: Vespertilionidae) in Britain in the mating season. *Journal of Zoology*, **240**, 767–773.

Barratt, E. M., Deaville, R., Burland, T. M., Bruford, M. W., Jones, G., Racey, P. A. and Wayne, R. K. 1997. DNA answers the call of pipistrelle bat species. *Nature*, **387**, 138–139.

Batten, L. A., Bibby, C. J., Clement, R., Elliott, G. D. and Porter, R. F. 1990. *Red Data Birds in Britain*. London, Poyser.

Brown, R. G., Kimmins, W. C., Mezei, M., Parsons, J., Pohajdak, B. and Bowen, W. D. 1996. Birth control for grey seals. *Nature*, **379**, 30–31.

Cook, R. M., Sinclair, A. and Stefánsson, G. 1997. Potential collapse of North Sea cod stocks. *Nature*, **385**, 521–522.

Cooke, A. S. 1994. Colonisation by muntjac deer (*Muntiacus reevesi*) and their impact on vegetation. In Massey, N. and Welch, R. C. (Eds.) *Monks Wood National Nature Reserve: the Experience of 40 Years 1953–1993*. Peterborough, English Nature, 45–61.

Cooke, A. S. and Farrell, L. 1995. Establishment and impact of Muntjac (*Muntiacus reevesi*) on two national nature reserves. In Mayle, B. (Ed.) *Muntjac Deer, their Biology, Impact and Management in Britain*. Farnham, The British Deer Society and the Forestry Commission, 48–62.

Daniels, M. J., Balharry, D., Hirst, D., Kitchener, A. C. and Aspinall, R. J. In press. Morphological and pelage characteristics of wild living cats in Scotland: implications for defining the wildcat. *Journal of Zoology*.

Eckert, K. L. and Eckert, S. A. (Eds.) 1996. Decline of the world's largest nesting assemblage of leatherback turtles. *Marine Turtle Newsletter* No. 74. Washington, Conservation International.

Evans, P. G. H. 1987. *The Natural History of Whales and Dolphins*. London, Academic Press.

Ferguson, A., Taggart, J. B., Prodöhl, P. A., McMeel, O., Thompson, C., Stone, C., McGinnity, P. and Hynes, R. A. 1995. The application of molecular markers to the study and conservation of fish populations, with special reference to *Salmo. Journal of Fisheries Biology*, **47**, 103–126.

Gardiner, K. J., Racey, P. A. and Hiby, L. 1994. Population management of seals: an evaluation of non-lethal methods of population control, with population modelling. Report to MAFF.

Green, J. and Green, R. 1997. *Otter Survey of Scotland 1990–1992*. London, The Vincent Wildlife Trust.

Grellier, K. In press. Bottlenose dolphins in the Sound of Barra? A pilot study. *Hebridean Naturalist*.

Groth, J. G. 1995. Genetics of crossbills (*Loxia*) in Scotland. Preliminary report to RSPB.

Hammond, P., Benke, H., Berggren, P., Collet, A., Heide-Jorgensen, M. P., Heimlich-Boran, S., Leopold, M. and Øien, N. 1995. The distribution and abundance of harbour porpoises and other small cetaceans in the North Sea and adjacent waters. Final report. LIFE 92.2/UK/027.

Hiby, L., Duck, C., Thompson, D., Hall, A. and Harwood, J. 1996. Seal stocks in Great Britain. *NERC News*, **34**, 20–22.

Hutchinson, P. and Mills, D.H. 1987. Characteristics of spawning-run smelt, *Osmerus eperlanus* (L.) from a Scottish river, with recommended actions for their conservation and management. *Aquaculture and Fisheries Management*, **18**, 249–258.

Hutson, A. M. 1993. *Action Plan for the Conservation of Bats in the United Kingdom*. London, The Bat Conservation Trust.

Jones, G. and van Parijs, S. 1993. Bimodal echolocation in pipistrelle bats: are cryptic species present? *Proceedings of the Royal Society of London*, B **251**, 119–125.

Kruuk, H. 1995. *Wild Otters – Predation and Populations*. Oxford, Oxford University Press.

Lambin, X., Telfer, S., Cosgrove, P. and Alexander, G. 1996. Survey of water voles and mink on the rivers Don and Ythan. Unpublished report to Scottish Natural Heritage. 33 pp.

Langton, T. E. S. and Beckett, C. L. 1995. Home range size of Scottish amphibians and reptiles. Scottish Natural Heritage Review No. 53.

Langton, T. E. S., Beckett, C. L., King, G., Dunmore, I. and Gaywood, M. 1996. Distribution and studies of marine turtles in Scottish waters. Scottish Natural Heritage Research, Survey and Monitoring Report No. 8.

Maitland, P. S. and Lyle, A. A. 1991. Conservation of freshwater fish in the British Isles: the current status and biology of threatened species. *Aquatic Conservation: Marine and Freshwater Ecosystems*, **1**, 25–54.

Maitland, P. S. and Lyle, A. A. 1992. Conservation of freshwater fish in the British Isles: proposals for management. *Aquatic Conservation: Marine and Freshwater Ecosystems*, **2**, 165–183.

Maitland, P. S., East, K. and Morris, K. H. 1983. Ruffe *Gymnocephalus cernua* (L.), new to Scotland, in Loch Lomond. *Scottish Naturalist*, **1983**, 7–9.

Mitchell, C. 1996 The 1995. National census of pink-footed and greylag geese in Britain. Slimbridge, Wildfowl and Wetlands Trust report to JNCC.

Nickson, Lord 1997. *Report of the Scottish Salmon Strategy Task Force*. Edinburgh, The Scottish Office.

Park, K. J., Altringham, J. D. and Jones, G. 1996. Assortative mating in the two phonic types of *Pipistrellus pipistrellus* during the mating season. *Proceedings of the Royal Society of London*, B **263**, 1495–1499.

Ramsay, P. 1997. *Revival of the Land. Creag Meagaidh National Nature Reserve*. Perth, Scottish Natural Heritage.

Reading, C. J., Buckland, S. T., McGowan, G. M., Gorzula, S., Jayasinghe, G., Staines, B. W., Elston, D. A. and Ahmadi, S. 1995. Status of the adder *Vipera berus* in Scotland. Scottish Natural Heritage Research, Survey and Monitoring Report No. 38.

Royal Society for the Protection of Birds 1996. *Birds of Conservation Concern in the United Kingdom, Channel Islands and Isle of Man*. Sandy, Royal Society for the Protection of Birds.

Rydell, J. and Swift, S. 1995. Observations of Nathusius' pipistrelle, *Pipistrellus nathusii* in northern Scotland. *Scottish Bats,* **3**, 6–7.

Scottish Natural Heritage 1994. *Red Deer and the Natural Heritage.* Perth, Scottish Natural Heritage, 70pp.

Sigfússon, A. 1996. A new system of bag reporting from Iceland. *Wetlands International Goose Specialist Group Bulletin,* **8**, 9–11.

Speakman, J. R., Racey, P. A., Hutson, A. M., Webb, P. I. and Burnett, A. M. 1991. Status of Nathusius' pipistrelle (*Pipistrellus nathusii*) in Britain. *Journal of Zoology,* **225**, 685–690.

Speakman, J. R., Racey, P. A., McLean, J. and Entwistle, A. C. 1993. Six new records of Nathusius' pipistrelle (*Pipistrellus nathusii*) for Scotland. *Scottish Bats,* **2**, 14–16.

Stewart, F. 1996. The effects of red deer on the regeneration of upland birch woodland. Ph.D. thesis, University of Aberdeen.

Strachan, C., Jefferies, D. J., Barreto, G. R., Macdonald, D. W. and Strachan, R. In press. The rapid impact of resident American mink in water voles: case studies in lowland England. *Symposium of the Zoological Society.*

Telfer, S. 1996. Distribution and demography of fragmented water vole *Arvicola terrestris* (L.) populations within the River Ythan catchment. M.Sc. thesis, University of Aberdeen.

The Scottish Office 1996. *Wild Geese and Agriculture in Scotland – a discussion paper.* Edinburgh, The Scottish Office Agriculture, Environment and Fisheries Department.

Thompson, P. M. 1992. The conservation of marine mammals in Scottish waters. *Proceedings of the Royal Society of Edinburgh,* B **100**, 123–140.

Tollitt, D. J. and Thompson, P. M. 1996. Seasonal and between-year variations in the diet of harbour seals in the Moray Firth, Scotland. *Canadian Journal of Zoology,* **74**, 1110–1121.

Velander, K. A. 1983. *Pine Marten Survey of Scotland, England and Wales 1980–1982.* London, Vincent Wildlife Trust.

Wilson, B. 1995. The ecology of bottlenose dolphins in the Moray Firth, Scotland: a population at the northern extreme of the species' range. Ph.D. thesis, University of Aberdeen.

Wynne, G., Avery, M., Campbell, L., Juniper, T., King, M., Smart, J., Steel, C., Stones, A., Stubbs, A., Taylor, J., Tydeman, C. and Wynde, R. 1993. *Biodiversity Challenge.* Sandy, Royal Society for the Protection of Birds.

Wynne, G., Avery, A., Campbell, L., Gubbay, S., Hawkswell, S., Juniper, T., King, M., Newberry, P., Smart, J., Steel, C., Stones, T., Stubbs, A., Taylor, J., Tydeman, C. and Wynde, R. 1995. *Biodiversity Challenge,* 2nd Edn. Sandy, Royal Society for the Protection of Birds.

10 THE GENETIC BIODIVERSITY OF SCOTTISH PLANTS

R. A. Ennos and E. P. Easton

Summary

1. Biodiversity at the genetic level is defined here as all forms of genetic variation within a taxon that affect the ecological attributes of individuals. Genetic biodiversity is important because it is a determinant of the ecological amplitude and the evolvability of species, and therefore underpins present and future biodiversity at the species and community levels.

2. The level and distribution of genetic biodiversity can only be assessed by using a genecological approach. Studies of genetic markers (isozymes, DNA variation) cannot be used to assess genetic biodiversity. However, they may provide valuable information for understanding its dynamics.

3. In the species-poor higher plant flora of Scotland, genetic biodiversity is expected to comprise a significant proportion of total biodiversity. Despite its importance very few measurements of the genetic biodiversity of Scottish plants have been made. Those that have been undertaken indicate the uniqueness of Scottish populations in terms of regional climatic adaptation, and the presence of considerable diversity between and within local populations.

4. To overcome our ignorance of genetic biodiversity a renaissance in low-technology studies is required, rather than the pursuit of increasingly sophisticated analysis of DNA variation.

10.1 Introduction

Biodiversity is a term that has been of enormous utility in directing political attention and a limited degree of funding to the general area of biological conservation. The success of biodiversity as a banner for conservation interests lies to a large extent in the flexibility and ambiguity of the term, allowing its use by non-specialists without the need to circumscribe its meaning. To be of value as a focus for scientific research, however, a rigorous definition is required, together with a scientific justification of its importance for the functioning of ecosystems. Our aim in this chapter is to provide both a definition of and a justification for studying biodiversity at the genetic level. We then outline an agenda for measuring and understanding the dynamics of genetic biodiversity, illustrating the ideas with work on higher plant species in Scotland.

10.2   Definition

Biodiversity will be defined here as the coexistence of distinct biological units that differ in their ecological attributes. Depending upon the biological unit chosen under this definition, biodiversity can be recognised and studied at a wide variety of scales. Landscape-, community- and particularly species-level biodiversity are most easily defined and classified. To a large extent, biodiversity programmes have become preoccupied with the cataloguing and classification of these types of variation.

It must be remembered, however, that biodiversity also occurs below the species level because of the presence of genetic variation within species. In accordance with the definition of biodiversity outlined above, biodiversity at the genetic level comprises all forms of genetic variation that affect the ecological attributes of individuals within a single taxon. In higher plants such attributes might include growth rate, phenology, competitive ability, and pest and disease resistance. Note that this definition excludes differences in the genetic information possessed by individuals that are of no ecological relevance either now or in the future to those individuals. Such genetic variation can be described as selectively or ecologically neutral.

10.3   Distribution of genetic biodiversity

Genetic biodiversity is present at two levels within any taxon: as variation between populations, and as variation among individuals within populations. Both forms of variation are biologically important. Variation between populations is associated with local adaptation of populations to their abiotic and biotic environments (Bradshaw, 1984). The extent of genetic biodiversity among populations is a major determinant of the ecological amplitude of species, the range of environments over which they can compete effectively (Bradshaw, 1984). Genetic biodiversity within a population on the other hand represents the raw material for future evolutionary change. Its abundance determines the evolutionary potential of that population, and its ability to respond to future environmental change (Fisher, 1930).

Thus the present and future ranges of species are substantially influenced by the extent of their genetic biodiversity. The greater the genetic biodiversity, the greater the potential range of the species and the greater the buffering against range reduction resulting from environmental change. As a consequence of these effects on species distribution, genetic biodiversity can be seen to underpin biodiversity at the species, community and landscape levels. There is therefore clear ecological justification for studies of genetic biodiversity because of its ultimate influence on the species composition of ecosystems and their dynamic behaviour.

10.4   Agenda for genetic biodiversity research

The extent and structure of genetic variation within species are not static, but are continually evolving. Appreciation and understanding of the dynamics of genetic biodiversity are the keys to conservation and management of this natural resource (Avise and Hamrick, 1996; Ennos *et al.*, 1997b). The agenda for research on genetic biodiversity must therefore have two aims. The first is clearly to measure and monitor the present levels and distribution of genetic biodiversity within taxa. The second, and arguably more important and demanding objective, must be to understand the processes that have led to the present distribution of genetic biodiversity and will be responsible for future changes in response to environmental perturbation. Genetic biodiversity research (along with other levels of biodiversity research) must be concerned with understanding processes as well as with providing descriptions and inventories.

10.5   Techniques for studying genetic biodiversity

Genetic variation in higher plants (and other taxonomic groups) can be studied by using two fundamentally different approaches. The first involves measurements on characteristics of the plant itself (phenotypic or genecological approach) (Heslop-Harrison, 1964). Plants from different populations or different families from the same population are grown under common environmental conditions. If differences are exhibited between them this indicates genetic differences between the populations or within the populations, respectively (Lawrence, 1984). If characters with important adaptive roles are measured under ecologically realistic conditions it is reasonable to regard the differences detected as adaptive genetic variation.

With the addition of modern quantitative genetic analysis to the straightforward genecological approach (which was pioneered more than two centuries ago (Langlet, 1971)), quantitative measurements of both adaptive differences between populations and the extent of adaptive variation within populations (their evolvability) can be measured for any chosen character (Mitchell-Olds and Rutledge, 1987; Houle, 1992). By using a range of environments for such tests the expression of the genotypes under different conditions (e.g. presence or absence of stress, pests, pathogens) can be explored to build up a complete picture of adaptive differences. The genotype of individuals is probed by using different environmental treatments to reveal differences that are expressed in the phenotype (Ennos, 1989).

These low-technology techniques are ideally suited to assessing the important aspects of biodiversity at the genetic level outlined above, namely regional and local adaptation of populations. Their major limitation is that the differences detected cannot readily be attributed to specific genes (so called quantitative trait loci, or QTLs), making modelling and monitoring of changes difficult. Although it is becoming technically possible with the help of modern DNA technology to identify QTLs in experimental crosses between genetically differentiated lines of model species such as tomato and *Arabidopsis thaliana* (Paterson *et al.*, 1991), their recognition in natural populations of less well-studied species is likely to remain an unattainable goal for the foreseeable future. In addition to this limitation of the genecological approach, it should also be remembered that the experiments required may be costly in terms of time, space and materials, although the scale of these difficulties will depend on the species being studied.

The alternative approach to assessing diversity at the genetic level is to use molecular analysis directly to detect variation in the genetic information possessed by populations or individuals (genotypic approach). For this purpose a whole suite of techniques is now available, allowing variation in all three plant genomes (nucleus, chloroplast and mitochondrion) to be measured (Avise, 1994). Superficially, this approach appears to overcome all the limitations inherent in traditional genecological research. Analysis can be conducted on tissue samples without the need to grow up large numbers of plants. The variation detected is attributable to single genetic differences and its distribution within and among individuals and populations is governed by processes, such as breeding system, gene flow and random genetic drift, that are well understood and easily modelled.

The problem with the approach, however, is that it measures variation in the genetic information possessed by individuals, not the effects of this variation on their phenotype and ecological attributes. Genetic diversity is assessed rather than biodiversity at the genetic level. The vast majority of genetic differences detected will have negligible effects on the phenotype or performance of an individual, and we have no means of recognising those rare genetic differences that are adaptively important. For all practical purposes, therefore, the genetic variation detected by this approach is best regarded as ecologically neutral.

The behaviour of ecologically neutral genetic variants (genetic markers) is quite different from that of variants that have effects on the phenotype and ecology of the plant. Population genetics theory predicts that the level of genetic-marker variation and the distribution of this variation within and among populations is unlikely to be correlated with levels or distribution of adaptive genetic differences, and cannot be used as a surrogate for measuring biodiversity at the genetic level (Ennos, 1996; Lynch, 1996). Experimental proof of this comes from parallel studies of genetic marker variation and genetic variation for a range of ecologically important characters in the Scottish endemic *Primula scotica* (Figure 10.1).

Isozyme analysis in 14 natural populations revealed variation in only one of 15 enzymes scored, and in a subsample of four populations no variation for randomly amplified polymorphic DNA (RAPD) markers was found (Glover and Abbott, 1995). It can be concluded that genetic-marker variation is low or non-existent both within and among populations. However, this analysis does provide evidence of fixed heterozygosity at the isozyme loci, suggesting a hybrid origin for the taxon.

In contrast, a genecological study using quantitative genetic analysis of 11

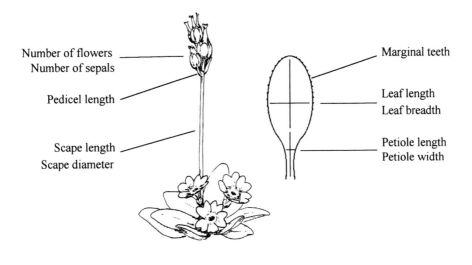

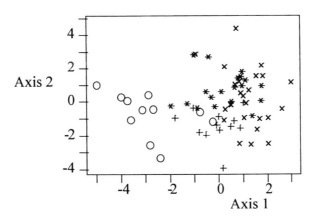

Figure 10.1   Genetic biodiversity among *Primula scotica* populations revealed by principal components analysis of quantitative characters measured in a common garden trial. Populations are: Dunnet Links (circles), Armadale Head (+), Sandside Head (*) and Ushat Head (x) (Ennos *et al.*, 1997a).

vegetative and flowering characters in four populations indicated significant genetic variation within populations for six of the 11 characters, and significant variation among populations for all characters measured (Ennos *et al.*, 1997a). The greatest differences were between populations occupying distinct cliff heath and dune grassland sites. It is clear that biodiversity at the genetic level is substantial both within and between populations of this species, even though there is an almost complete lack of detectable genetic-marker variation.

The implication is that studies of genetic markers cannot be used directly to assess the levels and patterns of adaptive genetic variation with which biodiversity studies are concerned. Failure to appreciate or state explicitly this conclusion is common throughout the conservation genetics literature and has led to a perception that genecological studies are somehow inferior to, and have been superseded by, molecular techniques. This is clearly untrue. If we wish directly to measure biodiversity at the genetic level the genecological approach is entirely appropriate. Our current understanding of the patterns and extent of adaptive variation in our native flora is woefully inadequate. If this situation is to be redressed it is essential to stimulate a revival in genecology to provide us with the relevant data to answer questions about biodiversity below the species level (Ennos, 1996).

Although the data above and those gathered by others (Giles, 1983; Cheverud *et al.*, 1994) clearly demonstrate that genetic markers are inappropriate for providing direct measurements of adaptive genetic variation and its distribution, genetic markers can none the less be exploited very successfully in a biodiversity context to provide us with insights into the dynamic processes that alter genetic biodiversity. As we will illustrate later, genetic markers can be used to study clonal spread, mating systems, gene flow and population history. All of these parameters must be quantified if the dynamic behaviour of genetic biodiversity is to be modelled and predicted.

10.6 Importance and assessment of genetic biodiversity in the Scottish flora

Western Europe as a whole, and Scotland in particular, has a species-poor higher plant flora. Repeated recent glaciation, the low species diversity of recolonising populations and the restricted time period during which a land bridge to Europe was open after glaciation have all contributed (Huntley and Birks, 1983). Moreover, the time since isolation of Britain from mainland Europe has been too short for a significant number of speciation events to have occurred. As a consequence there is only a handful of truly endemic higher plant taxa in Scotland, mostly of recent hybrid or polyploid origin.

This species-poor flora of Scotland invaded at a time when climate was more continental in nature than at present, the current oceanic conditions having developed after the loss of the land bridge and the isolation of populations of plants from their progenitors in continental Europe. This means that native Scottish species with European distributions have evolved for some 7,000 years in an oceanic climate distinct from that in the rest of their range. With low species diversity, and lack of competitors, those species that succeeded in reaching Scotland were often able to colonise a wider variety of habitats than would have been possible in a species-rich situation. Occupation of diverse habitats by single species was facilitated by the proximity of sites differing significantly in climatic, edaphic and topographical variables within the limited geographical area of Scotland.

This episode of colonisation and isolation of the Scottish flora is likely to have had two significant genetic consequences: regional genetic adaptation to oceanic conditions, and extensive local adaptation of widely distributed species within

Scotland. There is good evidence, where it has been sought, that regional adaptation of species to the oceanic climate of Scotland has occurred, giving climatic races distinct from their counterparts in continental Europe.

For instance, Scottish populations of Scots pine when tested against their continental European counterparts under Scottish conditions show at least 10% greater height growth and from 25% to 75% better survival (Worrell, 1992). In continental Europe, on the other hand continental populations have proved superior to Scottish sources, outperforming them in height growth by some 20% on average (Giertych, 1979). Similar evidence for genetic adaptation of native populations to Scottish conditions has been documented for a range of other tree species (Worrell, 1992).

Within Scotland, local adaptation to diverse abiotic and biotic environments can also be demonstrated. Classic genecological experiments on *Plantago maritima* by Gregor (1936) revealed independently evolved local genetic adaptation to gradients of salinity and exposure in coastal regions of Scotland. Trials in east Scotland of Scots pine populations from the east of Scotland showed that they were superior to populations from the west for height growth and resistance to local rust pathogens (Lines and Mitchell, 1964). Recent experiments at a single site in the east of Scotland demonstrated differences between native Scots pine populations for adaptive characters such as height growth, survival and phenology (Ennos *et al.*, 1997b).

The profound human influence exercised through grazing, burning, harvesting and cultivation regimes over prolonged periods has in many ways extended the range of conditions experienced by plant populations in Scotland and is likely to have encouraged local adaptation to the diversity of cultural environments so generated. Evidence for this comes from further studies by Gregor and Watson (1961) on *P. lanceolata* in which adaptive genetic differences in leaf length and other characters were demonstrated among populations taken from pastures subject to different intensities of grazing.

This evidence, though extremely limited and very patchy, suggests that in the Scottish situation biodiversity of higher plants at the genetic level, recognised as adaptive genetic variation, makes a significant contribution both to the distinctiveness of the Scottish flora and to total biodiversity within Scotland. The uniqueness of the Scottish flora could be considered to lie not so much in its community or species composition, but in its genetic adaptation to oceanic conditions. The species diversity of communities in a depauperate flora is enhanced where individual species display wide ecological amplitude. Where local genetic adaptation facilitates high ecological amplitude (as in the cases cited above), biodiversity at the genetic level can be seen as underpinning the biodiversity observed at species and community levels.

These considerations demonstrate that investigations of the levels of regional genetic adaptation to the unique climate of Scotland, and studies of the extent of local adaptation of populations, are needed if we are fully to document the biodiversity of the Scottish flora. They are as valid a focus for biodiversity research in Scotland as are surveys of species distribution and abundance. Moreover, they alone will provide the information needed for identifying well-adapted native seed sources for the restoration of native plant communities (Ennos, 1996; Ennos *et al.*, 1997a).

10.7   Dynamics of genetic biodiversity

Although the assessment of current levels and distribution of genetic biodiversity with the techniques described above is essential, it must be remembered that biodiversity at all levels is dynamic. Studies of biodiversity must be concerned with understanding the changes that have occurred in the past and with collecting information that will allow future changes to be predicted. Past changes in species distributions and abundance affect the abundance and patterning of selectively neutral genetic-marker variation. Studies of genetic markers, if properly interpreted, can therefore be used to infer past population history and behaviour. The history of the species can be read in the genes. In addition the processes that will facilitate and set the limits to genetic change in the future, notably breeding system and interpopulation gene flow, also leave their imprint on the arrangement of genetic markers within present-day populations.

These arguments indicate that, although genetic markers are inappropriate for estimating absolute levels of genetic biodiversity, they may be used indirectly to provide information on the dynamics of genetic and other levels of biodiversity, and the parameters that affect these dynamics. In this context genetic markers have a role to play in biodiversity research when used to elucidate processes of change. To illustrate the diverse forms of information that genetic marker studies can yield, a number of examples of their use in studies of Scottish plants are outlined below.

10.8   Use of genetic markers in biodiversity research

10.8.1   *Investigating the history of populations*

On a long timescale it is often of importance to understand the origin of Scottish plant populations following the last glaciation to provide the context for future conservation. Studies of monoterpene, isozyme and mitochondrial DNA markers have been invaluable in elucidating the origin of native Scots pine after glaciation. The evidence suggests multiple origins of the populations, possibly from two refugia in continental Europe (Ennos *et al.*, 1997b).

An appreciation of the recent history of threatened populations is often essential for the rational development of biodiversity conservation programmes. A good example is that of the sedge *Carex chordorrhiza*, which was for a long time known only from a single site at Altnaharra in Caithness. In 1978 a second population was discovered at Insh Marshes on Speyside (Page and Rieley, 1985). Given the rarity of the plant, it was of considerable interest to determine whether the population at Insh had been recently founded from the Altnahara locality, or whether it represented a case of under-recording. Isozyme analysis of plants revealed that at each site ten genotypes could be distinguished by using four polymorphic loci. Five genotypes were common to each site (R. A. Ennos *et al.*, unpublished). However, five genotypes were found at Altnaharra that were not present at Insh, and vice versa. The data also indicated considerable clonal spread at both sites. This evidence strongly suggests that the Insh marshes population has not been derived from that of Altnaharra but has been present independently for a considerable time. This is of interest from the point of view of species biodiversity because it doubles the number of known native sites occupied by *C. chordorrhiza*.

10.8.2   *Investigating the dynamics of asexual populations*

A number of threatened plant species in Scotland show very low levels of sexual reproduction and propagate almost exclusively by vegetative means. Questions then arise about the clonal dynamics of these species. Is clonal diversity lost within an area, leading to lack of genetic biodiversity, or are multiple clones retained, which may differ for ecologically important traits? These questions are of considerable importance in native aspen, *Populus tremula,* a dioecious tree that is of conservation interest in native woodlands, where it propagates almost exclusively by root suckers. Isozyme markers have proved ideal tools for identifying clones within aspen stands and for answering questions about clonal dynamics and diversity.

At Tomnagowhan wood on Speyside, some 21 clones could be distinguished and mapped within a 6 ha area (Figure 10.2). Ecologically important differences in phenology and growth rate were apparent between the clones (Easton, 1997). These results provide a benchmark against which to assess clonal diversity in other threatened populations, and suggest appropriate levels of genetic biodiversity to be included in newly established aspen woods. In contrast the woodland at Berridale on Hoy, Orkney, was found to contain only one clone and therefore no genetic biodiversity. Eleven other clones could be found within the whole island, however, suggesting that there may be sufficient genetic biodiversity on which to base a woodland restoration programme.

10.8.3   *Estimating mating systems and gene flow*

The ecologically neutral properties of most genetic markers mean that their behaviour is unaffected by natural selection. It is determined instead by an inter-action between the process of genetic drift, the exchange of genes between individuals (governed by the mating system) and the migration of genes between populations (governed by seed and pollen flow in plants) (Ennos, 1996). A survey of the distribution of genetic markers within and between individuals and populations can therefore be used to infer useful information concerning the mating system, gene flow and genetic drift. Such information is required if future changes in biodiversity at the genetic level are to be modelled and predicted.

A recent example of the inferences that can be drawn from genetic-marker data concerns a nationwide isozyme survey of native aspen populations in Scotland

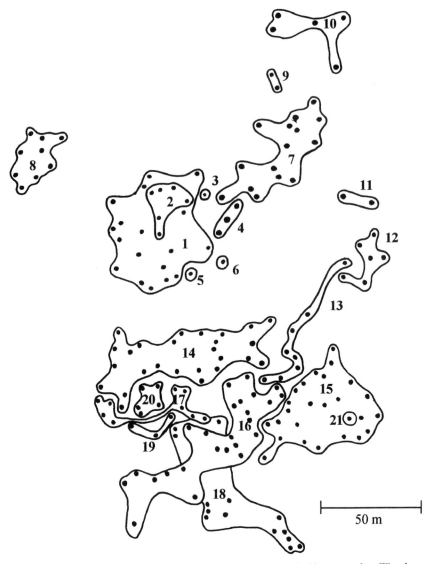

Figure 10.2   Diversity and distribution of aspen (*Populus tremula*) clones in Tomnagowhan Wood, as revealed by analysis of isozyme genetic markers (Easton, 1997).

(Easton, 1997). Results indicated that the populations had been formed by random mating. This is of interest because although populations are dioecious they rarely reproduce sexually in present-day Scottish populations. Genetic-marker differences were very small between populations (accounting for only 2% of total variation) indicating substantial gene flow among populations in the past. The inference is that present-day populations of aspen, although they are now reproducing almost exclusively by vegetative means, were founded in the past by a population actively engaged in sexual reproduction with few restrictions on gene flow by pollen and/or seed across Scotland. The change in the reproductive behaviour of aspen since colonisation may have been brought about by an alteration in climate from a continental to an oceanic regime after flooding of the English Channel around 7,000 years BP, removing the climatic cues needed by the tree for stimulation of flowering. The unprecedented abundance of aspen flowering in 1996 following the extremely hot and dry summer of 1995 in Scotland lends some credence to this theory (Easton, 1997). The implication of this result from the practical point of view is that establishment of new populations by sexual means should be attempted if we wish to mimic the original mode of foundation of native aspen populations.

10.9 Conclusions

The purpose of this chapter has been to clarify the meaning of biodiversity at the genetic level and to analyse how it may be assessed. We have concluded that biodiversity at the genetic level is synonymous with adaptive genetic variation. We have argued that, in the depauperate higher plant flora of Scotland, diversity at the genetic level is likely to constitute a significant component of total biodiversity underpinning variation at the species and community levels. We have also indicated that, to quantify this genetic biodiversity, appropriate genecological trials must be designed and performed. Analyses of genetic markers do not provide a short-cut, although when properly targeted they can make valuable contributions to understanding the history and dynamics of genetic biodiversity.

In recent years low-technology genecological research has been neglected in favour of increasingly sophisticated analysis of DNA variation. Our arguments suggest that if we are serious about understanding biodiversity at the genetic level a renaissance in the genecological approach is needed. Despite its shortcomings, it is the only approach that can reveal genetic variation that is directly relevant to ecologists (Ennos *et al.*, 1997a; Lynch, 1996). The study of biodiversity at the genetic level can then be developed as an integral part of biodiversity research, not as an expensive and possibly ecologically irrelevant extra.

Acknowledgements

Research on *Primula scotica*, *Carex chordorrhiza* and *Populus tremula* has been made possible by contracts from Scottish Natural Heritage and the Forestry Commission.

References

Avise, J. C. 1994. *Molecular Markers, Natural History and Evolution*. New York, Chapman and Hall.

Avise, J. C. and Hamrick, J. L. 1996. *Conservation Genetics. Case Histories from Nature*. New York, Chapman and Hall.

Bradshaw, A. D. 1984. Ecological significance of genetic variation between populations. In Dirzo, R. and Sarakhan, J. (Eds.) *Perspectives on Plant Population Ecology*. Sunderland, Massachusetts, Sinauer Associates, 213–228.

Cheverud, J., Routman, E., Jaquish, C., Tardif, S., Peterson, G., Belfiore, N. and Forman, L. 1994. Quantitative and molecular genetic variation in captive cotton-top tamarins (*Saguinus oedipus*). *Conservation Biology,* **8**, 95–105.

Easton, E. P. 1997. Genetic variation in Scottish aspen *(Populus tremula* L.). M. Phil. thesis, University of Edinburgh.

Ennos, R. A. 1989. Detection and measurement of selection – genetic and ecological approaches. In Brown, A. H. D., Clegg, M. T., Kahler, A. L. and Weir, B. S. (Eds.) *Plant Population Genetics, Breeding and Genetic Resources.* Sunderland, Massachusetts, Sinauer, 200–214.

Ennos, R. A. 1996. Utilising genetic information in plant conservation programmes. In Hochberg, M. E., Clobert, J. and Barbault, R. (Eds.) *The Genesis and Maintenance of Biological Diversity.* Oxford, Oxford University Press, 278–291.

Ennos, R. A., Cowie, N. R., Legg, C. J. and Sydes, C. 1997a. Which measures of genetic variation are relevant in plant conservation? A case study of *Primula scotica.* In Tew, T. E., Crawford, T. J., Spencer, J. W., Stevens, D. P., Usher, M. B. and Warren, J. (Eds.) *The Role of Genetics in Conserving Small Populations.* Peterborough, Joint Nature Conservation Committee, 73–79.

Ennos, R. A., Sinclair, W. T. and Perks, M. P. 1997b. Genetic insights into the evolution of Scots pine, *Pinus sylvestris* L., in Scotland. *Botanical Journal of Scotland,* **49**, 257–265.

Fisher, R. A. 1930. *The Genetical Theory of Natural Selection.* Oxford, Oxford University Press.

Giertych, M. 1979. Summary of results on Scots pine (*Pinus sylvestris* L.) height growth in IUFRO provenance experiments. *Silvae Genetica,* **28**, 136–152.

Giles, B. E. 1983. A comparison of quantitative and biochemical variation in the wild barley *Hordeum murinum. Evolution,* **38**, 34–41.

Glover, J. G. and Abbott, R. J. 1995. Low genetic diversity in the Scottish endemic *Primula scotica* Hook. *New Phytologist,* **129**, 147–153.

Gregor, J. W. 1936. Experimental Taxonomy II. Initial population differentiation in *Plantago maritima L.* of Britain. *New Phytologist,* **37**, 15–49.

Gregor, J. W. and Watson, P. J. 1961. Ecotypic differentiation: Observations and Reflections. *Evolution,* **15**, 166–173.

Heslop-Harrison, J. 1964. Forty years of genecology. *Advances in Ecological Research,* **2**, 159–247.

Houle, D. 1992. Comparing evolvability and variability of quantitative traits. *Genetics,* **130**, 195–204.

Huntley, B. and Birks, H. J. B. 1983. *An Atlas of Past and Present Pollen Maps for Europe 0–13,000 Years Ago.* Cambridge, Cambridge University Press.

Langlet, O. 1971. Two hundred years of genecology. *Taxon,* **20**, 653–722.

Lawrence, M. J. 1984. The genetical analysis of ecological traits. In Shorrocks, B. (Ed.) *Evolutionary Ecology.* (33rd Symposium of the British Ecological Society.) Oxford, Blackwell, 27–63.

Lines, R. and Mitchell, A. F. 1964. Results of some older Scots pine provenance experiments. In *Report on Forest Research for the Year Ended March 1964.* London, HMSO, 171–194.

Lynch, M. 1996. A quantitative-genetic perspective on conservation issues. In Avise, J. C. and Hamrick, J. L. (Eds.) *Conservation Genetics. Case Histories from Nature.* New York, Chapman and Hall, 471–501.

Mitchell-Olds, T. and Rutledge, J. J. 1987. Quantitative genetics in natural plant populations: a review of the theory. *American Naturalist,* **127**, 379–402.

Page, S. E. and Rieley, J. O. 1985. The ecology and distribution of *Carex chordorrhiza* L. fil. *Watsonia,* **15**, 253–259.

Paterson, A. H., Damon, S., Hewitt, J. D., Zamir, D., Rabinowitch, H. D., Lincoln, S. E., Lander, E. S. and Tanksley, S. D. 1991. Mendelian factors underlying quantitative traits in tomato: Comparison across species, generations and environments. *Genetics,* **127**, 181–197.

Worrell, R. 1992. A comparison between European continental and British provenances of some British native trees: growth, survival and stem form. *Forestry,* **65**, 253–280.

11 IMPLICATIONS OF CLIMATE CHANGE FOR BIODIVERSITY

A. D. Watt, P. D. Carey and B. C. Eversham

Summary

1. By AD 2050, it is predicted that in the UK there will be a doubling of the atmospheric concentration of CO_2, a mean increase in temperature of 1.6°C, a 10% rise in annual precipitation, and a 30% increase in the frequency of gales. This chapter discusses the implications of these changes for biodiversity.

2. By drawing on our knowledge of past influences of climate change on plants and animals, and our current understanding of the effects of climate on individual species and their interactions, it is concluded that predicted changes in atmospheric CO_2 and associated changes in climate are likely to have detrimental effects on biodiversity.

3. Although plants and animals have survived climate change in the past, anthropogenic reductions in the abundance, range and genetic diversity of species will seriously restrict their response to future climate change.

4. The implications of climate change for biodiversity are, therefore, such that new strategies for species conservation must be developed.

11.1 Introduction

Climate change is one of several potential threats facing biodiversity in Scotland and elsewhere. Although threats such as habitat loss, poisoning, 'collecting', invasion by exotic species and pollution are important, conservation agencies have extensive experience of dealing with most of them. However, anthropogenic climate change is a new phenomenon, and although we do not know exactly how the climate will change, or when these changes will occur, it is possible that they will have a profound effect on biodiversity.

The aims of this chapter are to discuss (a) the current predictions for future climate change and some of the uncertainties; (b) the influence of climate on plants and animals, with particular emphasis on the effects of climate on individual species and species interactions; (c) the different possible responses of species to climate change, with particular emphasis on predicting the influence of future climate change on species distributions; and (d) the management implications of future climate change.

11.2 Climate change predictions

There is increasing evidence that, as a result of human activity, mainly the burning of fossil fuels in the developed world, the global climate is changing (UKCCIRG, 1996). The following summary comes from the 1996 United Kingdom Climate Change Impacts Review Group Scenario (UKCCIRG, 1996). Globally, the climate in 2050 is predicted to be about 1.6°C warmer (1.1 – 2.4°C). In the UK, temperatures are likely to rise in a way similar to that predicted for the world overall, but it is thought that temperature will rise at a slower rate over the northwest UK than over the southeast, and slightly less in winter than in summer. Extremely warm seasons and years are expected to occur more frequently. For example, a summer like that of 1995 (currently an event that occurs one year in ninety) is predicted to occur one year in three. Other predictions are that, by 2050, atmospheric CO_2 will rise to be almost double its pre-industrial concentration, the sea level will be about 37 cm higher (26–50 cm), annual precipitation will increase by about 10% and average wind speeds will increase, with the frequency of storms increasing by 30% by 2050. Notably, most evidence suggests that these predicted changes imply a rate of climate change greater than ever experienced in the past.

It is worth emphasising that these are the current 'best' estimates for climate change in the UK and worth noting, for example, that the 'best' estimate two years ago suggested that the increase in temperature in the north of the UK would be more pronounced than in the south (Viner and Hulme, 1994).

11.3 The influence of climate on biodiversity: current knowledge

It has long been known that climatic variables, particularly temperature and precipitation, affect plants and animals. For example, there are temperature thresholds for the growth and development of plants and insects, for the emergence of buds, insects and hibernating mammals in the spring, and for flowering and seed production in plants (UKCCIRG, 1996). Between upper and lower temperature limits for growth, development or other aspects of performance, there are optimum temperatures at which performance is at a maximum (see Andrewartha and Birch, 1954). The relationship between performance and temperature varies between different stages of the same species (see Régnière and Turgeon, 1989); most species have complex life-histories, which mean that they are adapted to climatic *variability*, particularly inter-season variability. However, this does not mean that they are adapted to climate *change*.

Plants and animals may undergo periods of dormancy, hibernation or aestivation, or they may migrate in anticipation of the onset of unfavourable climatic conditions. However, they may also have periods in their life cycles (e.g. leaf flushing in plants and larval emergence in insects) when they are particularly vulnerable to adverse climatic conditions. Failure to tolerate the most severe climatic conditions puts plants and animals at a competitive disadvantage, but adoption of avoidance strategies makes them vulnerable to natural and artificial climate change.

Susceptibility to climate change, particularly among temperate species, is therefore more subtle than is often thought. Although several writers have emphasised the importance of changes in the frequency and magnitude of climatic extremes, rather than changes in climatic means, it must be remembered that plants and animals are more susceptible to extremes (and to climate change in general) in some seasons than others. Put simply, warm springs may have more impact than mild winters.

The relationship between performance and temperature varies between species of the same trophic level (see Hunt and Neal, 1988) and between species at different trophic levels, such as plants and their herbivores, insects and their predators and parasites (see Uvarov, 1931). It is, therefore, difficult to predict the effect of climate change on interspecies interactions such as competition, pollination and predation. Thus, within the climatically suitable area of a species, its range may be further limited by interactions with other species.

For example, the emergence of the larvae of many insects must synchronise with bud burst of their host plants. Dewar and Watt (1992) showed that climate warming would disrupt the synchrony between winter moth larvae and one of their host plants, Sitka spruce. However, Buse and Good (1997) showed that the emergence of winter moth larvae remained synchronous with bud burst of another of its host plants, oak, when the temperature was increased by 3°C. Randall (1982a,b) investigated the population dynamics of the moth *Coleophora alticolella* along an altitudinal transect in northern England. At the highest altitude, the abundance of *C. alticolella* is limited by low availability of larval food, seeds of the rush *Juncus squarrosus*, and at the lowest altitude by parasitism, which becomes more intense with decreasing altitude. That is, the altitudinal difference in climate does not have a great influence directly on the moth but it has a profound influence through the insect–parasitoid relationship at low altitudes (and high temperatures) and the insect–plant relationship at high altitudes (and low temperatures).

As well as interactions with other species, local habitat factors such as soil type or land use may additionally limit the range of plants and animals. However, despite the fact that the distribution of plants and animals is influenced by factors other than temperature, the magnitude of the effect of temperature can be seen by examining both the distribution of individual species and trends in species richness in relation to latitudinal variation in temperature.

At the individual species level, for example, the natural distribution of the small-leaved lime (*Tilia cordata*) is limited not by the effects of temperature on tree growth – individuals introduced to Scotland survive and grow well – but by the fact that seed production is inhibited below about 15°C (Pigott and Huntley, 1981). The northerly limit of the distribution of sites in the UK where seed production occurs consistently correlates well with August maximum temperature isotherms.

In terms of species richness, most groups of species show a decrease in diversity with increasing latitude. For example, in the UK there is a decline in the number of grasshoppers, crickets, dragonflies and butterflies with increasing latitude (Watt *et al.*, 1990). In global terms, orchids, non-oceanic birds, mammals, fish, stream invertebrates, marine invertebrates, ants, beetles and most other taxa are all more diverse in the tropics than in temperate latitudes, and this diversity is due to more than the greater diversity of distinctive habitats in the tropics (Stevens, 1989). It is clear that climate is a major determinant of overall species richness.

Further evidence of the importance of climate comes from historical distributions of plants and animals as shown by pollen deposits and fossil records. Pollen records show that the limit of several tree species has moved northwards in response to climate warming since the last glaciation (see Birks, this volume). The estimated rate of spread varies from species to species: in eastern North America following the retreat of the ice, jack and white pines spread northwards at 250–500 m yr^{-1}, whereas chestnut moved at 100 m yr^{-1} (Bennett, 1986).

The fossil record shows large-scale geographic changes in the distribution of Coleoptera and other insects in response to climate change during the latest glacial–interglacial cycle. Some species of beetle now found in northern Scandinavia occurred throughout the UK during previous 'cold' episodes. Others, such as *Bembidion octomaculatum*, were widespread in the UK during a brief period of high temperatures, but are now found much further south and rarely in the UK (although, interestingly, *B. octomaculatum* became established in Sussex in the late 1980s). Overall, species extinctions of Coleoptera and other insects on a global scale during this cycle appear to have been uncommon despite the rapidity of climate change at the end of the glacial era (Coope, 1995).

Increasing atmospheric CO_2 is predicted to have a beneficial effect on plant growth as a result of enhanced photosynthesis, reduced photorespiration and increased water-use efficiency (Bazzaz, 1990). Analysis of recent growth trends of trees suggests that rising atmospheric CO_2 is already having an effect, at least in Europe (Spiecker *et al.*, 1996; but see Graumlich, 1991). If trees do respond positively to elevated atmospheric CO_2, the consequences for biodiversity are not necessarily beneficial. Increased growth and associated changes in resource partitioning of trees may expose them to an increased risk of windthrow and drought (Possingham, 1993). In addition, forest and woodland understorey species with specific light and moisture requirements may suffer if the response of trees to increasing temperature and atmospheric CO_2 leads to an increased leaf-area index and darker forests (Possingham, 1993). Because plant species vary in their response to elevated CO_2, shifts in the composition of plant communities are also likely (Hunt *et al.*, 1991; Schwartz, 1992). Herbivores face a potential problem because elevated atmospheric CO_2 leads to a reduction in the food quality of their host plants (specifically a decline in plant nitrogen). Much research on insect herbivores, particularly pest species, has shown that it is unlikely that insect pests will become more serious as a result of increasing atmospheric CO_2 (Watt *et al.*, 1995, 1996; Docherty *et al.*, in press). However, the potentially negative conse-

PLATE 9 RAISED BOG

Typical raised bog, Flanders Moss, Stirling (Photo: L. Gill).

Green hairstreak butterfly (Callophrys rubi) feeding from flowers of Labrador tea (Ledum palustre) at Flanders Moss (Photo: M.B. Usher).

Commercial mechanised peat cutting on a lowland raised bog (Photo: L.V. Fleming).

PLATE 10   BLANKET BOG

Pools on blanket bog, Blar nam Faoileag (Photo: D.B.A. Thompson).

*Large heath
butterfly (Coenonympha tullia), widely
distributed on Scottish peatlands
(Photo: M.B. Usher).*

*The insectivorous intermediate bladderwort
(Utricularia intermedia, sensu lato),
Rannoch Moor (Photo: M.B. Usher).*

*Greenshank (Tringa nebularia): confined
as a breeding bird in Britain to bogs in
north and west Scotland (Photo: D.B.A
Thompson).*

quences of increasing atmospheric CO_2 for the diversity of insect herbivores has not yet been addressed.

11.4   Predicting the influence of climate change on biodiversity

Plants and animals will respond to climate change in one or more of four ways (Possingham, 1993): by tolerating the change in climate (without adaptation), by adapting (genetically), by changing distribution, or by becoming locally extinct.

The tolerance of some species to a wide range of climatic conditions is demonstrated by their large geographic ranges. For example, one of Scotland's best-known rare birds, the osprey (*Pandion haliaetus*), breeds in every continent apart from South America and Antarctica. There are many other examples of widely distributed organisms such as the barn owl (*Tyto alba*) and bracken (*Pteridium aquilinum*). In addition, the fossil record shows that the range of some species of plant has changed very little in response to past climate change, such as the juniper (*Juniperus osteospermum*) in the Great Basin area of the USA over 30,000 years (Nowak *et al.*, 1994). Such resilience to climate may be due to high genetic diversity within or between populations of these species (Begon *et al.*, 1996), and implies that the conservation of genetic diversity is not just an end in itself but an essential means of conserving species.

There has been little research on the way that species might respond to climate change by genetically based changes in tolerance or life-history parameters. Partridge *et al.* (1994) showed that *Drosophila* fruit flies adapt (in terms of survival, growth and development) to changes in temperature within five years (of continuous exposure). Species with much longer life-cycles may, however, not be able to adapt genetically to climate change. Billington and Pelham (1991) studied the capacity of Scottish populations of birch (*Betula pubescens* and *B. pendula)* to adapt to climatic warming by a change in the mean date of bud burst. They suggested that the heritability of bud burst was such that the predicted rate of change in temperature is too rapid for native populations of birch to evolve earlier dates of bud burst. A further problem with adaptation to climate change is that, in addition to global warming, other abiotic and biotic factors are likely to be changing rapidly, and plants and animals have a limited ability to adapt to multiple changes in their environments (Holt, 1990).

In the absence of, and perhaps in addition to, tolerance of climate change, species will move. There has been much recent research on predicting the bio-geographic response of plants and animals to climate change (see Carey and Brown, 1994; Sutherst *et al.*, 1995). Underlying this research is the hypothesis that individual species each have a multidimensional response to environmental factors including climatic factors, defined as the niche (Hutchinson, 1957). These physiological or potential niches are constrained in reality by interactions with other species so that each species has a realised niche (Austin, 1980, 1992). Determining the physiological and realised response to all climate variables for one, not to mention all, species is impracticable. However, the distribution limits of many species can be described adequately by correlating the current distribution with a few standard climate variables. For plants, measures of growing season, absolute minimum temperature and available moisture may adequately describe the realised niche or 'climate envelope' of a species. Thus, for example, the change in distribution of globeflower (*Trollius europaeus*) within the next 60 years can be predicted (Figure 11.1). This approach may be refined by including non-climatological factors (Carey *et al.,* 1995; Carey and Brown, 1994).

This type of approach may be used to predict the impact of climate change on the distribution of rare species. Elmes and Free (1994) recently published a summary of research on the effect of climate change on British Red Data Book (RDB) species. They showed that a temperature rise of 2°C would lead to no change in the distribution of 71% of RDB species, a reduction in range of 50% in 14% of RDB species and the possible loss from Britain of 15% of RDB species. Failure to estimate the ability of species to move in response to climate change may underestimate the impact of climate change in these analyses (see below).

11.5 Management strategies to reduce the impact of climate change on biodiversity

Our general understanding of the ecology of plants and animals, analyses of the distribution of individual species, and our understanding of the consequences of past changes in climate, all strongly suggest that, as a result of climate change, there will be a movement of species towards the poles and to higher altitudes (see Possingham, 1993). Species restricted to the poleward end of continents and montane species will, therefore, be most at risk (Peters and Darling, 1985; Holten and Carey, 1992; Crawford, 1995): they will have nowhere else to go. Thus, although we are likely to see an increase in biodiversity in Scotland as species move northwards, the geographical position of Scotland and the presence of so many montane species mean that there is also likely to be a loss of species, such as ptarmigan (*Lagopus mutus*) and mountain bearberry (*Arctostaphylos alpinus*), some of which are typical or rare and which are therefore currently afforded high conservation status.

However, this view is rather biased towards the effects of climate on individual species and the assumption that species can respond, and have time to respond. It must be emphasised that species interact and that the net response to climate change will depend upon the interaction between species of the same and different trophic levels (Peters, 1988). Although the responses of particular species to changes in climate are unknown, it is probable that different species will respond in different ways and to different degrees; this will lead to changes in the balance between competing species, plants and herbivores, and insects and their predators (Lawton, 1995). This may result in local extinctions of species if currently benign insect herbivores become damaging 'pests' when they spread faster than their predators and parasitoids.

Differences in the response of species to climate change is likely to result in the formation of different plant and animal communities with novel interactions (between competitors, plants and herbivores, herbivores and predators, etc.)

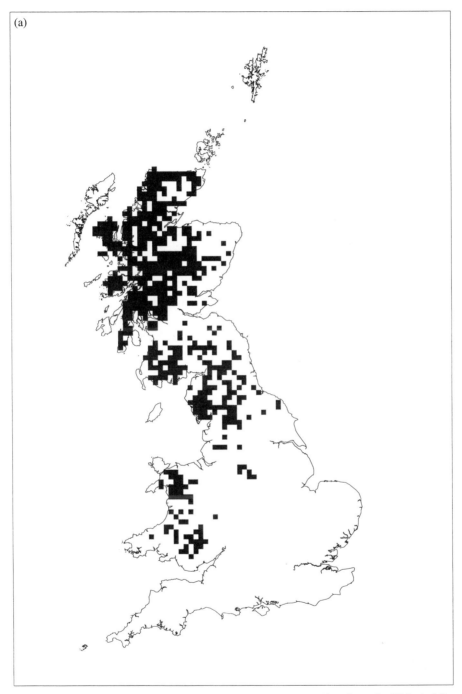

Figure 11.1   (a) Known distribution of globeflower (*Trollius europaeus*) (data from the Biological Records Centre).

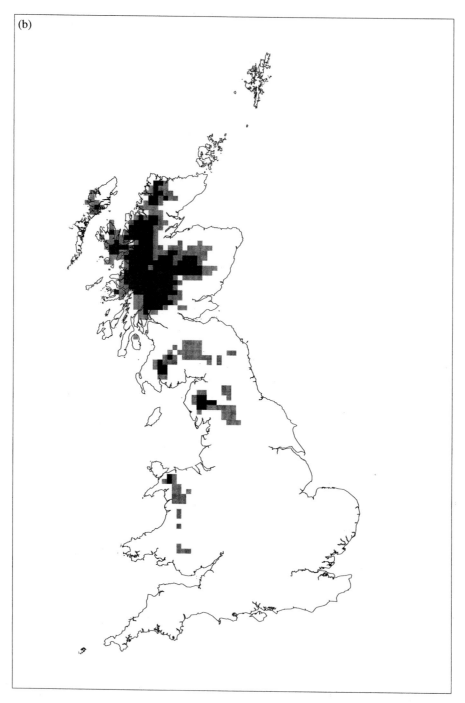

Figure 11.1   (b) Predicted current distribution using January minimum temperature, July maximum temperature and annual precipitation as independent variables in a logistic regression.

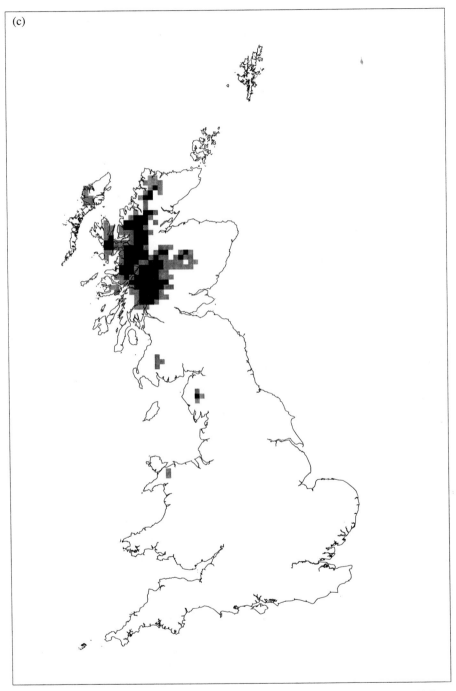

Figure 11.1 (c) Predicted distribution in 2050 using the same model as (b) but imposing the UK transient climate scenario for 2050.

(Peters, 1988). Some of these interactions could threaten biodiversity, resulting in loss of mutualists (such as specialist pollinators) and the emergence of new diseases and increasing numbers of epidemics (Holmes, 1996). Again, local extinctions are likely.

In the face of a direct or indirect decline in the suitability of the local environment as a result of climate change, the ability of species to disperse to new habitats is critical and is dependent on the presence of new habitats or environments, the distance to those new habitats or environments, and the dispersal ability of each species. The speed of climate change in the next 100 years will severely limit the response of species with poor rates of dispersal (Peters, 1990; Fowler and Loodinkins, 1992).

In terms of management, one additional aspect compounds the threat of climate change itself: the erosion of natural habitats and the restriction of many species to protected areas or other refuges. The negative impacts of climate change are made worse because protected areas may cease to be suitable environments for the species that they exist to protect. In addition, protected areas (or other suitable areas) with the right climatic conditions after climate change either may not exist, or may be too far for the threatened species to disperse to readily (Peters and Darling, 1985; McNeely, 1994; Lawton, 1995). In other words, past changes in climate occurred when the landscape was relatively intact. Movement of many species in response to future climate change will be severely restricted by the increasingly fragmented nature of Scottish landscapes.

We clearly need to know more through research about the likely impact of climate change. There are current studies across altitudinal and latitudinal transects; experimental studies based on the manipulation of existing environments and the creation of artificial ones; and the development of models to predict the response of species to climate change (Beerling and Woodward, 1996), especially under the NERC Terrestrial Initiative for Global and Environmental Research (see Cummins *et al.* (1995) for a summary of this research programme).

As well as providing information on the future response of species to climate change, some of this research is already producing evidence of the effects of current climate change. For example, Chapin *et al.* (1995) detected changes in species diversity in untreated plots over a 9-year period in the Arctic. Research in Mexico, the US and Canada on the checkerspot butterfly (*Euphydras editha*) has shown that populations of this species more frequently become extinct at the southern end of its range (Parmesan, 1996), evidence for the first time across the entire range of a species that it is currently shifting in response to climate change. There is also evidence that spring events are occurring earlier: the appearance of some migrating animals and the leafing dates of common species of plant have been getting earlier in the calendar year over the past 250 years, since the end of the mini ice age, in eastern England (Sparks and Carey, 1995). The many studies of recent range changes of species in Britain are tentative in their implication of climate change as a cause, but some have shown a temperature-related response: the flight period of the hedge brown butterfly (*Pyronia tithonus*) is increasing as it spreads north (Pollard, 1991). This species occurred in Scotland in the 19th century and seems likely to return within the next two decades.

We need to know where plants and animals would move, if free to do so in a future climate, so that species conservation efforts can be concentrated where they will be effective. There is little point in trying to conserve species where their environment is no longer suitable; more importantly, there is a need to identify areas where new protected areas (or conservation efforts in general) should be targeted. From the point of view of conservation, the most worrying aspect of climate change is that the current distribution of protected areas may be inappropriate in the future.

It is time we started to plan to protect the biodiversity of Scotland in the future but we are not currently in a position to identify accurately existing or new areas which might serve as future protected areas. However, we can take practical steps by: increasing the number of protected areas, particularly in areas of the country where they do not currently exist, and focusing on areas with altitudinal variation; linking protected areas as far as possible with wildlife 'corridors' or 'stepping stones' (for all major habitats); developing protocols to translocate species that cannot disperse fast enough; and promoting conservation outside protected areas. As the Scotland of the future may be very different from that of the present, such steps may require the development of novel and far-sighted approaches if biodiversity is to be effectively conserved.

Acknowledgements

We thank our colleagues in the Institute of Terrestrial Ecology and elsewhere for encouragement and advice during our research on the implications of climate change for biodiversity, particularly Melvin Cannell, Mark Hill, and John Good for constructive comments on the manuscript. The support of the NERC TIGER and BBSRC BAGEC programmes is acknowledged.

References

Andrewartha, H. G. and Birch, L. C. 1954. *The Distribution and Abundance of Animals*. Chicago, University of Chicago Press.

Austin, M. P. 1980. Searching for a model for use in vegetation analysis. *Vegetatio*, **42**, 11–21.

Austin, M. P. 1992. Modelling the environmental niche of plants: implications for plant community response to elevated CO_2 levels. *Australian Journal of Botany*, **40**, 515–525.

Bazzaz, F. A. 1990. The response of natural ecosystems to the rising global CO_2 levels. *Annual Review of Ecology and Systematics*, **21**, 167–196.

Beerling, D. J. and Woodward, F. I. 1996. *In situ* exchange responses of boreal vegetation to elevated CO_2 and temperature: first season results. *Global Ecology and Biogeography Letters*, **5**, 117–127.

Begon, M., Harper, J. L. and Townsend, C. R. 1996. *Ecology*. Oxford, Blackwell Science.

Bennett, K. D. 1986. On the rate of spread and population increase of forest trees during the postglacial. *Philosophical Transactions of the Royal Society of London*, B **314**, 523–531.

Billington, H. L. and Pelham, J. 1991. Genetic variation in the date of bud burst in Scottish birch populations: implications for climate change. *Functional Ecology*, **5**, 403–409.

Buse, A. and Good, J. 1997. Synchorization of larval emergence in winter moth (*Operphtera brumata* L.) and budburst in pedunculate oak (*Quercus robur* L.) under stimulated climate change. *Ecological Entomology*, **21**, 335–343.

Carey, P. D. and Brown, N. J. 1994. The use of GIS to identify sites that will become suitable for a rare orchid, *Himantoglossum hircinum* L., in a future changed climate. *Biodiversity Letters*, **2**, 117–123.

Carey, P. D., Preston, C. D., Hill, M. O., Usher, M. B. and Wright, S. M. 1995. An environmentally defined biogeographical zonation of Scotland designed to reflect species distributions. *Journal of Ecology*, **83**, 833–845.

Chapin, F. S., Shaver, G. R., Giblin, A. E., Nadelhoffer, K. J. and Laundre, J. A. 1995. Responses of Arctic tundra to experimental and observed changes in climate. *Ecology*, **76**, 694–711.

Coope, G. R. 1995. The effects of Quaternary climatic changes on insect populations: lessons from the past. In Harrington, R. and Stork, N. E. (Eds.) *Insects in a Changing Environment*. London, Academic Press, 29–48.

Crawford, R. M. M. 1995. Plant survival in the high Arctic. *Biologist*, **42**, 101–105.

Cummins, C. P., Beran, M. A., Bell, B. G. and Oliver, H. R. 1995. The TIGER programme. *Journal of Biogeography*, **22**, 897–905.

Dewar, R. C. and Watt, A. D. 1992. Predicted changes in the synchrony of larval emergence and budburst under climatic warming. *Oecologia*, **89**, 557–559.

Docherty, M., Salt, D. T. and Holopainen, J. K. In press. The impacts of climate change on pollution on forest pests. In Watt, A. D., Stork, N. E. and Hunter, M. D. (Eds.) *Forests and Insects*. London, Chapman and Hall.

Elmes, G. W. and Free, A. 1994. *Climate Change and Rare Species in Britain*. London, HMSO.

Fowler, D. P. and Loodinkins, J. A. 1992. Breeding strategies in a changing climate and implications for biodiversity. *Forestry Chronicle*, **68**, 472–475.

Graumlich, L. J. 1991. Subalpine tree growth, climate, and increasing CO_2: an assessment of recent growth trends. *Ecology*, **72**, 1–11.

Holmes, J. C. 1996. Parasites as threats to biodiversity in shrinking ecosystems. *Biodiversity and Conservation*, **5**, 975–983.

Holt, R. D. 1990. The microevolutionary consequences of climate change. *Trends in Ecology and Evolution*, **5**, 311–315.

Holten, J. I. and Carey, P. D. 1992. *Responses to Climate Change of Natural Terrestrial Ecosystems in Norway*. Trondheim, NINA.

Hunt, R., Hand, D. W., Hannah, M. A. and Neal, A. M. 1991. Response to enrichment of 27 herbaceous species. *Functional Ecology*, **5**, 410–421.

Hunt, R. and Neal, A. M. N. 1988. *Annual Report of the Unit of Comparative Plant Ecology*. Sheffield, UCPE.

Hutchinson, G. E. 1957. Concluding remarks. *Cold Spring Harbor Symposium on Quantitative Biology*, **22**, 415–427.

Lawton, J. H. 1995. The response of insects to environmental change. In Harrington, R. and Stork, N. E. (Eds.) *Insects in a Changing Environment*. London, Academic Press, 3–26.

McNeely, J. A. 1994. Protected areas for the 21st-century – working to provide benefits to society. *Biodiversity and Conservation*, **3**, 390–405.

Nowak, C. I., Nowak, R. S., Tausch, R. J. and Wigand, P. E. 1994. Tree and shrub dynamics in northwestern Great Basin woodland and shrub steppe during the Late Pleistocene and Holocene. *American Journal of Botany*, **8**, 265–277.

Parmesan, C. 1996. Climate and species' range. *Nature*, **382**, 765–766.

Partridge, L., Barrie, B., Fowler, K. and French, V. 1994. Thermal evolution of pre-adult life-history traits in *Drosophila melanogaster*. *Journal of Evolutionary Biology*, **7**, 645–663.

Peters, R. L. 1988. The effect of global climatic change on natural communities. In Wilson, E. O. (Ed.) *Biodiversity*. Washington, D.C., National Academy Press, 450–461.

Peters, R. L. 1990. Effects of global warming on forests. *Forest Ecology and Management*, **35**, 13–33.

Peters, R. L. and Darling, J. D. S. 1985. The greenhouse effect and nature reserves. *Bioscience*, **35**, 707–717.

Pigott, C. D. and Huntley, J. P. 1981. Factors controlling the distribution of *Tilia cordata* at the northern limits of its geographical range III. Nature and causes of seed sterility. *New Phytologist*, **87**, 817–839.

Pollard, E. 1991. Changes in the flight period of the hedge brown butterfly *Pyronia tithonus* during range expansion. *Journal of Animal Ecology*, **60**, 737–748.

Possingham, H. P. 1993. Impact of elevated atmospheric CO_2 on biodiversity: Mechanistic population-dynamic perspective. *Australian Journal of Botany*, **41**, 11–21.

Randall, M. G. M. 1982a. The dynamics of an insect population throughout its altitudinal distribution – *Coleophora alticolella* (Lepidoptera) in northern England. *Journal of Animal Ecology*, **51**, 993–1016.

Randall, M. G. M. 1982b. The ectoparasitization of *Coleophora alticolella* (Lepidoptera) in relation to its altitudinal distribution. *Ecological Entomology*, **7**, 177–185.

Régnière, J. and Turgeon, J. J. 1989. Temperature-dependent development of *Zeiraphera canadensis* and simulation of its phenology. *Entomologia Experimentalis et Applicata*, **50**, 185–193.

Schwartz, M. W. 1992. Potential effects of global climate change on the biodiversity of plants. *Forestry Chronicle*, **68**, 462–471.

Sparks, T. H. and Carey, P. D. 1995. The responses of species to climate over 2 centuries – an analysis of the Marsham phenological record, 1736–1947. *Journal of Ecology*, **83**, 321–329.

Spiecker, H., Mielikäinien, K., Köhl, M. and Skovsgaard, J. 1996. *Growth Trends in European Forests*. Berlin, Springer.

Stevens, G. C. 1989. The latitudinal gradient in geographical range: how so many species coexist in the tropics. *American Naturalist*, **133**, 240–256.

Sutherst, R. W., Maywald, G. F. and Skarrat, D. B. 1995. Predicting insect distributions in a changed climate. In Harrington, R. and Stork, N. E. (Eds.) *Insects in a Changing Environment*. London, Academic Press, 59–91.

United Kingdom Climate Change Impacts Review Group 1996. *Review of the Potential Effects of Climate Change in the United Kingdom*. London, HMSO.

Uvarov, B. P. 1931. Insects and climate. *The Transactions of the Entomological Society of London*, **79**, 1–247.

Viner, A. and Hulme, M. 1994. *The Climate Change Impacts Link Project: Providing Climate Change Scenarios for Impact Assessment in the UK*. Norwich, Climatic Research Unit, University of East Anglia.

Watt, A. D., Lindsay, E., Leith, I. D., Fraser, S. M., Docherty, M., Hurst, D., Hartley, S. E. and Kerslake, J. 1996. The effects of climate change on the winter moth, *Operophtera brumata*, and its status as a pest of broadleaved trees, Sitka spruce and heather. *Aspects of Applied Biology*, **45**, 307–316.

Watt, A. D., Ward, L. K. and Eversham, B. C. 1990. Effects on animals: invertebrates. In Cannell, M. G. R. and Hooper, M. D. (Eds.) *The Greenhouse Effect and Terrestrial Ecosystems*. London, HMSO, 32–37.

Watt, A. D., Whittaker, J. B., Docherty, M., Brooks, G., Lindsay, E. and Salt, D. T. 1995. The impact of elevated atmospheric CO_2 on insect herbivores. In Harrington, R. and Stork, N. E. (Eds.) *Insects and Environmental Change*. London, Academic Press, 197–217.

PART THREE
SUSTAINABLE MANAGEMENT FOR BIODIVERSITY

PART THREE

SUSTAINABLE MANAGEMENT FOR BIODIVERSITY

Some 20% of the total land area of Scotland is designated either as Site of Special Scientific Interest, National Nature Reserve or National Scenic Area (Anon., 1995). Such protected areas are central to the conservation of the natural heritage of Scotland, even if their effectiveness may sometimes be difficult to demonstrate (see Fleming, this volume). However, it is increasingly being recognised that, in any country, the designation of protected areas can only be a partial solution to the conservation of biodiversity (Cassells, 1995). Indeed, the majority of species and habitats, including many that are scarce or threatened, occur outside reserves in the landscape 'matrix'. Much of the Scottish landscape has been subjected to human use for millennia, and continues to be an important source of economic revenue for many rural communities. The key challenge is therefore to identify methods of land management which maintain economic productivity, while maximising the value to wildlife. This objective is central to the current concept of sustainable development, which has been the subject of a variety of recent policy initiatives, including Agenda 21 (Holdgate, 1996).

Four natural resources have historically been particularly important in Scotland: fish, forests, game and agriculture. In this section, each of these resources is considered individually in different chapters. Maitland highlights the high economic and conservation value of Scotland's freshwater fish stocks, but notes that they are undervalued and under threat. Fisheries management, and the recent expansion of fish farming, have partly been responsible for the decline in stocks of native fish. To develop management approaches that are genuinely sustainable, major shifts in attitude are required, accompanied by the development and implementation of suitable policies. The management of marine resources has been discussed by Matthews *et al.* (this volume).

In the case of forests, the situation is perhaps more optimistic. Newton and Humphrey note that there has been a major shift in forest policy during recent years, and the need to manage forests for biodiversity is now central to the overall forest design and management process. Efforts at restoration of Scottish native forests are at an all-time high. Despite a wealth of current initiatives, and the formation of new partnerships between different organisations with an interest in the future of Scottish forests, significant challenges remain. In particular, difficult decisions will need to be made where management strategies with different biodiversity objectives conflict.

In terms of the impact on the landscape, the management of land for game is of exceptional importance in Scotland. As noted by Thompson *et al.*, some 80% of the total upland area of Scotland is managed for grouse shooting, and some 13–19% of the total land area of the country is managed primarily for red deer. Although management of the game species themselves is now relatively well understood, the impacts of such management on other components of biodiversity are far less well known. Sustainable management for game requires an understanding of such impacts so as to ensure that biodiversity within managed areas is maintained and enhanced.

Agricultural land will perhaps be the most important challenge to biodiversity conservation within Scotland in the coming years. Many recent declines in habitat quality (such as hedgerows) and in the abundance of particular species (e.g. songbirds; see Racey, this volume) have been attributed to the continual intensification of agriculture throughout this century. The international policy initiatives that have encouraged the development of sustainable forest management practices are likely to result in increased pressure for a similar evolution in farming practice. It is hoped that this will ensure that biodiversity is maintained in agricultural land. Foster *et al.* indicate that farmland may be an important habitat for certain groups of species, such as ground beetles. Such research highlights the current lack of information about the species associated with agricultural land, a point reiterated by French *et al.* in their account of a recent research initiative. The provision of such information is clearly vital if sustainable management practices are to be developed. Scotland already provides some internationally important examples of how agricultural land management may be combined with conservation objectives, such as the partnerships between conservation organisations and crofting communities in the Western Highlands and Islands and the management of grazing land for geese on Islay.

All of the chapters in this section highlight the need for further research to assist land managers in the decision-making process, both by the provision of improved information and by practical tools. The outputs of such research will need to be coupled with the further development of appropriate policies, and the provision of suitable mechanisms to ensure their successful implementation. A number of additional challenges remain if the sustainable management of Scotland's natural resources is to be achieved in the coming millennium. Most importantly, landowners, land managers and planning authorities will need to collaborate closely to ensure that different land uses are integrated in a manner conducive to conservation of biodiversity. Such approaches require consideration of different land uses at the landscape scale, and an understanding of how the availability of different habitats influences the distribution of individual organisms. The understanding of spatial patterns and processes in the landscape, the theme of another recent conference held in Scotland (Simpson and Dennis, 1996), is therefore fundamental to sustainable management for biodiversity throughout rural Scotland.

References

Anon. 1995. *The Natural Heritage of Scotland: an Overview*. Perth, Scottish Natural Heritage.

Cassells, D. S. 1995. Considerations for effective international cooperation in tropical forest conservation and management. In Sandbukt, O. (Ed.) *Management of Tropical Forests: Towards an Integrated Perspective*. University of Oslo, Centre for Development and the Environment, 357–375.

Holdgate, M. 1996. *From Care to Action: Making a Sustainable World*. London, Earthscan.

Simpson. I. A. and Dennis, P. (Eds.) 1996. *The Spatial Dynamics of Biodiversity: Towards an Understanding of Spatial Patterns and Processes in the Landscape*. University of Stirling, The International Association for Landscape Ecology (UK Region).

12 SUSTAINABLE MANAGEMENT FOR BIODIVERSITY: FRESHWATER FISHERIES

P. S. Maitland

Summary

1. The freshwater fish fauna of Scotland includes 35 native species (one of which is extinct) and 5 alien species. The 'native' species include several that have only recently been introduced from England.

2. The main factors currently affecting native stocks and their diversity include land use, pollution, engineering developments and the management of freshwater fisheries and fish farms.

3. Scotland's freshwater fish stocks are a highly valuable resource, at present under-valued and under threat. The existing protection, conservation and management of freshwater fish populations is fragmented and greatly biased in favour of game species.

4. Sustainable management of Scotland's fish fauna for biodiversity has yet to be seriously addressed. On the contrary, fisheries management and fish farming have themselves caused significant losses of, and damage to, individual stocks of native fish.

5. To reverse this process and save what remains of the truly native stocks of freshwater fish in Scotland, a radical change of policy and attitude is necessary. With a positive approach it is not too optimistic to believe that fishermen and native fish can both have a sustainable future.

12.1 Introduction

The valuable native fish populations of Scotland have suffered substantial losses over the past two hundred years. The concept of sustainable management of Scotland's freshwater fish fauna for biodiversity is an admirable target for the future, but one that has not been seriously addressed in the past. In order to reverse this process and save what remains of the truly native stocks of freshwater fish in Scotland, a radical change of policy and attitude in relation to the management of freshwater fish and fisheries is necessary. Such a policy has recently been proposed by Maitland (1996); this chapter reviews the role of that policy in reversing the damaging trends that are affecting Scotland's valuable freshwater fish and fisheries.

Biodiversity, in the context of the Scottish ichthyofauna, is defined here as the range of native fish species naturally present in each of Scotland's fresh waters, including the full span of genetic diversity and phenotypic expression contained within all the populations concerned.

12.2   Fish habitat

The fresh waters of Scotland are a major resource, comprising over 31,000 lochs and lochans and 5,000 rivers and burns. These include some of the most spectacular waters in Great Britain as well as the largest: by surface area, Loch Lomond (71 km^2); by length, Loch Awe (41 km); by depth, Loch Morar (310 m); and by volume, Loch Ness (7,452 million m^3). The River Tay is the largest river by flow (194 m^3 s^{-1}).

The demands on this resource are enormous and often conflicting. Each person requires about 100 litres a day and many reservoirs for water supply have been built in the hills. Industry too has huge demands. Wastes from domestic sewage and industry are passed to rivers to be carried to the sea. In addition, many adjacent land uses and activities, such as the application of fertilisers, herbicides and pesticides used in agriculture and forestry, may affect nearby watercourses. Land drainage means that water runs off faster, resulting in higher (and then lower) flows in associated rivers. Fish farming has recently created new demands. Hydro-electric schemes have harnessed most larger mountainous rivers. Finally, water is of major importance for recreation: humans picnic beside it, birdwatch over it, paddle and bathe in it, boat on it and fish in it.

Thus, there are problems, both for humans and for wildlife. Many waters have been so misused that they are unfit for either. Over-enrichment from agricultural fertilisers causes algal blooms on lowland lochs; when these blooms reach a maximum and die, they may use up all the available oxygen and fish kills result. Domestic and industrial pollution can eliminate all aquatic plants and animals. These stretches of extreme pollution also act as a barrier to migrating fish, which cannot occupy clean stretches upstream. Weirs and dams have the same effect. Aerial pollution has acidified a number of hill lochs, eliminating fish. Other problems are created by the introduction of new plant or fish species or by humans trampling vegetation and disturbing wildlife.

With so many waters in Scotland there should be enough to go round, but present national planning is inadequate and there are many problems and conflicts to be resolved (Maitland *et al.*, 1994). What is needed is a national framework and policy for the lochs and rivers of Scotland that considers their distribution, quality and value in relation to current demands. Some are clearly essential to a major need, say, hydro-electricity or water supply, and this must be the overriding factor, although lesser uses may fit in with this. Other waters may be of such wildlife importance that their conservation needs are paramount. Several of the largest waters are so important nationally (e.g. Loch Lomond) that each requires an individual management plan to reconcile the needs of potential users to the overwhelming importance of maintaining the quality and value of that water. In all these cases, users must be prepared to accept compromises and even give up claims somewhere, if not on one water then on another.

12.3   Native fish

The freshwater fish fauna of Scotland is substantially impoverished compared with the communities found further south in Europe. Nevertheless, 41 out of the 56 species found in the British Isles as a whole are found here (Table 12.1) and the number is very gradually increasing as more species appear from the south. Taking the starting point of the fish communities of Scotland as the closing stages of the last ice age, it is clear that euryhaline fish, many of which come into fresh water to spawn, had no difficulty in invading new waters as the ice receded. Thus sturgeon (*Acipenser sturio*), shads, sparling (*Osmerus eperlanus*), sea bass (*Dicentrarchus labrax*), gobies and mullets must have occurred in Scottish estuaries for thousands of years (Maitland, 1977).

Apart from these species, the only fish that were able to colonise truly fresh waters as the ice receded were also those with marine affinities and capable of existing in the ice lakes and glacial rivers that prevailed at that time. At most there were then probably only about 12 species, most notable among which were lampreys, Atlantic salmon (*Salmo salar*), brown trout (*S. trutta*), Arctic charr (*Salvelinus alpinus*), powan (*Coregonus lavaretus*), vendace (*C. albula*), eel (*Anguilla anguilla*), sticklebacks and flounder (*Platichthys flesus*). By about AD 1790 only another five species had been added to the Scottish fauna: pike (*Esox lucius*), minnow (*Phoxinus phoxinus*), roach (*Rutilus rutilus*), stone loach (*Barbatulus barbatulus*) and perch (*Perca fluviatilis*). Some of these may have moved north naturally; others were transferred by humans. Ninety years later (1880) another five species had been translocated and were known to be established in Scotland: brook charr (*Salvelinus fontinalis*) (from North America), grayling (*Thymallus thymallus*), tench (*Tinca tinca*), bream (*Abramis brama*) and chub (*Leuciscus cephalus*) (all from England). By 1970 another eight species (rainbow trout (*Oncorhynchus mykiss*), carp (*Cyprinus carpio*), goldfish (*Carassius auratus*), gudgeon (*Gobio gobio*), rudd (*Scardinius erythrophthalmus*), orfe (*Leuciscus idus*), dace (*L. leuciscus*) and bullhead (*Cottus gobio*)) were known to have established viable populations in Scotland; humans appear to have been responsible for the introduction of all of them. The latest species in this saga is the ruffe (*Gymnocephalus cernuus*), discussed below.

Thus the present freshwater fish fauna of Scotland is a mixture of natural immigrants from the sea and from further south along with many more recent fish that have been brought in by humans from England, continental Europe and even North America. The situation is by no means stable and other arrivals are likely in future years. These will certainly add to the 'diversity' of Scottish fish

Table 12.1   A check list of freshwater fishes in Scotland.

Species	Origin	Status	Value
Sea lamprey *Petromyzon marinus*	native	decline	—E
River lamprey *Lampetra fluviatilis*	native	decline	—E
Brook lamprey *Lampetra planeri*	native	decline	—
Common sturgeon *Acipenser sturio*	native	decline[a]	—E
Allis shad *Alosa alosa*	native	decline	—E
Twaite shad *Alosa fallax*	native	decline	—E
Atlantic salmon *Salmo salar*	native	decline	SE
Brown trout *Salmo trutta*	native	decline	SE
Rainbow trout *Oncorhynchus mykiss*	alien	increase	SE
Arctic charr *Salvelinus alpinus*	native	decline	SE
Brook charr *Salvelinus fontinalis*	alien	increase	SE
Powan *Coregonus lavaretus*	native	stable	—E
Vendace *Coregonus albula*	native	extinct[b]	—E
Grayling *Thymallus thymallus*	(native)	increase	SE
Sparling *Osmerus eperlanus*	native	decline	SE
Pike *Esox lucius*	native	increase	SE
Common carp *Cyprinus carpio*	alien	increase	SE
Crucian carp *Carassius carassius*	(native)	increase	—E
Goldfish *Carassius auratus*	alien	decline	—
Gudgeon *Gobio gobio*	(native)	increase	—E
Tench *Tinca tinca*	(native)	increase	SE
Common bream *Abramis brama*	(native)	increase	SE
Minnow *Phoxinus phoxinus*	native	increase	—
Rudd *Scardinius erythrophthalmus*	(native)	increase	SE

continued

Table 12.1 — *continued.*

Species	Origin	Status	Value
Roach *Rutilus rutilus*	native	increase	SE
Chub *Leuciscus cephalus*	(native)	increase	SE
Orfe *Leuciscus idus*	alien	decline	—E
Dace *Leuciscus leuciscus*	(native)	increase	SE
Stone loach *Noemacheilus barbatulus*	native	stable	—
European eel *Anguilla anguilla*	native	decline	SE
Three-spined stickleback *Gasterosteus aculeatus*	native	stable	—
Nine-spined stickleback *Pungitius pungitius*	native	decline	—
Sea bass *Dicentrarchus labrax*	native	stable	SE
Perch *Perca fluviatilis*	native	increase	SE
Ruffe *Gymnocephalus cernuus*	(native)	increase	—
Common goby *Pomatoschistus microps*	native	stable	—
Thick-lipped mullet *Chelon labrosus*	native	stable	SE
Thin-lipped mullet *Liza ramada*	native	stable	SE
Golden mullet *Liza aurata*	native	stable	SE
Bullhead *Cottus gobio*	(native)	increase	—
Flounder *Platichthys flesus*	native	stable	SE

Origin: (native) means not originally Scottish but introduced by humans from England.

Status: reflects any change in the number of Scottish populations this century.

Value (commercial): S, fished in Scotland; E, fished elsewhere in Europe; —, not fished commercially in Scotland.

[a] *Likely to become extinct worldwide soon.*

[b] *Extinct in Scotland.*

communities – but not to their natural biodiversity – and may bring with them threats to native species in the form of disease, competition and predation.

As well as their scientific importance (Maitland, 1985), the economic, social and conservation value of many Scottish native species is often not realised. The importance of Atlantic salmon and both sea trout and brown trout is, of course, well known and there are a few small fisheries for eels. However, for example, although there are no fisheries in Scotland for sea lamprey (*Petromyzon marinus*) or river lamprey (*Lampetra fluviatilis*), both these species are highly valued as commercial species in Portugal and Finland, respectively, where they fetch high prices. Similarly, both pike and perch, which many anglers and fishery managers regard as vermin, command high prices in parts of Europe (e.g. Switzerland). Thus the total worth of Scotland's freshwater fish resource has never been properly evaluated and is much higher than commonly appreciated.

12.4   Alien fish

The present situation with regard to controlling the movement of fish by humans into and within Scotland is unsatisfactory. It was thought, until recently, that many fish, especially cyprinids, had already reached their northern limits in Britain and were unlikely to disperse much further. This is clearly not the case, however, as the successful moves (i.e. transfers by humans) made recently by several species (e.g. gudgeon, chub, dace and ruffe) have resulted in the establishment of thriving new populations well to the north of their previous areas of distribution. For example, 25 years ago only native fish species were found in Loch Lomond (Maitland, 1972); since then, five alien species have become established there (Maitland and East, 1989). All of these new stocks appear to have resulted from deliberate introductions by humans, either intentionally to initiate new populations or by discarding excess livebait at the end of fishing.

Although present controls on the movement of fish are probably adequate in England and Wales, this is not the case in Scotland. The Wildlife and Countryside Act 1981 covers alien species but not fish native to Great Britain. If the introduction of fish species (like ruffe) native to England but not to Scotland is to be controlled then such fish need to be specified by the Secretary of State under Section 1 of the Import of Live Fish (Scotland) Act 1978. This should be done now for the southern species concerned, which include barbel (*Barbus barbus*), silver bream (*Blicca bjoerkna*), bleak (*Alburnus alburnus*) and spined loach (*Cobitis taenia*). There are, of course, a number of exotic species already established in England (but not in Scotland) and these include bitterling (*Rhodeus sericeus*), Danube catfish (*Silurus glanis*), largemouth bass (*Micropterus salmoides*), pumpkinseed (*Lepomis gibbosus*), rock bass (*Ambloplites rupestris*) and pikeperch (*Stizostedion lucioperca*).

A further problem, of course, relates to the movement of fish within Scotland. The southern species that have become established here relatively recently can be moved around the country and introduced to new catchments without hindrance. The recent population explosion of four of these species in Loch Lomond and the River Endrick within a few years of their introduction there (Maitland and East, 1989) shows how quickly they can establish themselves. This problem can only be tackled when there is a nationwide catchment-based system of fishery management.

12.5 Management: the problems

Although in a few areas of Scotland there is excellent management for game fish at a local level, elsewhere there continues to be poor or even damaging management. There is no comprehensive monitoring of key fish stocks across the range of Scotland's fish fauna, nor even any realisation of the full economic value of this diverse resource. The educational, social and cultural importance of our fish fauna is diminishing as biodiversity is lost, as, for example, in the case of the vendace (see below). Many aspects of essential fish habitat (e.g. spawning grounds, riparian cover and shade, backwaters) are declining, not only through agricultural practices, etc., but also through damaging 'management' of some fisheries (Maitland, 1992). The *status quo* is unacceptable.

The existing protection, conservation and management of freshwater fish populations in Scotland is fragmented and greatly biased in favour of game species, especially Atlantic salmon. The present system is inadequate in that it does not meet all the needs of anglers, riparian owners or fish, particularly species other than salmonids. It is apparent that this system has, in the past, failed to safeguard the status of hundreds of populations of freshwater fish (including salmonids) in many parts of the country from numerous hazards such as the introduction of parasites and diseases, overfishing, barriers to migration, habitat loss, etc. (Campbell *et al.*, 1994). There is currently concern about the status of many stocks of all native salmonids: Atlantic salmon, brown and sea trout, and Arctic charr. One freshwater fish species, the vendace, has become extinct in Scotland this century (the most recent vertebrate to do so) and several other species (e.g. allis shad (*Alosa alosa*), twaite shad (*A. fallax*) and sparling) have declined significantly.

A substantial number of angling activities and management practices can cause damage to both the aquatic and the terrestrial components of water bodies (Maitland and Turner, 1987). The removal of 'unwanted' species may not only eliminate native fish species but also completely alter the ecology of a water. Stocking with non-native stocks or the introduction of alien fish species may also create problems and change the ecological character of a site. Diseases and parasites may be introduced directly by anglers and their gear or indirectly along with stocked fish. Groundbaiting is of little concern in Scotland at the moment, but may need to be monitored at certain sites in the future, as coarse fishing increases in popularity. The control of fish predators is a very contentious issue and substantial further research is required here. Habitat management of various kinds is widely practised by anglers and can cause substantial damage at sensitive sites. Physical access, including the use of boats, may cause damage to terrestrial and aquatic vegetation as well as disturbance to many forms of wildlife.

The freshwater cage-rearing of salmonids is a recent development; an increasing number of lochs in Scotland are being used for this purpose. The main species involved are Atlantic salmon and rainbow trout, although brown trout and, more recently, Arctic charr are also farmed. Several ecological problems have arisen from the use of such cages: there is concern over the use of pharmaceutical chemicals (which may affect other organisms), the enriching effect of large amounts of waste feed and fish faeces, the introduction of fish diseases, the impact of farmed fish which are released or escape to the wild, and the ecological threat to a number of waters that have great scientific importance.

There is increasing fisheries pressure on lochs in Scotland; more and more are being pushed towards 'put-and-take' systems. These are probably a disaster for native salmonids and also depend on fish farms (with their attendant problems) for their stock. In some instances, the introduction of fish for angling may be preceded by poisoning out any native stock.

It is incumbent upon the fish-farming industry that it moves towards solving its own environmental problems. To indicate that freshwater cage culture is crucial to salmon and trout farming in Scotland is not enough. There should be much more research on the use of land-based units and discrete systems using smaller volumes of water, which can be controlled and treated so as to virtually eliminate the dangers of disease, pollution and escape of farmed fish. This is not such a radical suggestion as it may seem. There is a complete ban on freshwater cage rearing in some parts of Europe (e.g. Bavaria). In other countries, cage rearing is permitted provided that the trophic state of the lake concerned is not altered.

Overall, the main difficulty with many aspects of fishery management in Scotland is that there is no nationwide organisation for freshwater fishes and fisheries in Scotland comparable to that in England and Wales and in most other western countries. The closest equivalent in Scotland are the Salmon District Fishery Boards, which are only concerned with Atlantic salmon and sea trout. In several parts of Scotland there is no Board and thus, in such areas, it is possible to introduce Atlantic salmon from any source. Such an introduction in most other catchments would require to be sanctioned by the local Salmon District Fishery Board under the Salmon Act (1986). Nowhere is there adequate state control or scientific management of fish species other than Atlantic salmon. Although the former Fishery Board for Scotland studied sea trout many years ago (Nall, 1955), it is only recently (because of catastrophic decline in some rivers) that attention has turned back to this species (Department of the Marine, 1993; West Highland Sea Trout and Salmon Group, 1995).

12.6   Management: the science

One of the major problems with fisheries of all kinds is that their management is rarely based on scientific principles or conceptions of conservation and sustainability. The usual approach is a short-term one of maximising yield regardless of the consequences to the fish populations concerned. As a result, many fisheries (and, consequently, local fish-based economies) have collapsed, especially those where there is multinational ownership of the fishery (Wise, 1984; Keen, 1988). To avoid such catastrophes, substantial effort was directed, in the 1950s, to developing quantitative theories of fisheries management; this led to the development of 'surplus yield' models (Ricker, 1954) and the foundation for 'cohort analysis' models (Beverton and Holt, 1957).

For many years since then, the scientific philosophy preached has been that of 'maximum sustainable yield' (MSY) but this concept has fallen into some disrepute after failing to ensure sustainable use in some fisheries (Larkin, 1977). In practice, the Maximum Economic Yield (MEY) is a more realistic concept. Nevertheless, MSY is a convenient framework for considering fishery management problems (Gulland, 1989) because it serves three distinct functions: a description of the status of fish stocks in relation to exploitation, a definable objective of management, and a measure of the success of stock management. Many workers have attempted to improve the existing models; much more sophisticated statistical tools are now available (see, for example, Beddington and Rettig, 1984; Walters, 1986; Getz and Haight, 1989).

The problem in most fisheries is really two-fold. Firstly, it is now known that fish populations (and fish production) can undergo large annual variations. The Beverton and Holt model assumes constant recruitment and is useful only under equilibrium conditions. Secondly, the statistics (e.g. the Catch Per Unit Effort: CPUE) that are essential to successful management policies are usually lacking. By implication, therefore, successful management of freshwater fish populations for their sustainable use requires regular statistics (Hocutt and Stauffer, 1980) on the populations of each species, in particular: the annual CPUE in the fishery; the age structure in the catch and in the population; and the growth rate of each year class.

Most fisheries should have a close season of some kind (usually during the spawning season) when the above data, for the previous season, can be analysed. Results may then be available for the start of the next fishing season and catch quotas for the whole fishery can be set at limits that will allow a successful fishery to be sustained.

In conclusion, the successful management of freshwater fish populations to allow their sustainable use must rely on a sensible mixture of annual scientific information about each species and the extrapolation of this, by means of appropriate statistical models, into the allowable catch for the following year. There must also be equitable policies in relation to the ownership of fresh waters and fisheries, and a realisation that, in the long term, the status of freshwater fish populations is dependent not only on the quality of the water in which they live but also on the land use and other activities of humans in the catchment that it drains.

12.7 Management: new proposals

The proposed new structure (Maitland, 1996) has been designed to take the best of the present systems and involve minimal change to existing organisations. Its main features are as follows.

1. Scotland should be covered by a series of catchment-based statutory Regional Fishery Boards, replacing the present Salmon District Fishery Boards.
2. Separate Regional Fishery Trusts would be created in parallel with each Regional Fishery Board.

3. Each Regional Fishery Board would have its own Protection Order, the procedures for which should be simplified to speed them up. In addition, further legislation is needed to give freshwater fish more protection nationally.
4. Both the Regional Fishery Boards and the Regional Fishery Trusts should have associations to represent them at national level.
5. A Central Fisheries Unit, with which both the Regional Fishery Boards and the Regional Fishery Trusts would interrelate (the former on a statutory basis, the latter on a voluntary one), should be based in a geographically suitable location.
6. The Regional Fishery Boards would be funded more or less as at present by local levies on proprietors.
7. The Scottish Anglers National Association (SANA) could be the recognised body representing anglers and angling clubs in Scotland.
8. The Institute of Fisheries Management (IFM) could be the recognised body for fishery managers in Scotland.

12.8 Discussion

The problem of inadequate national fisheries organisation and management in Scotland is not new; the problems have been outlined from time to time by various authors (e.g. Mills, 1989). Over a century ago, Maitland (1887) noted that if his studies 'persuade District Fishery Boards that their sphere of usefulness is wider than they have hitherto held, I shall be amply rewarded'. Yet, 100 years later, in spite of the new Salmon Act (1986), even salmonid fisheries still have inadequate protection and management, as noted by the UK's Salmon Advisory Committee (1991):

> There is ... a need for a stronger and more comprehensive framework for Scotland and we therefore recommend that the present regulatory structure there should be urgently reviewed in this context and amended if appropriate.

In Scotland there has never been a national organisation that has had the authority to protect all species of freshwater fish; it is significant that the the most recent vertebrate to become extinct in Scotland this century is a fish (the vendace). This extinction was not just the disappearance of a slightly obscure fish and a reduction in biodiversity but the loss of part of our Scottish culture, with local social and economic consequences.

> The vendace was of old held in high repute as a delicacy, and two local clubs were wont to hold an annual feast at Lochmaben ... There were two clubs – the Vendace Club and the St. Magdalene Vendace Club ... After fishing the lochs for vendace in the usual way, they held a feast and a meeting for Border games ... this was rather a big event ... (Service, 1902).

Similarly, with the substantial decline this century in the stocks of sparling (sometimes known as cucumber smelt), the loss is not only to Scottish wildlife, but also to human pleasure and experience.

> One mild November morning I enjoyed the best dish that I can recollect ever to have eaten at

breakfast. ... on Solway shore to watch the nets drawn ... a fragrance as of cucumbers and violets diffuses itself ... we had the sparling piping hot on the table before half an hour ... No one can have a notion of the subtle toothsomeness of this fish who has not savoured it in an inn parlour close at hand (Maxwell, 1922).

The latest in this sequence of losses of biodiversity is the Arctic charr. Having survived and evolved in isolation in many Scottish lochs for 10,000 years, the recent introduction of alien North American stocks for charr farming and the subsequent movement and mixing of both native and alien stocks by charr farmers is destroying one of the last remaining vestiges of the original pristine Scottish ichthyofauna (Maitland and Lyle, 1992). In spite of pressure from the author and others to save this part of Scotland's heritage, the Scottish Office has remained obdurate and apparently content to preside over the continued decline in the biodiversity of native Scottish fish stocks.

In this modern era of high stress and unemployment, the national importance of the recreational value of healthy populations of freshwater fish to the Scottish people must not be underestimated. In addition to angling, there is also a socio-economic, as well as a scientific, element within the recent proposals by the author (Maitland, 1996): it is hoped that more jobs will be created for fishery biologists and fishery wardens within Scotland as a whole. The value of Scotland's freshwater fisheries has never been accurately measured but clearly runs into many millions of pounds, much of which is spent on a local basis, not only on fishing permits and lets of beats, but also on accommodation, food, fishing tackle and so on. Thus fish and fisheries are an important element of the tourist industry. The broad objectives of management for biodiversity remain the same as in the Scottish statutes of 500 years ago: to put down practices that 'destroy the breed of fish, and hurt the commoun profite of the realme'. The much-needed restructuring of the way in which fish populations and fisheries in Scotland are managed should, therefore, benefit both fish and people.

References

Beddington, J. R. and Rettig, R. B. 1984. *Approaches to the Regulation of Fishing Effort*. Rome, Food and Agricultural Organisation of the United Nations.

Beverton, R. J. H. and Holt, S. J. 1957. *On The Dynamics of Exploited Fish Populations*. London, Ministry of Agriculture, Fisheries and Food.

Campbell, R. N., Maitland, P. S. and Campbell, R. N. B. 1994. Management of fish populations. In Maitland, P. S., Boon, P. J. and McLusky, D. S. (Eds) *The Fresh Waters of Scotland*. Chichester, Wiley, 489–513.

Department of the Marine 1993. *Report of the Sea Trout Working Group*. Abbotstown, Fisheries Research Centre.

Getz, W. M. and Haight, R. G. 1989. *Population Harvesting: Demographic Models of Fish, Forest and Animal Resources*. Princeton, Princeton University Press.

Gulland, J. A. 1989. *Fish Stock Assessment: a Manual of Basic Methods*. Chichester, Wiley/FAO.

Hocutt, C. H. and Stauffer, J. R. 1980. *Biological Monitoring of Fish*. Lexington, Lexington Books.

Keen, E. A. 1988. *Ownership and Productivity of Marine Fishery Resources*. Blacksburg, McDonald and Woodward.

Larkin, P. A. 1977. An epitaph for the concept of MSY. *Transactions of the American Fisheries Society*, **107**, 1–11.

Maitland, J. R. G. 1887. *A History of Howietoun*. Stirling, Guy.

Maitland, P. S. 1972. Loch Lomond: man's effects on the salmonid community. *Journal of the Fisheries Research Board of Canada*, **29**, 849–860.

Maitland, P. S. 1977. Freshwater fish in Scotland in the 18th, 19th and 20th Centuries. *Biological Conservation*, **12**, 265–278.

Maitland, P. S. 1985. Criteria for the selection of important sites for freshwater fish in the British Isles. *Biological Conservation,* **31**, 335–353.

Maitland, P. S. 1992. *Fish and Angling in SSSIs in Scotland.* Report to Scottish Natural Heritage, Edinburgh.

Maitland, P. S. 1996. *Review of Policies Concerning Freshwater Fish in Scotland.* Aberfeldy, WWF Scotland.

Maitland, P. S., Boon, P. J. and McLusky, D. J. (Eds.) 1994. *The Fresh Waters of Scotland.* Chichester, Wiley.

Maitland, P. S. and East, K. 1989. An increase in numbers of Ruffe, *Gymnocephalus cernua* (L.), in a Scottish loch from 1982 to 1987. *Aquaculture and Fisheries Management,* **20**, 227–228.

Maitland, P. S. and Lyle, A. A. 1992. Conservation of freshwater fish in the British Isles: proposals for management. *Aquatic Conservation,* **2**, 165–183.

Maitland, P. S. and Turner, A. K. (Eds.) 1987. *Angling and Wildlife Conservation.* Grange-over-Sands, Institute of Terrestrial Ecology.

Maxwell, H. 1922. *Memories of the Months,* 7th Series. London, Maclehose.

Mills, D. 1989. Conservation and management of Brown Trout, *Salmo trutta,* in Scotland: an historical review and the future. *Freshwater Biology,* **21**, 87–98.

Nall, G. H. 1955. Movements of salmon and sea trout, chiefly kelts, and of brown trout tagged in the Tweed between January and May, 1937 and 1938. *Freshwater and Salmon Fisheries Research,* **10**, 1–19.

Ricker, W. E. 1954. Stock and recruitment. *Journal of the Fisheries Research Board of Canada,* **11**, 559–623.

Salmon Advisory Committee 1991. *Factors Affecting Natural Smolt Production.* London, MAFF Publications.

Service, R. 1902. The vertebrates of Solway – a century's changes. *Transactions of the Dumfries-shire and Galloway Natural History and Antiquarian Society,* **17**, 15–31.

Walters, C. J. 1986. *Adaptive Management of Renewable Resources.* New York, Macmillan.

West Highland Sea Trout and Salmon Group 1995. *Report and Action Plan.* Pitlochry, West Highland Sea Trout and Salmon Action Group.

Wise, M. 1984. *The Common Fisheries Policy of the European Community.* London, Methuen.

13 FOREST MANAGEMENT FOR BIODIVERSITY: PERSPECTIVES ON THE POLICY CONTEXT AND CURRENT INITIATIVES

A. C. Newton and J. W. Humphrey

Summary

1. Scotland has suffered a massive decline in forest area during the past 5,000 years, primarily as a result of human activities. The precise impact of deforestation on Scotland's biodiversity is unknown, but is undoubtedly negative.

2. A number of national and international policies are now in place aimed at conserving and restoring native forest ecosystems and their associated biodiversity, and at managing plantation forests to enhance their value as a habitat for wildlife.

3. Efforts at restoration and creation of native woodland in Scotland have never been greater. This results from current policy initiatives, and the recent formation of partnerships between different organisations and landowners. The impact of such initiatives on biodiversity is likely to be positive, although the precise effects are difficult to predict.

4. Current research indicates that coniferous plantation forests are a valuable habitat for wildlife, but forest managers will be required to meet the contrasting needs of different groups of species. This is illustrated by results of recent research on the impact of introducing birch into coniferous plantations.

5. There is unprecedented interest in restoring and enhancing the biodiversity associated with Scottish forests. To capitalise on this interest, further development of policy and associated legislation will be required. Research is required to identify more precisely the impacts of forest management on different components of biodiversity. Effective tools for decision-making need to be made available to forest managers.

13.1   Introduction

As forests historically covered at least 75% of Scotland's land area, it is arguable that the majority of terrestrial species native to Scotland are forest-dwelling organisms. The massive decline in the area of native forest resulting from human activity over the past few millennia must have been associated with pronounced declines in the abundance and distribution of many such species, leading to an overall impoverishment of biodiversity. In the context of biodiversity conservation, two main challenges face forest managers at the present time: the protection and restoration of native forest areas, and the development of management approaches for non-native plantation forests which enhance their value as a habitat for native species. Although it does not attempt to be comprehensive, this chapter describes a number of initiatives that are currently being implemented to achieve these two objectives. Examples are given to illustrate how efforts at woodland management and expansion in Scotland have been encouraged by a policy framework that promotes the development of appropriate partnerships and action plans, and by the provision of incentives. The importance of research is illustrated by recent results, which indicate the practical challenges facing forest managers in achieving policy commitments relating to biodiversity.

13.2   History and current status of forests in Scotland

The postglacial colonisation of Scotland by different tree species has been thoroughly documented (Birks, this volume). Although estimates vary with respect to the maximum extent of forest cover which developed in the Holocene, it is clear that forest covered most of the land surface of Scotland at around 5,700 years ago, declining subsequently throughout Scotland primarily as a result of human activity (Bennett, 1995). By the early years of this century, forest cover in Scotland had declined to less than 5% of the total land area (Anon., 1995b). The past 80 years have been characterised by a steady expansion in the area of Scottish forests, mainly as a consequence of the afforestation of upland areas with conifers. By the 1990s, forests were estimated to cover approximately 12% of Scotland's land area, divided more or less equally between private and state ownership (Anon., 1995b).

Although the total forest area in Scotland has more than doubled this century, the area of native woodland has more than halved during the same period, resulting in a shift in composition from about 75% native to about 75% introduced species (Worrell and Callander, 1996). Approximately half of the forest area composed of native species may be classified as 'ancient' or 'semi-natural', where natural processes such as tree colonisation, growth and mortality predominate (Table 13.1). Scotland may therefore be considered to have a relatively limited area of native woodland, and a relatively high proportion of forest area devoted to exotic species, compared with other European countries. Total forest area and composition also vary markedly between different regions of Scotland (Table 13.1). Recent inventories have indicated that nearly 90% of Scotland's native woodlands are

Table 13.1   The area and composition of forests in different regions of Scotland (figures in kilohectares). Forests of less than 0.25 ha in area are excluded; percentages are of region land area. Based on Anon. (1995b) after Roberts *et al.* (1993).

Region	Land area	Ancient woodland area	%	Semi-natural woodland area	%	Plantation area	%	Total area	%
Borders	467	0.5	0.1	5.5	1.1	62	13.3	68	14.6
Central	263	3.5	1.3	2.5	0.9	30	11.4	36	13.7
Dumfries and Galloway	637	3.2	0.5	3.8	0.6	128	20.1	135	21.2
Fife	131	2.9	2.2	0.1	0.1	10	7.6	13	9.9
Grampian	871	6.1	0.7	11.9	1.4	113	12.9	131	15.0
Highland	2,530	40.0	1.6	11.0	0.4	189	7.5	240	9.5
Lothian	176	1.2	0.7	2.8	1.6	10	5.7	14	7.9
Strathclyde	1,353	19.0	1.4	8.0	0.6	182	13.5	209	15.4
Tayside	750	5.7	0.7	7.3	1.0	60	8.0	73	9.7
Total	**7,708**	**82.1**	**1.1**	**52.9**	**0.7**	**786**	**10.2**	**920**	**11.9**

located in the Highlands, where they account for over 30% of the total forest area (MacKenzie and Callander, 1995, 1996).

Native woodlands make a vital contribution to the biodiversity of Scotland. The western deciduous woodlands are of international importance for the bryophyte communities that they support (Birks, this volume). The native pine forests of Scotland are floristically, structurally and genetically distinct from those of Continental Europe (Rodwell and Cooper, 1995) as a result of their different climate and historical development. Many species characteristic of Scottish native forests, such as twinflower (*Linnaea borealis*), pinewood hoverfly (*Callicera rufa*) and Scottish crossbill (*Loxia scotica*), are very rare or absent in other parts of Britain (Anon., 1995b) or even mainland Europe. The value of non-native forests for biodiversity is much less obvious. To a large extent this reflects a lack of information of the extent to which native species are able to colonise and maintain themselves in plantation forests.

13.3   Policy context

Forestry has undergone something of a world-wide revolution in recent years, as a consequence of a number of key policy initiatives. At the international scale, the United Nations Conference on Environment and Development (UNCED) in 1992 was a landmark event at which the statement of Forest Principles and Chapter 11 of Agenda 21 were developed, which aimed to 'contribute to the management, conservation and sustainable development of forests' (Upton and Bass, 1995). This conference led to a proliferation of national and international initiatives, attempting to define and assess sustainable forest management. In Europe, the most significant such development was the Helsinki Declaration of

1993, at which countries committed themselves to implementing guidelines for the sustainable management of their forests. Although there has been a great deal of debate on the precise meaning of the word 'sustainable' (Upton and Bass, 1995), it is widely accepted that maintenance of biodiversity is central to the concept. The Helsinki Guidelines, for example, state that ' "sustainable management" means the stewardship and use of forests and forest lands in a way ... that maintains their biodiversity ...'.

In response to these international policy developments, a number of initiatives has been developed in the UK which specifically relate to biodiversity and forest management. For example, the document *Sustainable Forestry: the UK Programme* (Anon., 1994) details how the UNCED and Helsinki commitments will be met in the UK, and highlights the need for research into the 'measurement and cost-effective enhancement of biodiversity', and into the impacts of forestry practices on biodiversity. The UK Biodiversity Action Plan, which aims to conserve native species and the quality and range of wildlife habitats (Anon., 1995a), lists two forest types in Scotland as priority habitats, namely upland oakwoods and native pinewoods. The plans for both habitats involve bringing existing remnants under 'favourable management', and the significant expansion of these habitats through regeneration and establishment of new native woodlands (Anon., 1995a). The Plan sets targets for new areas of around 10,000 ha and 25,000 ha for upland oakwood and native pinewood, respectively, to be established by 2005 (Anon., 1995a).

The implementation of such policy initiatives largely depends upon the voluntary participation of forest owners and managers, supported by a limited amount of legislation. For example, tree felling may be prevented under the 1967 Forestry Act (and its subsequent amendments) if it would seriously damage biodiversity (Hodge *et al.*, in press). Tree felling and regeneration is generally carried out in the context of forest management plans, which are approved by the Forestry Authority. For state forests this is achieved by Forest Design Plans; for privately owned forests, similar plans are required under the Woodland Grant Scheme (WGS) (Hodge *et al.*, in press). In addition, a UK Forestry Standard is currently being developed; this will set minimum design and management standards for a range of forest types in the UK, including a number of recommendations specifically relating to biodiversity conservation. These recommendations build upon the guidelines for nature conservation (Forestry Commission, 1990), which include a list of recommended forestry operations. New management guidelines for biodiversity are being planned (Hodge *et al.*, in press).

The provision of financial incentives for forest establishment and management is a vitally important tool for achieving these policy objectives. Most important of these is the WGS, which provides incentives to forest owners to enhance biodiversity by grant payments for planting and regenerating areas of native woodland. The grants allow for areas of up to 20% to remain unplanted and to be managed as open areas, in line with the Nature Conservation guidelines and the draft UK Forestry Standard. Special grants are now available to provide benefits for

PLATE 11 THE COAST

Petalwort (Petalophyllum ralfsii), a liverwort of sand dunes, threatened throughout its European range (Photo: L. Gill).

Flower-rich coastal grasslands, Sumburgh Links, Shetland (Photo: L. Gill).

Typical machair, North Uist (Photo: M.B. Usher).

Adder's tongue (Ophioglossum vulgatum), a fern typical of unim- proved grassland, including dune slacks (Photo: M.B. Usher).

PLATE 12 INTER-TIDAL

Saltmarsh development at the head of a sea loch, Harris
(Photo: D.B.A Thompson).

A bed of eelgrass (Zostera marina) exposed by spring tides at the mouth of
the River Tay (Photo: M.B. Usher).

Eelgrass (Zostera marina) in flower (Photo: M.B. Usher).

biodiversity at specific sites, and to help implement the habitat and species action plans developed under the UK Biodiversity Action Plan (Hodge *et al.*, in press).

As a result of changes in policy, the associated changes in financial climate, and the current age class structure in the forest estate, forestry in Scotland is undergoing a major transformation. For much of this century, afforestation in Scotland has mostly involved the establishment of coniferous plantations for commercial timber production, primarily with exotic tree species. In recent years, most afforestation has involved the creation of new native woodlands, with much less emphasis on timber production. The recent policy shifts towards 'multi-purpose' and 'sustainable' forestry, and the development of 'Action Plans' for specific forest habitats and some forest species, have placed the maintenance and enhancement of biodiversity at the centre of management activity in many Scottish forests, even in some that were originally established primarily for timber production. The following sections detail some of the current initiatives aimed at improving the status of biodiversity within Scottish forests, and highlight some of the challenges that still need to be overcome if the policy objectives relating to biodiversity conservation and enhancement are to be achieved.

13.4   Current initiatives

13.4.1   *Native woodlands*

Efforts at creation and enhancement of native woodland within Scotland have never been greater than at the present time. This is illustrated by the case of Scottish native pinewoods. In the first five years after the WGS was introduced in 1989, some activity was initiated in 58 of the 77 genuinely native pinewoods listed on a recent inventory, involving planting or encouragement of natural regeneration in 5% of the total area (Gill, 1995). In addition, 519 applications for new native pinewoods were approved during the same period, involving a total area of 11,000 ha to be established through both natural regeneration and planting, using trees of appropriate native origin (Gill, 1995). Under the Biodiversity Action Plan, a total of 36,000 ha are scheduled to be established or prepared for regeneration by 2005 under the WGS (Anon., 1995a). Earlier grant schemes met with more limited uptake; some 300 ha of pinewood were established either by planting or by natural regeneration in the decade prior to initiation of the WGS (Gill, 1995).

One of the key recent developments has been the formation of partnerships between interested organisations, to promote the restoration of native forests within Scotland (see, for example, Aldhous, 1995). For example, the Caledonian Partnership project, focusing on restoration of native pinewoods at Glen Affric, involves collaboration between Highland Birchwoods, Scottish Natural Heritage (SNH), the Forestry Commission (FC), the Scottish Wildlife Trust and the Institute of Terrestrial Ecology. In a similar way, the Cairngorms Partnership are developing a project for the expansion of major pinewood remnants in Strathspey and Mar. A network of area-based Native Woodland Initiatives has been established by a variety of organisations around Scotland, such as Scottish Native Woodlands, Tayside Native Woodlands and Highland Birchwoods, some of which have received financial support from the Millennium Fund through the recently created Millennium Forest for Scotland project. With a budget of some £6 million, this project could potentially make a very significant contribution to restoration of Scottish forests in coming years.

Such initiatives are being supported by a variety of research activities, such as the Forestry Commission's Native Woodland Research Programme (see, for example, Nixon and Cameron, 1994; Nixon and Clifford, 1995). Guidance is now available for determining the appropriate woodland composition based on information relating to site type (Pyatt and Suarez, 1997) and the National Vegetation Classification (Rodwell, 1991; Rodwell and Patterson, 1994) (Figure 13.1). Practical guidance for the management of native woodlands is also available, in the form of the Forestry Practice Guides produced by the Forestry Authority.

The current large-scale restoration and expansion of native woodland in Scotland will undoubtedly have major benefits for biodiversity by increasing and improving the habitat for forest-dwelling organisms. However, the precise impact of such activities on biodiversity is difficult to predict. The composition of a newly established woodland will have a pronounced influence on the biodiversity that subsequently develops, as tree species differ markedly in terms of the number of species that associate with them. For example, the number of species of herbivorous insects and ectomycorrhizal fungi differs by more than an order of magnitude between tree species, the highest values being associated with *Salix* spp. (willow) and *Betula* spp. (birch), respectively (Figure 13.2). For other species groups, such as birds, the structure of the forest stand is generally more important than the tree species composition in determining which species are likely to colonise (Fuller, 1997).

Although the establishment of appropriate tree species is a logical first step in forest restoration, attention will need to be paid to the requirements of other components of biodiversity if forest ecosystems are to be restored in their entirety (Ferris-Kaan, 1995), including the functional processes that will ensure their long-term viability. Forest design aspects such as spatial structure and planting density will clearly have a major effect on which organisms are able to colonise forest areas as they mature (Good *et al.*, 1996). The area of the woodland established, and the proximity to sources of colonists, will influence the diversity that develops in any particular woodland, as predicted by island biogeography theory (Usher, 1995). Such concepts have led to the proposal that ecologically isolated woodlands should be linked through the establishment of connecting forest habitats (Peterken *et al.*, 1995) to assist species colonisation and conserve metapopulation structure.

One of the main problems facing woodland restoration initiatives within Scotland relates to the genetic aspects of biodiversity. Massive deforestation has undoubtedly resulted in the genetic impoverishment of many native species, although the precise effects are impossible to quantify. Habitat fragmentation, dysgenic selection resulting from logging, and genetic 'pollution' resulting from

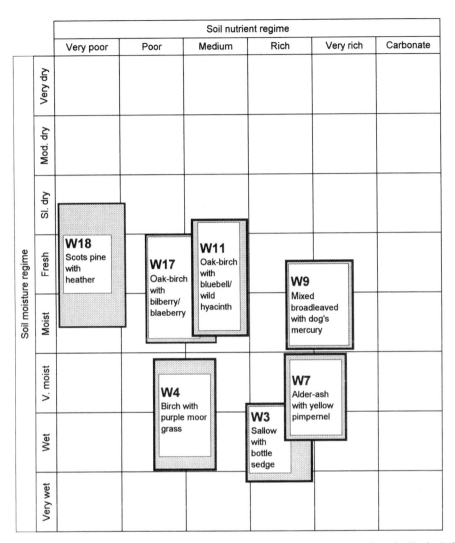

Figure 13.1   National Vegetation Classification (NVC) woodland types superimposed on the Ecological Site Classification (ESC) Soil Quality grid. Major soil types can also be stratified according to these soil variables (adapted from Pyatt and Suarez, 1997).

introduction of non-native genotypes, may all have had an adverse effect on genetic resources of native species (Ennos *et al.*, in press). The extent of such effects is not widely appreciated. For example, planting of non-indigenous sources of oak (*Quercus* spp.) has been so widespread in Scotland that almost all native populations are likely to have been affected (Ennos *et al.*, in press). The problem of identifying suitable locally adapted genotypes for planting is compounded by the almost total lack of research on genetic variation in native species other than Scots pine (*Pinus sylvestris*) (Ennos *et al.*, in press); organisms other than trees have received even less attention, although the same processes apply.

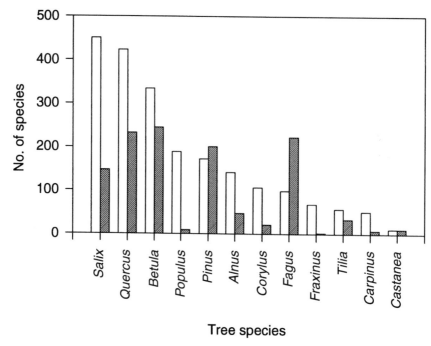

Figure 13.2   The number of species of herbivorous insects (open bars) and ectomycorrhizal fungi (hatched bars) associated with different native British tree species (data from Claridge and Evans (1990) and Newton and Haigh (in press), respectively).

13.4.2   *Plantation forests*

Most of the present forest cover in Scotland is composed of plantations, primarily of introduced coniferous species such as Sitka spruce (*Picea sitchensis*). Many of these forests were established on previously deforested sites, typically upland grassland or moorland, and therefore have inherited little forest biodiversity (Hodge *et al.*, in press). Traditionally, coniferous plantations have a poor reputation as a habitat for wildlife. However, it is being increasingly recognised that stands allowed to grow to maturity can support diverse assemblages of lower plants, vertebrates and invertebrates, including a number of endangered species (Hodge *et al.*, in press). The current challenge is to develop management strategies for these forests which enhance their value as habitat for native species, the focus of a research programme recently initiated by the Forestry Commission.

13.4.2.1   The Forestry Commission's Biodiversity Research Programme

In 1994, The Forestry Commission initiated a Biodiversity Research Programme (BRP) (Hodge *et al.*, in press), which aims to:

1. develop monitoring protocols and collect baseline information on species, structural and habitat diversity in secondary forests;
2. identify biodiversity indicators for secondary forests at the stand (ecological unit) and landscape (assemblages of units) scales;

3. identify and recommend practical standards by which to appraise biodiversity in secondary forests;
4. identify and recommend silvicultural systems and management practices that maintain and enhance biodiversity in secondary forests; and
5. provide decision support for forest managers.

These objectives are addressed through a number of research tasks, which are integrated into a decision process framework (Figure 13.3).

The first and second objectives have resulted in a programme of biodiversity assessments, which are currently being carried out at 16 sites representing a selection of main commercial crop types within three broad climate zones (Table 13.2). Where possible, data from plantation stands are compared with those from semi-natural stands on similar site types (i.e. oak and Scots pine). To encompass the range of stand conditions generated by the patch clearfell system, a chrono-sequence approach has been adopted (see, for example, Pollard, 1993), where

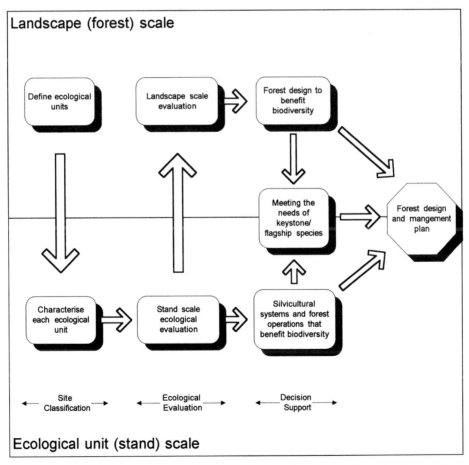

Figure 13.3 Process framework for the Biodiversity Research Programme illustrating linkages between projects and tasks (adapted from Hodge *et al.*, in press).

Table 13.2 Location of Biodiversity Assessment Sites of the Biodiversity Research Programme initiated by the Forestry Commission.

Climate zones	Crop type	Site location
Uplands	Sitka spruce	Clunes and South Laggan, Lochaber; Knapdale, Argyll
	Oak (semi-natural)	Taynish, Argyll; Beasdale/Moidart, Lochaber
Foothills	Scots pine	Glen Affric, Inverness; Glenmore, Strathspey
	Sitka spruce	Kielder, Northumberland; Glentress, Peeblesshire
	Norway spruce	Forest of Dean, Gloucestershire; Fineshade, Northamptonshire
Lowlands	Scots pine	Thetford, Norfolk; New Forest, Hampshire
	Corsican pine	Thetford, Norfolk; Clipstone, Nottinghamshire
	Oak (semi-natural)	New Forest, Hampshire; Alice Holt Forest, Surrey

Climate zones are defined by soil moisture deficit and accumulated temperature.

homogeneous stands of different age are selected on similar soils, site histories, topography and aspect (Ferris-Kaan *et al.*, in press). Assessments are carried out within a 1 ha plot established in each structure class, and are selected to encompass the three roles of biodiversity in forest ecosystems: composition (e.g. numbers of species), functioning (e.g. decomposer chains, nutrient cycling, etc.) and structure (e.g. canopy and understorey layers).

Results of this research are now becoming available (see Ferris-Kaan *et al.*, in press). For example, initial data indicate that the number of fungal species associated with plantations of Sitka spruce (*Picea sitchensis*) may be comparable with that in plantations or semi-natural forest of Scots pine (*Pinus silvestris*) (Figure 13.4). During the survey, a number of threatened fungi (i.e. listed on a provisional Red Data List (Ing, 1992)) have been recorded in plantations, including the saprotrophs *Mycena purpureofusca* and *M. rubromarginata* under Sitka spruce, and the ecto-mycorrhizal species *Suillus flavidus* and *Rozites caperata* under Scots pine. These preliminary data indicate no consistent effect of stand age on the number of fungal species, but suggest that differences between sites may be pronounced (Figure 13.4).

The next stage of this research is to attempt to establish relations between different components of biodiversity, with particular emphasis on testing structural parameters as indicators of diversity in other groups. Deadwood is potentially one such indicator, as the presence of large volumes of both standing and fallen deadwood, particularly in older stands, has been correlated in other studies with a wide range of species groups (for a summary see Ratcliffe, 1993). Coarse woody debris recorded within the Scottish biodiversity assessment sites displayed considerable variation in volume between sites, crop types, and stand stage (Table 13.3). In general, the older stand stages had significantly higher volumes of deadwood than the younger stages for both pine and spruce, although to some extent this is a consequence of the plot selection criteria. The value of deadwood as an indicator of biodiversity in these forest types remains to be tested: regression

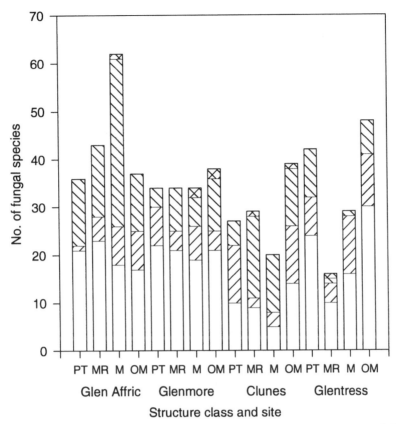

Figure 13.4 The number of fungal species recorded at four Biodiversity Assessment Research Sites as part of the Biodiversity Research Programme initiated by the Forestry Commission Research Agency. For details of sites, see Table 13.2. Abbreviations: PT, pre-thicket; MR, mid-rotation; M, mature; OM, overmature. Open bars, litter-colonising saprotrophs; right-hatched bars, wood-colonising saprotrophs; left-hatched bars, ectomycorrhizal species; cross-hatched bars, parasitic species.

analysis of the preliminary fungal data described above indicated no relation between deadwood volume and the number of saprotrophic fungal species.

In addition to collection of baseline data, the BRP is seeking to identify silvicultural systems, stand management practices and forest operations that improve the biodiversity of commercial forests. This requires an understanding of how natural and management processes affect the composition and structure of forest ecosystems. The research aims to identify ways in which these processes can be modified to generate the desired future forest structure and composition. However, as with native woodland, decisions have to be made about the forest structure and composition required. Value judgements are necessary concerning which species are to be managed for (Table 13.4). Additional strategies are required, which are complementary to the historic emphasis on conserving rare or endangered species and habitats.

Table 13.3 Preliminary deadwood volumes (sum of fallen and standing deadwood) recorded in Scottish Biodiversity Assessment plots.

Species	Site	Deadwood volume ($m^3\ ha^{-1}$)			
		Pre-thicket	Mid-rotation	Mature	Overmature
Scots pine	Glen Affric	0	4.2	34.9	28.6*
Scots pine	Glenmore	0.4	53.5	8.2	96.3*
Sitka spruce	Clunes	7.8	39.7	55.6	124.0
Sitka spruce	Knapdale	116.2	11.6	88.7	316.8
Sitka spruce	Glentress	11.9	7.1	26.1	26.8

Definition of structure classes: pre-thicket (crop height 2–4 m; age 10 years); mid-rotation (crop height 10–20 m; age 25 years); mature (crop height 20–25 m; age 50 years); overmature (crop height variable, with some canopy degeneration and creation of understorey layers; age over 60 years).

*Ancient semi-natural Scots pine sites.

13.4.2.2   Strategies for improving biodiversity in commercial forests

A strategy for improving biodiversity in plantation forests should combine elements of all four approaches to valuing biodiversity, and their management implications (Table 13.4). However objectively this is attempted, the necessary information is often lacking, and management decisions will often require a 'trade-off' between different objectives. This is illustrated by a recent joint SNH-FC initiative, which aimed to assess the ecological benefits of introducing birch into spruce plantations (Humphrey *et al.*, in press a). Studies were undertaken on plants (Wallace, in press), lichens (Orange, in press) and invertebrates (Watt *et al.*, in press; Barbour *et al.*, in press) in second-rotation Sitka spruce stands containing different proportions of self-sown birch (predominantly *Betula pubescens*). These stands were compared with pure spruce stands, clearfells, ancient semi-natural birch stands, and unplanted habitats.

In general, semi-natural birch stands were found to support a range of semi-natural vascular plant and bryophyte communities (including rare Atlantic bryophtes in wetter climate zones) and a range of invertebrate species including some rarities and many that are host-specific to birch. Although closed-canopy spruce crop stages have poorly developed plant, bryophyte and lichen communities, they were found to support a large diversity and abundance of invertebrates from groups such as the Hymenoptera, Collembola, Psocoptera and Acarina (Watt *et al.*, 1997), which thrive in dark, damp conditions. Clearfell sites are also important habitats for Carabidae, and when included with other stages of a spruce crop support a higher carabid diversity than semi-natural birchwoods. In contrast, spruce provides a poor habitat for phytophagous insects such as the Lepidoptera (Barbour *et al.*, in press).

Allowing birch to colonise spruce stands will benefit some species groups, but not others. At the landscape scale, birch groups in spruce stands could act as habitat 'stepping stones' for the dispersal of organisms and as a valuable habitat

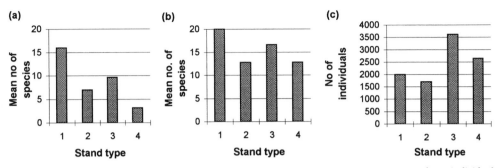

Figure 13.5 Number of (a) vascular plant species, (b) bryophyte and lichen species and (c) individual invertebrates recorded in semi-natural birchwoods (1), birch – Sitka spruce mixtures (2), birch groups in spruce (3) and pure spruce stands (4). Adapted from Humphrey *et al.* (in press b) using data from Wallace (in press), Orange (in press) and Watt *et al.* (1997, in press).

type in their own right (Humphrey *et al.*, in press b). Groups of birch tend to support more species than admixtures (Figure 13.5), suggesting that not every spruce stand should be managed to contain equal proportions of birch, but that larger pockets of birch should perhaps be encouraged in those stands where they can confer the most benefit (i.e. to provide linkages between existing semi-natural birchwood fragments). As the requirements of different groups of species sometimes conflict, management decisions may be difficult. This is particularly true in the case of rare or threatened species with exacting requirements, for example the Kentish glory moth (*Endromis versicolora*), which requires birch below 3 m height for oviposition (see Young and Rotheray, this volume).

Management decisions may be easier to make where there is a consensus that a particular species or habitat type is of high conservation value. One such example is peatland habitat, which is now recognised to be of great importance for biodiversity conservation in many upland forests (Anderson, 1997; Anderson *et al.*, 1995). The 1992 EC Habitats Directive listed raised bogs, blanket bogs, transition mire and bog woodlands as natural habitats requiring special protective measures. Current challenges facing forest managers relate to the design of forest edges to minimise drying of bog surfaces, and the creation of bog woodland habitat. The current emphasis on restructuring in many Scottish plantation forests provides an unprecedented opportunity for increasing the diversity of habitats within forest areas, as well as increasing the structural diversity within forest stands.

13.5   Conclusions and forward look

After centuries of neglect, there is currently unprecedented interest in restoring and enhancing the native woodland of Scotland, and increased appreciation of the international importance of these woodlands for biodiversity. As the extensive coniferous plantations established during the 20th century mature, their value for biodiversity is also being appreciated, and opportunities for improving their value as habitat for wildlife are being recognised. There is therefore considerable cause

Table 13.4 Four approaches to valuing biodiversity as a basis for forest management decisions (based in part on Burton *et al*, 1992; Ratcliffe, 1977; Ratcliffe and Peterken, 1995 and Ratcliffe, 1995).

Name of approach	Details of approach	Management implications	Current initiatives
Generalist	Involves increasing habitat and species diversity in a general sense. Values all species equally, regardless of whether they are rare or threatened. This is the ethos most associated with biodiversity conservation, in other words 'saving all the pieces' (Ratcliffe, 1993).	The biodiversity potential of any particular habitat can be judged in terms of the potential number of ecological niches (for species) which are available naturally on any give site type. Although it may be possible to increase species diversity by introducing species, this may reduce biodiversity at the landscape scale by compromising the distinctiveness of individual habitats. Similarly, individual tree selection systems, or small patch felling, may increase structural diversity at the stand scale, but if repeated across the whole forest, would ultimately result in a reduction in structural diversity at the landscape scale.	Reviews of the impacts of forest management practices on biodiversity have been made by Mitchell and Kirby (1989), Thornber *et al.* (1993) and Patterson (1993). The BRP contains a number of long-term studies on the impacts of forest operations and management on aspects of biodiversity (see, for example, Humphrey *et al.*, 1995; Humphrey and Coombs, in press). Research on forest herbivores and their impacts on forest systems is underway (see, for example, Ratcliffe and Mayle, 1992; Gill, 1992). Current research also aims to develop recommendations for target deer densities in different forest types to achieve improvements in biodiversity and sustainability of both herbivore and forest production (Humphrey and Gill, in press).
Naturalist	Involves increasing 'naturalness' (see, for example, Peterken, 1987) or 'authenticity'. Management interventions should be based as far as possible on an understanding of ecological processes in natural forests.	Silvicultural systems could be based on the effects of natural disturbance (e.g. in the frequency, periodicity of disturbance events and resultant landscape patterns). However, selecting the appropriate 'natural model' is proving problematic for Scotland. Prevailing disturbance regimes of Scandinavian forests are very different (i.e. less disturbance due to wind, more disturbance due to fire).	These issues are currently being addressed by a literature review and scoping study (Quine and Humphrey, 1996).

Specialist	Attaches special value to rare, often endangered species or habitats that require special management consideration, but in themselves do not necessarily contribute significantly to overall biological diversity.	Involves introducing rare species or encouraging their spread, or encouraging the development of rare habitats. Requires information on the ecological requirements of individual species, and the processes influencing population or habitat status, for the impacts of management to be evaluated.	Priority forest habitats and species have been indentified for special protection as part of the UK Biodiversity Action Plans, and the EU Species and Habitats Directive. The BRP aims to identify appropriate stand conditions and management for encouraging the spread of rare vascular plants (currently focused on pinewood species), and to construct a database of habitat requirements for rare forest-dwelling organisms.
Keystone species	Approach involves encouraging keystone species, which are those having a major role in the functioning of forest ecosystems.	It is necessary to consider forest management at both the landscape and stand scales. The aim is maintenance of ecosystem function.	The Habitat-Predator-Prey project of the BRP seeks to test emerging design and management plans for their capacity to support these species. Current work is focused on the modelling of field voles (*Microtus agrestis*) and tawny owl (*Strix aluco*) population dynamics in relation to habitat variability, and on modelling the habitat requirements of red squirrels (Hodge *et al*, in press).

for optimism that the long-term trend of decline in the biodiversity associated with forest ecosystems in Scotland is finally being reversed. However, such optimism must be tempered with the realisation of the enormity of the challenges still to be faced.

The recent development of partnerships between organisations with an interest in forest management is a welcome and essential development. However, this is perhaps only the beginning of a process; greater collaboration between institutions is required for conservation objectives to be fully realised. The implementation of policy relating to biodiversity will require the continued provision of appropriate incentives, as well as practical guidelines, for forest managers. New legislation may be required to enable the implementation of policy, and integration of policies and planning mechanisms needs to be promoted between forestry and other land uses (Worrell and Callander, 1996). It has also been suggested that a strategic framework is required to guide the development of forestry in Scotland, and to provide a basis for long-term planning (Worrell and Callander, 1996).

There is a clear need for further research. Very little is known concerning the ability of species to colonise plantations established on previously deforested sites, whether of native or exotic species. In new native woodlands being established, will other species colonise naturally, or will artificial introductions be required? How is the colonisation process of different groups of organisms influenced by landscape design and woodland management? How will the biodiversity of coniferous plantations develop as they mature? To what extent may exotic conifer plantations act as a reservoir of native biodiversity? Further efforts at quantifying the biodiversity associated with both native and non-native woodlands are required to resolve these questions. Research on genetic aspects of biodiversity is a particularly high priority; without such basic information, many current restoration initiatives run the risk of failure (Ennos *et al.*, in press). In addition, information is required on the impacts of management on different components of biodiversity so as to integrate successfully management of forests for conservation with management for other objectives, such as timber production, landscape value and recreation.

Acknowledgements

The authors thank the following people, who gave assistance in the drafting of this paper and allowed their work to be summarised and referenced: Russell Anderson, David Barbour, Richard Ferris-Kaan, John Good, Simon Hodge, Gary Kerr, Chris Nixon, Alan Orange, Gordon Patterson, Graham Pyatt, Hillary Wallace and Allan Watt. The members of the Fungus Group of South-East Scotland, and Lynn Davy in particular, are thanked for their assistance in collection of field data on fungal diversity.

References

Aldhous, J. R. (Ed.) 1995. *Our Pinewood Heritage.* Farnham, Forestry Authority.

Anderson, A. R. 1997. Forestry and Peatlands. In Parkyn, L., Stoneman, R. E. and Ingram, H. A. P. (Eds.) *Conserving Peatlands.* Oxford, CAB International, 234–244.

Anderson, A. R., Pyatt, D. G. and White, I. M. S. 1995. Impacts of conifer plantations on blanket bogs and prospects of restoration. In Wheeler, B. D., Shaw, S. C., Fojt, W. J. and Roberston, R. A. (Eds.) *Restoration of Temperate Wetlands,* Wiley, Chichester, 533–548.

Anon. 1994. *Sustainable Forestry: the UK Programme.* HMSO, London. 32pp.

Anon. 1995a. *Biodiversity: the UK Steering Group Report.* Vol. 2, *Action Plans.* London, HMSO.

Anon. 1995b. *The Natural Heritage of Scotland: An Overview.* Perth, Scottish Natural Heritage.

Barbour, D., Watt, A. D. and McBeath, C. In press. The diversity and larval feeding status of moths in spruce forests. In Humphrey, J. W., Holl, K. and Broome, A. (Eds.) *Birch in Spruce Plantations: Management for Biodiversity.* Forestry Commission Technical Paper.

Bennett, K. D. 1995. Post-glacial dynamics of pine (*Pinus sylvestris* L.) and pine woods in Scotland. In Aldhous, J. R. (Ed.) *Our Pinewood Heritage.* Farnham, Forestry Authority, 23–39.

Burton, P. J., Balisky, A. C., Coward, L. P., Cumming, S. G. and Kneeshaw, D. D. 1992. The value of managing for biodiversity. *The Forestry Chronicle,* **68,** 225–237.

Claridge, M. F. and Evans, H. F. 1990. Species-area relationships: relevance to pest problems of British trees? In Watt, A. D., Leather, S. R., Hunter, M. D. and Kidd, N. A. C. (Eds.). *Population Dynamics of Forest Insects.* Andover, Intercept, 59–69.

Ennos, R. A., Worrell, R. and Malcolm, D. C. In press. The genetic management of native species in Scotland. *Forestry.*

Ferris-Kaan, R. (Ed.) 1995. *Ecology of Woodland Creation.* Chichester, Wiley.

Ferris-Kaan, R., Peace, A. J. and Humphrey, J. W. In press. Assessing structural diversity in managed forests. In *Proceedings of the Monte Verita Conference on Assessment of Biodiversity for Improved Forest Planning.* Ascona, Switzerland, 7–11 October, 1996.

Forestry Commission 1990. *Forest Nature Conservation Guidelines.* London, HMSO.

Fuller, R. J. 1997. Native and non-native trees as factors in habitat selection by woodland birds in Britain. In Ratcliffe, P. R. (Ed.) *Native and Non-native in British Forestry.* Edinburgh, Institute of Chartered Foresters, 131–140.

Gill, J. G. S. 1995. Policy framework for the native pinewoods. In Aldhous, J. R. (Ed.) *Our Pinewood Heritage.* Farnham, Forestry Authority, 52–59.

Gill, R. M. A. 1992. A review of damage by mammals in North Temperate Forests: 1. Deer. *Forestry,* **65,** 145–170.

Good, J. E. G., Humphrey, J. W., Cough, D., Willis, R. W. M., Nixon, C. J., Harris, I., Roden, P., Norris, D. A. and Thomas, T. H. 1996. *Upland Woodlands in Wales. Report on Phase II.* Unpublished report to the Countryside Council for Wales.

Hodge, S. J., Patterson, G. and McIntosh, R. In press. The approach of the British Forestry Commission to the conservation of forest biodiversity. In *Proceedings of the Monte Verita Conference on Assessment of Biodiversity for Improved Forest Planning.* Ascona, Switzerland, 7–11 October, 1996.

Humphrey, J. W. and Coombs, E. L. In press. Effects of forest management on understorey vegetation in a *Pinus sylvestris* plantation in NE Scotland. *Botanical Journal of Scotland.*

Humphrey, J. W. and Gill, R. M. A. (Eds.) In press. *Grazing as a Management Tool in European Forest Ecosystems.* Forestry Commission Technical Paper.

Humphrey, J. W., Glimmerveen, I. and Mason, W. L. 1995. The effects of soil cultivation techniques on vegetation communities and tree growth in an upland pine forest. I: Vegetation responses. *Scottish Forestry,* **49,** 198–205.

Humphrey, J. W., Holl, K. and Broome, A. (Eds.) In press, a. *Birch in Spruce Plantations: Management for Biodiversity.* Forestry Commission Technical Paper.

Humphrey, J. W., Mason, W. L., Holl, K. and Patterson, G. S. In press, b. Birch and Biodiversity: options for management in upland spruce plantations. In Humphrey, J. W., Holl, K. and Broome, A. (Eds.) *Birch in Spruce Plantations: Management for Biodiversity.* Forestry Commission Technical Paper.

Ing, B. 1992. A provisional Red Data List of British fungi. *Mycologist,* **6,** 124–8.

MacKenzie, N. A. and Callander, R. F. 1995. *The Native Woodland Resource in the Scottish Highlands.* Forestry Commission Technical Paper No. 12. London, HMSO.

MacKenzie, N. A. and Callander, R. F. 1996. *The Native Woodland Resource in the Scottish Lowlands.* Forestry Commission Technical Paper No. 17. London, HMSO.

Mitchell, P. L. and Kirby, K. J. 1989. *Ecological Effects of Forestry Practices in Long Established Woodland and Their Implications for Nature Conservation.* Oxford Forestry Institute Occasional Papers No. 39. Oxford, Oxford Forestry Institute.

Newton, A. C. and Haigh, J. M. In press. Diversity of ectomycorrhizal fungi in Britain: a test of the species-area hypothesis, and the role of host specificity. *New Phytologist.*

Nixon, C. J. and Cameron, E. 1994. A pilot study of the age structure and viability of the Mar Lodge pinewoods. *Scottish Forestry,* **48**, 22–27.

Nixon, C. J. and Clifford, T. 1995. The age and structure of native pinewood remnants. In Aldhous, J. R. (Ed.) *Our Pinewood Heritage.* Farnham, Forestry Authority, 177–185.

Orange, A. In press. Lichens in upland spruce plantations. In Humphrey, J. W., Holl, K. and Broome, A. (Eds.) *Birch in Spruce Plantations: Management for Biodiversity.* Forestry Commission Technical Paper.

Patterson, G. S. 1993. *The Value of Birch in Upland Forests for Wildlife Conservation.* Forestry Commission Bulletin No. 109. London, HMSO.

Peterken, G. F. 1987. Natural features in the management of upland conifer forests. *Proceedings of the Royal Society of Edinburgh,* B **93**, 223–230.

Peterken, G. F., Baldock, D. and Hampson, A. 1995. *A Forest Habitat Network for Scotland.* Scottish Natural Heritage Research, Survey and Monitoring Report No. 44.

Pollard, D. W. F. 1993. An introduction to the Forest Ecosystem Dynamics Program. In Marshall, V. (Ed.) *Proceedings of the Forest Dynamics Workshop,* February 10–11, 1993, Pacific Forestry Centre, Victoria, British Columbia. Forestry Canada and British Columbia, Ministry of Forests, 1–3.

Pyatt, D. G. and Suarez, J. C. 1997. *An Ecological Site Classification for Forestry in Great Britain with special reference to Grampian, Scotland.* Forestry Commission Technical Paper No. 20. Edinburgh, Forestry Commission.

Quine, C. P. and Humphrey, J. W. 1996. The potential role of wind in contributing to the floral and structural diversity of planted coniferous forests in Britain. In Simpson, I. A. and Dennis, P. (Eds.) *The Spatial Dynamics of Biodiversity: Towards an Understanding of Spatial Patterns and Processes in the Landscape.* Stirling, The International Association for Landscape Ecology (UK Region), 183–186.

Ratcliffe, D. A. 1977. *Nature Conservation Review.* Cambridge University Press, Cambridge.

Ratcliffe, P. R. 1993. *Biodiversity in Britain's Forests.* Edinburgh, Forestry Authority.

Ratcliffe, P. R. 1995. Ecological diversity in managed forest. In Ferris-Kaan, R. (Ed.) *Managing Forests for Biodiversity.* (Forestry Commission Technical Paper No. 8.) Edinburgh, Forestry Commission, 3–7.

Ratcliffe, P. R. and Mayle, B. A. 1992. *Roe Deer Biology and Management.* Forestry Commission Bulletin 105. London, HMSO.

Ratcliffe, P. R. and Peterken, G. F. 1995. The potential for biodiversity in British upland spruce forests. *Forest Ecology and Management,* **79**, 153–160.

Roberts, A. J., Russell, C., Walker, G. J. and Kirby, K. J. 1993. Regional variation in the origin, extent and composition of Scottish woodland. *Botanical Journal of Scotland,* **46**(2), 167–189.

Rodwell, J. S. (Ed.) 1991. *British Plant Communities.* Vol. 1, *Woodland and Scrub.* Cambridge, Cambridge University Press.

Rodwell, J. S. and Cooper, E. A. 1995. Scottish pinewoods in a European context. In Aldhous, J. R. (Ed.) *Our Pinewood Heritage.* Farnham, Forestry Authority, 4–22.

Rodwell, J. S. and Patterson, G. S. 1994. *Creating New Native Woodlands.* Forestry Commission Bulletin 112. London, HMSO.

Thornber, K. A., Legg, C. J. and Malcolm, D. C. 1993. *The Influence of Stand Manipulation on the Biodiversity of Managed Forests: A Review.* Unpublished report to the Forestry Commission Research Division.

Upton, C. and Bass, S. 1995. *The Forest Certification Handbook.* Earthscan Publications, London.

Usher, M. 1995. Species richness and the application of island biogeography theory to farm woodlands. In Ferris-Kaan, R. (Ed.) *Managing Forests for Biodiversity.* (Forestry Commission Technical Paper No. 8.) Edinburgh, Forestry Commission, 22–27.

Wallace, H. In press. Ground flora associated with birch in Scottish spruce plantations. In Humphrey, J. W., Holl, K. and Broome, A. (Eds.) *Birch in Spruce Plantations: Management for Biodiversity.* Forestry Commission Technical Paper.

Watt, A. D., Barbour, D. A. and McBeath, C. 1997. *The Invertebrate Fauna Associated with Birch in Spruce Forests.* Scottish Natural Heritage Research, Survey and Monitoring Report No. 82.

Watt, A. D., Barbour, D. A. and McBeath, C. In press. The abundance, diversity and management of arthropods in spruce forests. In Humphrey, J. W., Holl, K. and Broome, A. (Eds.) *Birch in Spruce Plantations: Management for Biodiversity.* Forestry Commission Technical Paper.

Worrell, R. and Callander, R. 1996. *Native Woodlands and Forestry Policy in Scotland. A Discussion Paper.* Aberfeldy, WWF Scotland.

14 THE CONTRIBUTION OF GAME MANAGEMENT TO BIODIVERSITY: A REVIEW OF THE IMPORTANCE OF GROUSE MOORS FOR UPLAND BIRDS

D. B. A. Thompson, S. D. Gillings, C. A. Galbraith, S. M. Redpath

and J. Drewitt

Summary

1. Game management is practised widely across the uplands and lowlands of Britain. Some of the more recent game conservation proposals are summarised.

2. There is an ongoing debate about the importance of grouse moor management in the uplands. This paper considers differences in the breeding bird assemblages associated with grouse moors, as opposed to non-grouse moors, across five regions in upland Britain.

3. In all regions, 10 km squares holding grouse moors tend to have a higher number of breeding species than non-grouse moor squares. In the Highlands, more species are at higher abundance in non-grouse moor squares, whereas elsewhere the converse applies. At the tetrad (2 km square) level, game species tend to be more abundant in grouse moor squares, and birds of prey more abundant in non-grouse moor squares.

4. Comparisons between recorded breeding distributions in 1968–72 and 1988–91 showed regional differences. In the Highlands, grouse moor squares are associated with an expansion in the recorded breeding range of merlin (*Falco columbarius*) and peregrine (*Falco peregrinus*), and a decrease in the distributions of raven (*Corvus corax*) and Eurasian golden plover (*Pluvialis apricaria*). Elsewhere, grouse moor squares are associated with an expansion between the two periods for peregrine, and smaller range contractions for red grouse (*Lagopus scoticus scoticus*) and ring ouzel (*Turdus torquatus*).

5. Grouse moor management practices may not account for these differences. Suggestions are made for further work and analyses to tease out the relevant factors, and some proposals are made regarding the future direction of grouse moor management in the uplands.

14.1   Introduction

> Systematic destruction of birds and beasts of prey began with the rise of the organised management of small game. The creation of artificial populations of gamebirds began many years earlier but the active *preservation* of game really took off from around the middle of the 19th century. The widespread rearing of pheasants and the development of the breech-loaded shotgun in 1853 created the motivation and the tools with which to destroy the birds on the gamekeeper's list of targets, and as the number of gamekeepers grew so the population of avian predators fell. (Simon Holloway, 1996.)

The above quotation is a recent example of the mixture of factual and emotive writing encountered frequently on matters relating to game and their management. For the purposes of this paper, game are considered to be wild mammals, birds or fish now or formerly hunted, shot or caught for sport. The quarry list of game in Britain includes 16 species of bird, 8 species of mammal, and over 10 species of fish. These include some of our most abundant species of birds (e.g. pheasant (*Phasianus colchicus*)), species which are regarded as 'pests' in many parts (e.g. wood pigeon (*Columba palumbus*) and rabbit (*Oryctolagus cuniculus*)), and species which present some of the greatest challenges for land and waterway managers (red deer (*Cervus elaphus*), salmon (*Salmo salar*)). At least one organisation in Britain is devoted solely to the conservation and management of game: the Game Conservancy Trust (Game Conservancy Trust, 1995). The Deer Commission for Scotland (DCS) is the only statutory body charged with advising the UK Government specifically on the management of a group of game species – deer – under the Deer (Scotland) Act (1996). Under the Salmon Act (1986), Salmon District Fishery Boards control some measures affecting salmon and sea trout (*Salmo trutta*) in parts of Scotland.

14.1.1   Game conservation in Britain

Several recent reviews have considered the status and/or conservation needs of game mammals (see, for example, Harris *et al.*, 1995), gamebirds (see, for example, Potts, 1990; Coulson *et al.*, 1992; Gibbons *et al.*, 1993; Tucker and Heath, 1994; Hudson, 1995) and fish (Maitland and Lyle, 1991; Maitland and Campbell, 1992; Maitland, this volume). *Biodiversity: the UK Steering Group Report* (Anon., 1995) contained 116 action plans for priority threatened species which were subsequently endorsed by Government. Three of these plans are are for game species. The substantial number of proposed actions range from protection of existing habitat resources (capercaillie (*Tetrao urogallus*)), to provision of advice on specific land management actions (grey partridge (*Perdix perdix*)), to proposals to develop new agri-environmental schemes (brown hare (*Lepus europaeus*)).

14.1.2   Extent of game management

The extent of game management practised varies according to country and landscape. In the lowlands, there appear to be differences between Scotland and the rest of Britain. Piddington (1980) estimated that shooting (largely of pheasants) took place on 58% of agricultural properties smaller than 400 ha, and on 88% of those larger than 400 ha. Cobham Resource Consultants (1985) found that 67%

of members of the Country Landowners Association retained woodlands for shooting, and 56% planted new woodlands for shooting. Robertson (1997) reviewed these and other studies and suggested that pheasant shooting occurs in approximately 60% of lowland woods. However, work in Scotland into the Farm Woodland Premium Scheme concluded that game management was much less important (Crabtree, 1996). This study found that of those farmers asked to indicate their planting objectives, only 16% said the objective 'to encourage game for sport' was very important, 32% stated it as important, and 52% stated it as unimportant. Instead, the provision of shelter, improvements to the landscape, and encouragement of wildlife were listed as the three most important objectives.

So far as other game management practices are concerned, deer forests, which are managed primarily for their red deer populations, cover 13–19% of Scotland (Reynolds and Staines, in press). However, the overall range extent of red deer in Scotland is considerably greater, amounting to around 3.25 million hectares, 40% of Scotland's land area (Callander and Mackenzie, 1991). There is also considerable variation in the estimates of moorland managed for red grouse, principally through muirburn and other keepering activities. McGilvray (1995) estimates the area of grouse moor in Scotland to be 1.87 million hectares (23% of land surface), whereas the area of heather moorland is estimated to be around 3.08 million hectares (39% of Scotland's land surface) (Macaulay Land Use Research Institute, 1993; Ward *et al.*, 1995). Hudson (1995) states that there are 746 British estates with grouse shooting, occupying 3.8 million hectares (80% of the total upland land area).

These figures indicate that game management, not least for grouse in the uplands, may have a major influence on biodiversity. However, it is difficult to tease out this influence because so many factors covary with the presence or absence and scale of grouse moors. This chapter focuses on associations between grouse moors and the assemblage of breeding birds. On the basis of data collected by volunteers for the British Trust for Ornithology for the Breeding Bird Atlas studies (Gibbons *et al.*, 1993) we have contrasted regions and areas managed primarily for red grouse as opposed to other land uses. We are posing a simple question: do those parts of upland Britain which are managed predominantly for red grouse have the largest species number or most abundant complement of breeding birds?

14.2 Grouse moors and upland birds

Several papers have already considered the importance of grouse moors for upland breeding birds. None of these is conclusive so far as indicating whether or not a given area of moorland gains through having active management for grouse (see, for example, Haworth and Thompson, 1990; Ratcliffe, 1990; Brown and Bainbridge, 1995; Hudson, 1995; Thompson and Miles, 1995; Thompson *et al.*, 1995a,b; Etheridge *et al.*, 1997). However, at least two conclusions have emerged from these papers and others reviewed therein. First, illegal persecution of birds of prey practised on many grouse moor estates has resulted in a marked decline in the assemblage and abundance of birds of prey, and remains a serious problem (see also Royal Society for the Protection of Birds, 1997). Second, if many moorland estates in Scotland had not been managed for red grouse in the 1970s and 1980s they would have been forested, resulting in widescale diminutions of the characteristic upland birds.

Usher and Thompson (1993), Brown and Bainbridge (1995) and Thompson *et al* (1995b) provide prescriptions for grouse moor management which should benefit biodiversity in the uplands. Game management practices in the British uplands include legal control of pest species such as crows (*Corvus corone*) and foxes (*Vulpes vulpes*), reduced pesticide and herbicide use, supplementary feeding, and habitat manipulation. If we assume that most grouse moors are managed according to these guidelines then we can begin to weigh up the costs and benefits of grouse-moor management in the uplands. However, some important pieces of information are lacking, notably: (a) quantitative information on habitat selection for the more common and even well-studied species; (b) the actual nature of different moorland management practices, and their impacts on the birds both on and off moorland; and (c) a straightforward comparison of bird numbers and productivity between grouse moor and non-grouse moor areas.

At the broad level of the British uplands, Gibbons *et al.* (1995) and Ratcliffe (1997) have shown that breeding buzzards (*Buteo buteo*) and ravens tend to be absent from upland areas, principally in the east, where there are grouse moors. Gibbons *et al.* (1995) showed that the distribution of grouse moors strongly limits the distribution of these two species, even when variables such as climate, landcover, topography and sheep management were taken into account. Nevertheless, they did not conclude that grouse-moor-related persecution was the factor limiting distributions because several other factors, such as the nature of moorland management, food supplies and nest site availability, might have accounted for some or much of the variation (see, for example, Austin *et al.*, 1996; Ratcliffe, 1997).

14.3   Grouse moor areas in Britain

14.3.1   *Regional variation*

Ratcliffe (1997) provides a more up-to-date distribution map for grouse moors in Britain than that used by Gibbons *et al.* (1995). The latter was based on Hudson's (1992) map, which did not include all grouse moors, and excluded moors in Wales. We supplemented Ratcliffe's map on the basis of discussions with colleagues (Figure 14.1). We then divided the uplands (according to Bunce *et al.*, 1991) into five regions, and within each of these compared the distribution and abundance of upland birds (*sensu* Thompson *et al.*, 1988) in upland 10 km squares predominantly managed for red grouse ('grouse moor', GM, squares), or for other purposes ('non-grouse moor', non-GM, squares). The GM squares tended to have more than 50% cover of managed grouse moors and less than 10% cover of commercial forests or broadleaf woodland (most had under 5%). The non-GM squares consisted largely of rough grassland, heath, bog and/or commercial forests.

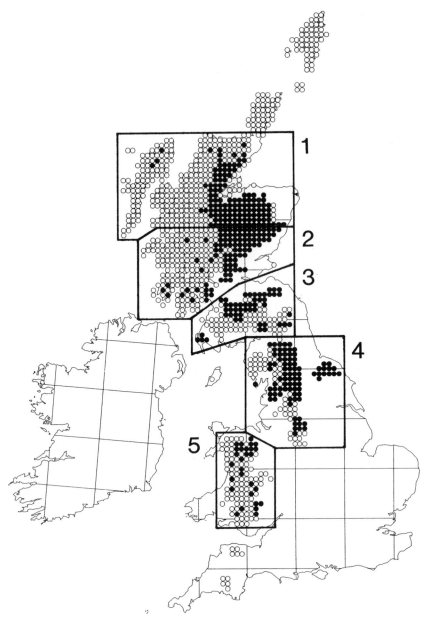

Figure 14.1   Distribution of upland 10 km squares in Great Britain, divided into 5 regions, contrasting where the squares are predominantly managed actively for red grouse (filled circles) with moors which are managed for other land uses (open circles). See text for details.

In regions 1 (N Highlands and W Isles) and 2 (S Highlands) there are additional, potentially important differences between the grouse moor and non-grouse moor squares: the eastern ones are mainly managed for grouse and these tend to have the more productive, gently sloping hills (see, for example, Ratcliffe and

Thompson, 1988; Ratcliffe, 1990). In S Scotland (region 3) and Wales (region 5), however, there are no such noticeable differences between the GM and non-GM squares, although in S Scotland substantial parts of non-grouse moor squares are forested. In N England (region 4) the eastern uplands are flatter than the rugged, and less fertile mountains of the Lake District in the west.

14.3.2 *Questions about bird assemblages associated with grouse moors*

The bird data used in this chapter are drawn from the Breeding Bird Atlas studies organised by the British Trust for Ornithology (BTO), collected during 1968–72 and 1988–91 (Gibbons *et al.*, 1993). These data are used to address three questions.

1. In a comparison of squares predominantly with and without active grouse moor management, does species richness and prevalence (extent of squares occupied) differ?
2. Does the abundance of birds differ between these types of squares?
3. For the most characteristic moorland species, did overall changes in distribution between 1968–72 and 1988–91 differ between GM and non-GM squares

Each of these questions is treated separately below. However, we emphasise at the outset that the comparisons are made at a coarse scale. The grouse moor squares contain many other habitats, not just managed moorland, and some of the non-grouse moor squares contain some grouse moor habitats, as well as others also present in the grouse moor squares. Any differences between the two types of upland are not necessarily attributable to differences in management.

14.4 Species richness and prevalence differences according to predominance of grouse moor management

In all five regions shown in Figure 14.1, species richness (number of species) was greater in grouse moor squares. Across all sample squares, the mean species richness for non-grouse moor squares varied between 26.3 (region 5) and 31.0 species (region 4); for grouse moor managed squares, mean species richness ranged between 26.4 (region 5) and 33.0 species (region 3). The differences were significant for regions 1, 3 and 4 (Wilcoxon 2-sample test, $p < 0.005$, $p < 0.001$ and $p < 0.05$, respectively).

The most prevalent species on grouse moor managed squares are meadow pipit (*Anthus pratensis*), wren (*Troglodytes troglodytes*), willow warbler (*Phylloscopus trochilus*), pied wagtail (*Motacilla alba*), curlew (*Numenius arquata*) and lapwing (*Vanellus vanellus*). On the non-grouse moor managed squares, the most prevalent species are meadow pipit, wren, pied wagtail and crow. Several of these species are not typical of grouse moor tracts, but rather of farmland, scrub or woodland fringes, or fragments of the uplands. The red grouse was among the most prevalent ten species only in the N Highlands and W Isles (region 1) in grouse moor managed squares. Kestrels (*Falco tinnunculus*) and buzzards are listed among the most prevalent species in non-grouse moor managed squares in the S Highlands (region 2), S Scotland (region 3) and N England (region 4). In Wales, however, the buzzard is among the most prevalent species in both grouse moor and non-grouse moor managed squares.

14.5   Abundance of species in squares with or without predominant grouse moor management

Table 14.1 presents a difference between the Scottish Highlands and the rest of Britain. In the Highlands (regions 1 and 2), non-grouse moor squares have more species at significantly higher densities (22 species) than in grouse moor squares (12 species at highest densities). The species significantly more abundant in grouse moor squares include curlew, kestrel and red grouse. The non-grouse moor squares have more common sandpiper (*Actitis hypoleucos*), stonechat (*Saxicola torquata*), snipe (*Gallinago gallinago*), twite (*Acanthis flavirostris*) and wren. However, many of these differences are probably accounted for by east-west differences in distribution due to factors other than grouse-moor-related land management (Gibbons *et al.*, 1993).

In S Scotland and N England some clearer grouse-moor-related differences seem to occur. In S Scotland, non-grouse moor squares are associated only with a significantly higher abundance of tree pipits (*Anthus trivialis*), with the grouse moor squares having the highest abundance of 11 species (again including curlew, kestrel and red grouse, and notably lapwing, oystercatcher (*Haematopus ostralegus*), redshank (*Tringa totanus*), skylark (*Alauda arvensis*) and meadow pipit). More than in any other region, differences in S Scotland probably reflect differences in grouse moor management, resulting in the retention of more heather, more grasslands (and inbye) and less commercial forestry associated with grouse moor estates. In N England, curlew, lapwing, redshank and red grouse are again more abundant in grouse moor squares, and crows are more abundant in non-grouse moor squares, perhaps reflecting differences in control measures, which are applied particularly rigorously in N England (Hudson, 1992). In Wales, only curlew and kestrel are more abundant in grouse moor squares. Indeed, both appear more abundant in grouse moor squares across much of upland Britain: this deserves closer examination.

Overall, whereas non-grouse moor squares in the Highlands tend to have more species at a higher abundance (presumably in part for the geographical reasons mentioned earlier), elsewhere the grouse moor squares have more species at significantly higher abundance compared with non-grouse moor squares. These results are limited by census methods used (see, for example, Gibbons *et al.*, 1993; Thompson *et al.*, 1996) and the first-order type of comparisons being made (cf. Gibbons *et al.*, 1995). However, we also compared grouse moors and non-grouse moors for the tetrad squares (2 km x 2 km) visited. Here, we compared the mean counts for tetrads across relevant 10 km squares, an approach which tends to favour comparisons between the rarer species. In all cases where there are significant differences (Wilcoxon 2-sample test), game species are more abundant in grouse moor squares (black grouse (*Tetrao tetrix*) in N and S Highlands; ptarmigan (*Lagopus mutus*) in S Highlands) and birds of prey or large predators are more abundant in non-grouse moor squares (golden eagle (*Aquila chrysaetos*) and raven

Table 14.1   Summary of significant differences in abundance (during 1988–91) between grouse moor and non-grouse moor 10 km squares.

Region	Abundance	
	Grouse moor	Non-grouse moor
Region 1: N Highlands and W Isles		
Common sandpiper	0.35	**0.43** ***
Curlew	**0.43**	0.20***
Grey Wagtail	**0.26**	0.11 ***
Kestrel	**0.23**	0.10 ***
Mistle thrush	**0.36**	0.11 ***
Meadow pipit	0.80	**0.90** ***
Red grouse	**0.38**	0.21 **
Redshank	0.10	**0.16** ***
Skylark	0.46	**0.63** ***
Stonechat	0.09	**0.30** **
Snipe	0.20	**0.32*****
Song thrush	**0.47**	0.38 *
Twite	0.08	**0.26** ***
Wheatear	0.43	**0.61** ***
Whinchat	0.14	**0.18** ***
Willow warbler	**0.63**	0.38 ***
Crow	0.48	**0.56** **
Region 2: S Highlands		
Cuckoo	0.21	**0.34** ***
Common sandpiper	0.36	**0.37** *
Curlew	**0.52**	0.30 ***
Kestrel	**0.30**	0.12 **
Mistle thrush	**0.36**	0.22 **
Meadow pipit	0.81	**0.84** *
Red grouse	**0.38**	0.09 ***
Redshank	**0.14**	0.13 **
Stonechat	0.06	**0.23** **
Snipe	0.17	**0.19** *
Song thrush	0.43	**0.46** *
Tree pipit	0.17	0.28 *
Twite	0.08	**0.15****
Whinchat	0.20	**0.32** **
Whitethroat	0.08	**0.23** **
Wren	0.55	**0.68** **
Willow warbler	0.62	**0.66** *
Region 3: S Scotland		
Common sandpiper	**0.38**	0.27 *
Curlew	**0.82**	0.58 ***
Kestrel	**0.46**	0.38 *
Lapwing	**0.64**	0.35 ***
Mallard	**0.55**	0.40 **
Meadow pipit	**0.85**	0.76 *
Oystercatcher	**0.56**	0.36 *

continued

Table 14.1 – *continued*

Region	Abundance	
	Grouse moor	*Non-grouse moor*
Pied wagtail	**0.70**	0.58 **
Red grouse	**0.37**	0.14 **
Redshank	**0.28**	0.13 **
Skylark	**0.77**	0.62 **
Tree pipit	0.11	**0.18 ***
Region 4: N England		
Curlew	**0.76**	0.52 ***
Lapwing	**0.66**	0.46 ***
Red grouse	**0.41**	0.12 ***
Redshank	**0.24**	0.13 *
Stonechat	0.003	**0.01 ***
Snipe	**0.31**	0.21 ***
Wheatear	0.34	**0.45 ***
Crow	0.66	**0.85 ***
Region 5: Wales		
Curlew	**0.46**	0.27 ***
Kestrel	**0.35**	0.22 **

Differences are tested by Wilcoxon 2-sample tests, with significance given by * $p < 0.05$, ** $p < 0.01$, *** $p < 0.001$. The highest abundance is given in bold. Values are the means of abundance scores (proportion of tetrads visited in which the species was recorded) for 10 km squares in each category.

in N Highlands; buzzard, raven and tawny owl (*Strix aluco*) in S Highlands; and buzzard in N England).

14.6 Overall changes in distribution between 1968–72 and 1988–91 in relation to predominance of grouse moor management

Table 14.2 lists the species most characteristic, in our assessment, of grouse moors in Britain. For each of the ten species we compared changes in presence or absence at the 10 km square level for present-day grouse moor and non-grouse moor squares.

As in Table 14.1, the patterns for the Scottish Highlands contrasted with the rest of Britain. Red grouse and black grouse distributions have contracted, notably outwith grouse moor areas south of the Highlands. In the Highlands, the curlew's distribution appears to have changed little, but the golden plover's distribution has contracted throughout Britain except, it seems, in non-grouse moor areas in the Highlands. There have been no marked recorded changes in the meadow pipit's distribution over the 20-year period, but the distribution of the ring ouzel (*Turdus torquata*) has contracted throughout the British uplands, particularly outwith grouse moor squares in S Scotland and N England.

Table 14.2   Changes in distribution between 1968–72 and 1988–91 (percentage change in numbers of 10 km squares occupied by birds characteristic of grouse moor (GM) and non-grouse moor (N-GM) squares).

	Percentage change									
	N Highlands		S Highlands		S Scotland		N England		Wales	
	GM	N-GM	GM	N-GM	GM	N-GM	GM	N-GM	GM	N-GM
Hen harrier	+55	+20	+29	+64	+9	+8	+50	−19	+86	+67
Golden eagle	+8	+2	+14	+2	+25	+5	—	−13	—	—
Merlin	+73	+26	+60	+29	0	+8	+9	+6	+16	−14
Peregrine	+54	0	+52	+29	+230	+77	+356	+86	+600	+125
Red grouse	+1	+4	−4	−11	−5	−24	−1	−23	0	−16
Black grouse	−2	−23	−16	+6	−24	−16	−6	−48	−14	−34
Golden plover	−10	+6	−28	+6	−22	−14	−5	−32	−5	−10
Curlew	−5	+3	−9	−10	0	−4	0	0	0	0
Ring ouzel	−22	−19	−16	−25	−18	−49	−21	−32	−16	−19
Meadow pipit	0	+2	0	0	0	−1	0	0	0	−2

The recorded distributions of birds of prey have expanded (presumably following low initial survey counts for peregrine, hen harrier (*Circus cyaneus*) and merlin in the late 1960s post-pesticide era). Grouse moor squares are associated with greater range expansions for peregrine throughout Britain, for merlin in the Highlands and Wales, and for hen harriers in the N Highlands, N England and Wales. The last result is perhaps surprising given the notable absence of hen harriers in many suitable areas of moorland or rough grassland in N England (see, for example, Thompson *et al.*, 1995a; Etheridge *et al.*, 1997). Nevertheless, the BTO data indicate that in that region the recorded number of 10 km squares with breeding hen harriers rose from 20 to 30 for grouse moor squares, and fell from 16 to 13 for non-grouse moor squares (here, however, 'breeding' included attempts at breeding as well as successful breeding).

14.7   Discussion

14.7.1   *Comparisons between grouse moor and non-grouse moor areas*

Our analyses of BTO survey data have provided some surprising results. If we consider the results beyond the Highlands first, the grouse moor squares tend to have a higher species richness, and many more species at significantly higher abundance, than do other upland squares. Grouse moor squares also tend to have less marked range contractions of red grouse, black grouse and ring ouzel, and greater range expansions of birds of prey. These results suggest that, at the broad 10 km square level and given the limitation of the survey methods (notably for the larger, rarer birds (see, for example, Ratcliffe, 1997)), grouse moor squares in the south tend to have more upland birds. Our own experiences of these areas suggest that these grouse moor areas have more active, favourable habitat management (and therefore retention of heather, control of crows and foxes, and maintenance of some wet flushes). Where these moors adjoin areas of woodland, rough grassland and even arable farmland, the diversity and abundance of birds can be even higher. However, those areas which are mainly grassy sheep walk, or acidic heath and bog (not managed for grouse) or even predominantly upland commercial forests are likely to have much lower numbers of many upland bird species. Nevertheless, factors other than grouse moor management (such as upland farming practices) may well account for differences between the two types of square. This should be examined more closely.

In the Highlands, the patterns (and explanations) were somewhat different. The tendency for the grouse moors to be mainly in the eastern half (Figure 14.1) may mean that factors such as wetness, topography, geology and afforestation account for many of the recorded differences. Here, the non-grouse moor squares had higher densities of more species than grouse moor squares, and experienced an increase in the range of golden plover (and of black grouse and hen harriers in the S Highlands). Grouse moor squares had a somewhat greater expansion of merlin and peregrine (and hen harrier in the N Highlands), but a decline in the recorded breeding range of ravens. There does seem to be a view that the 'classic' eastern grouse moors of the Highlands are poor for upland birds other than red grouse (see, for example, Brown and Stillman, 1993; Whitfield and Tharme, in press). This may apply to the large continuous moors with little additional diversity coming from agricultural and woodland mosaics – which tend to be distant from or much lower than these moors – or agricultural fringe habitats. Around the lower fringes of these grouse moors, particularly in glens and close to roads, there may be a greater diversity and abundance of moorland birds, but lower numbers of black grouse and red grouse (owing to higher heather losses around the moorland fringes). Scale is probably the most important factor here: large swathes of uniformly burnt heather moorland hold few upland species but high densities of red grouse. However, large-scale mosaics of well-managed grouse moor mixed with other habitats seem to hold high numbers of breeding bird species at high abundance. The question, therefore, is to what extent does grouse moor management *per se* account for the benefits? Given that so many species which are not characteristic of grouse moors appear to be more abundant in grouse moor squares, we suggest that factors other than grouse moor management may actually prove more important in explaining the differences.

14.7.2 *Biodiversity and game management: some scenarios*

In the lowlands of Britain, game management appears to be growing in popularity and sophistication. The potential and known impacts of massive releases of game (e.g. annual releases in Great Britain of 20 million pheasants and 1 million red-legged partridges (*Alectoris rufa*)) have been reviewed by Robertson (1997). More work is needed on this topic because the ramifications for the wider lowland environment may be considerable. There seems to be much scope for increasing the sporting amenity interests of lowland farmlands, and these seem to be more fully developed to the south of Scotland.

Management and design of lowland woods will be crucial to the management of game mammals such as roe deer (*Capreolus capreolus*) and sika deer (*Cervus nippon*). For these mammals, control measures are expensive, but there are important benefits through reductions in commercial damage to trees. There would seem to be some scope for increasing a sporting-related cull of these mammals (see, for example, Reynolds and Staines, in press). The capercaillie has become so rare that it is currently subject to a voluntary ban on hunting.

There are at least three strands to developing the game-biodiversity benefits for the uplands. First, for salmon and trout there are already well-tried habitat design and management measures which can benefit both (see, for example, Maitland, this volume). Second, red deer management continues to be the dominant game-orientated land use in the Scottish uplands. The cull of deer has increased in the past 25 years from 24,000 animals (1973) to almost 58,000 animals (in 1993–94): approximately 6–12% of hinds and 10–17% of stags are culled annually. The Deer (Scotland) Act (1996) provides new provisions for the natural heritage, so that deer may be controlled to prevent damage to woodland, moorland and mountain habitats. The trans-estate management approach of Deer Management Groups is vital to manage a deer resource which varies spatially, seasonally and even daily in its use of different habitats.

Thirdly, we have tried to draw out the main adverse impacts and benefits of grouse moor management. The adverse impacts include absence of or reductions in birds of prey, 'arrested' ecological succession with little scrub or natural wood-land, and access restrictions which may prevent or spoil appreciation or rec-reational use of the area. Benefits include the retention of a high diversity of growth phases of heather, control of pests such as rabbits, and maintenance of heather moorland, a rare European habitat with high proportions of rare or priority species, notably moorland birds.

So far as overall biodiversity is concerned, two factors influence benefits: spatial scale and past and present management. The former is crucial, but difficult to determine. The size and range of mosaics of grouse moor and other landscapes probably determine the overall bird interest. Hence, a small grouse moor with degenerating heather cover may nonetheless have high densities of nesting merlin and hen harriers (because of abundant small mammal and passerine prey on surrounding hill grasslands and plantations), whereas a large expanse of moor may be poor because there are few other habitats where particular feeding requirements can be met.

One of the difficulties in contrasting populations of birds of prey on grouse moors across the country is their historical destruction: one cannot be certain what has been removed, and where and when (see, for example, Etheridge *et al.*, 1997). Without persecution, species such as hen harriers, ravens and golden eagle would be considerably more widespread and abundant. Many moors have been so heavily grazed that shooting butts now lie amongst acidic grasslands, yet moor managers expect equal or higher grouse 'bags' from remaining areas of heather moorland. It is arguable that in Victorian or Edwardian times, when large bags were obtained from many grouse moors, the moors were larger, less fragmented and in better condition (through less sheep and deer grazing and more rotational burning), and all predators were ruthlessly controlled on a large scale. Is it reasonable to expect such large grouse bags today?

The future of driven grouse shooting as a sporting enterprise probably lies in its diversification, such as provision of more black game through scrub and woodland restoration, associated lowland game shooting, and the wider experience of hunting and enjoyment of wildlife in the wilds. Is it possible that the grouse shooting market will change, and clients will pay relatively more for smaller 'bags'? Currently, driven grouse moors need autumn densities of over 60 grouse km^{-2} to yield at least 25–30 birds shot km^{-2} in order to make a net annual profit. To achieve this, an estate employs one keeper per 20–40 km^2 in Scotland (one per 10–20 km^2 in N England (Hudson, 1995)). There are, therefore, two options open to estates: (i) seek to increase even further the grouse densities on remaining moors; or (ii) reduce the expectations for high grouse densities, and find alternative means of generating the equivalent income per unit area.

The social environment within which sporting estates now operate is different from that of 100, or even 50 years ago. It is unlikely that the public will tolerate any consideration given to so-called 'predator control' measures to reduce the impacts of birds of prey on grouse, when other moorland management needs are not being addressed (and large-scale persecution continues). Diversification of enterprise and extension of moorland cover, on the other hand, could offer a broader sense of enjoyment for hunters, of the sort experienced in Scandinavia. More cooperation between moorland owners in sharing land management experiences, and with other parties in sharing broader aspirations for the uplands, would also be beneficial to moorland interests (Phillips and Watson, 1995). Otherwise there will be a growing polarisation in outcome: many moors will be afforested, abandoned, or given over to intensive hill sheep grazing, thus reducing some biodiversity interests. A few owners will persist in pursuing large grouse bags for driven shooting whilst minimising every potential loss of grouse, often on moors which have already deteriorated, or continue to do so. A healthier regard for the biodiversity of grouse moors, as part of the upland landscape, therefore seems to offer both moor owners and conservationists more hope for the future. Moorland owners and conservationists do still seem to share a common agenda, so long as management is well practised and there is no killing of birds of prey.

Finally, the BTO data analysis needs to be developed. Three questions are at the forefront here: (i) are other landscape and/or biological factors accounting for suggested differences between grouse moor and non-grouse moor squares; (ii) are there relationships between the actual extent of grouse moors and bird species richness and abundance; and (iii) at what spatial scales are different moorland management practices influencing each species? A key pointer from our analysis is that factors other than grouse moor management *per se* may benefit breeding birds in areas associated with 'grouse moors'. This merits closer study.

Acknowledgements

We are grateful to: Derek Ratcliffe for providing an up-to-date distribution map of grouse moors; the Heather Trust, the Game Conservancy Trust, and other colleagues for refining the grouse moor map; volunteers who collected the data for the BTO atlases; Rob Fuller, Phil Whitfield, David Baines, Roger Burton, Andy Brown, Ian Bainbridge, Peter Hudson, Mark Avery, Stuart Housden, Simon Thirgood and Pete Robertson for discussions and comment; and Juliet Vickery for editorial support. The views expressed in this paper should not be read as being policy of SNH, JNCC, BTO or ITE.

References

Anon. 1995. *Biodiversity: the UK Steering Group Report.* London, HMSO.

Austin, G. E., Thomas, C. J., Houston, D. C. and Thompson, D. B. A. 1996. Predicting the spatial distribution of buzzard *Buteo buteo* nesting areas using a Geographical Information System and remote sensing. *Journal of Applied Ecology*, **33**, 1541–1550.

Brown, A. F. and Bainbridge, I. P. 1995. Grouse moors and upland breeding birds. In Thompson, D. B. A., Hester, A. J. and Usher, M. B. (Eds.) *Heaths and Moorland: Cultural Landscapes.* Edinburgh, HMSO, 51–66.

Brown, A. F. and Stillman, R. A. 1993. Bird-habitat associations in the eastern Highlands of Scotland. *Journal of Applied Ecology*, **30**, 31–42.

Bunce, R. G. H., Howard, D. C., Clarke, R. T. and Lane, M. 1991. *ITE Land Classification: Classification of all 1 km Squares in Great Britain.* London, Department of the Environment.

Callander, R. F. and Mackenzie, N. A. 1991. *The management of wild red deer in Scotland.* Scotland, Rural Forum.

Cobham Resource Consultants 1985. *Countryside Sports: their Economic Significance.* Reading, A Standing Conference on Countryside Sports.

Coulson, J. C., Fielding, C. A. and Goodyer, S. A. 1992. The management of moorland areas to enhance their nature conservation interest. JNCC report no. 134. Peterborough, Joint Nature Conservation Committee.

Crabtree, B. (Ed.) 1996. *Evaluation of the Farm Woodland Premium Scheme.* Edinburgh, Scottish Office Agriculture, Environment and Fisheries Department.

Etheridge, B., Summers, R. W. and Green, R. H. 1997. The effects of illegal killing and destruction of nests by humans on the population dynamics of the hen harrier *Circus cyaneus* in Scotland. *Journal of Applied Ecology*, **34**, 1081–1105.

Game Conservancy Trust 1995. *The Game Conservancy Review of 1995.* Fordingbridge, Game Conservancy Trust.

Gibbons, D. W., Reid, J. B. and Chapman, R. A. (Comp.) 1993. *The New Atlas of Breeding Birds in Britain and Ireland: 1988–1991.* London, Poyser.

Gibbons, W., Gates, S., Green, R. E., Fuller, R. J. and Fuller, R. M. 1995. Buzzards *Buteo buteo* and ravens *Corvus corax* in the uplands of Britain: limits to distribution and abundance. *Ibis*, **137** (Suppl. 1), 75–84.

Harris, S., Morris, P., Wray, S. and Yalden, D. W. 1995. *A Review of the Status of British Mammals.* Peterborough, Joint Nature Conservation Committee.

Haworth, P. F. and Thompson, D. B. A. 1990. Factors associated with the breeding distribution of upland birds in the south Pennines, England. *Journal of Applied Ecology*, **27**, 562–577.

Holloway, S. 1996. *The Historical Atlas of Breeding Birds in Britain and Ireland, 1875–1900.* London, Poyser.

Hudson, P. J. 1992. *Grouse in Space and Time: the Population Biology of a Managed Gamebird.* Fordingbridge, Game Conservancy Trust.

Hudson, P. J. 1995. Ecological trends and grouse management in upland Britain In Thompson, D. B. A., Hester, A. J. and Usher, M. B. (Eds.) *Heaths and Moorland: Cultural Landscapes*. Edinburgh, HMSO, 282–293.

Macaulay Land Use Research Institute (1993). *The Land Cover of Scotland 1988. Final Report*. Aberdeen, Macaulay Land Use Research Use Institute.

Maitland, P. S. and Campbell, R. N. 1992. *Freshwater Fishes*. London, Harper Collins.

Maitland, P. S. and Lyle, A. A. 1991. Conservation of freshwater fish in the British Isles: the current status and biology of threatened species. *Aquatic Conservation,* **1**, 25–54.

McGilvray, J. 1995. An economic study of grouse moors. Fordingbridge, Game Conservancy Trust.

Phillips, J. and Watson, A. 1995. Key requirements for management of heather moorland. In Thompson, D. B. A., Hester, A. J. and Usher, M. B. (Eds.) *Heaths and Moorland: Cultural Landscapes*. Edinburgh, HMSO, 344–361.

Piddington, H. R. 1980. Shooting and fishing in land-use: a study of economic, conservation and recreation aspects. Unpublished report. Cambridge, University of Cambridge, Department of Land Economy.

Potts, G. R. 1990. Numbers, harvest, status and management recommendations for game species in the UK. *The Game Conservancy Annual Review*, **33**, 30–35.

Ratcliffe, D. A. 1990. *Bird Life of Mountain and Upland*. Cambridge, Cambridge University Press.

Ratcliffe, D. A. 1997. *The Raven*. London, Poyser.

Ratcliffe, D. A. and Thompson, D. B. A. 1988. The British uplands: their ecological character and international significance. In Usher, M. B. and Thompson, D. B. A. (Eds.) *Ecological Change in the Uplands*. Oxford, Blackwell, 9–36.

Reynolds, P. and Staines, B. In press. Red deer management in Scotland. In Bolton, M. (Ed.) *Conservation and Use of Wildlife Resources*. London, Chapman and Hall.

Robertson, P. 1997. Naturalised introduced game birds in Britain. In Holmes, J. S. and Simons, J. R. (Eds.) *The Introduction and Naturalisation of Birds*. London, HMSO, 63–69.

Royal Society for the Protection of Birds 1997. *Birds of Prey in the UK: Back from the Brink*. Sandy, Royal Society for the Protection of Birds.

Thompson, D. B. A., MacDonald, A. J., Marsden, J. H. and Galbraith, C. A. 1995a. Upland heather moorland in Great Britain: a review of international importance, vegetation change, and some objectives for nature conservation. *Biological Conservation,* **71**, 163–178.

Thompson, D. B. A., MacDonald, A. J. and Hudson, P. J. 1995b. Upland moors and heaths. In Sutherland, W. J. and Hill, D. A. (Eds.) *Managing Habitats for Conservation*. Cambridge, Cambridge University Press, 292–326.

Thompson, D. B. A. and Miles, J. 1995. Heaths and moorland: some conclusions and questions about environmental change. In Thompson, D. B. A., Hester, A. J. and Usher, M. B. (Eds.) *Heaths and Moorland: Cultural Landscapes*. Edinburgh, HMSO, 362–388.

Thompson, D.B.A., Stroud, D.A. and Pienkowski, M.W. (1988). Afforestation and upland birds: consequences for population ecology. In Usher, M. B. and Thompson, D. B. A. (Eds.) *Ecological Change in the Uplands*. Oxford, Blackwell, 237–259.

Thompson, D. B. A., Watson, A., Rae, S. and Boobyer, G. 1996. Recent changes in breeding bird populations in the Cairngorms. *Botanical Journal of Scotland (Special Issue on Environmental History of the Cairngorms)*, **48**, 99–110.

Tucker, G. A. and Heath, M. 1994. *Birds in Europe: their Conservation Status*. Cambridge, Birdlife International.

Usher, M. B. and Thompson, D. B. A. 1993. Variation in the upland heathlands of Great Britain: conservation importance. *Biological Conservation,* **66**, 69–81.

Ward, S. D., MacDonald, A. J., and Matthew, E. M. 1995. Scottish heaths and moorland: how should conservation be taken forward? In Thompson, D. B. A., Hester, A. J. and Usher, M. B. (Eds.) *Heaths and Moorland: Cultural Landscapes*. Edinburgh, HMSO, 319–333.

Whitfield, D. P. and Tharme, A. P. In press. Priority forest areas and moorland birds in the Cairngorms. Scottish Natural Heritage Review Report.

15 BIODIVERSITY ON SCOTTISH AGRICULTURAL LAND

D. D.French, R. G. H.Bunce and R. P.Cummins

Summary

1. A study is being undertaken to assess the relative importance of agricultural practices, pollution and climate change in determining biodiversity on Scottish agricultural land.

2. Existing data have been collated and land use has been divided into five strata to assist analysis.

3. Land cover composition, management changes and distributions of plant and animal species and communities have been assessed for the five strata.

4. There is a greater proportion of more diverse land cover types in Scotland than in England; analysis of the biological data will show whether this is reflected in greater biodiversity.

5. The arable stratum is likely to be most affected by changes in climate and management, with moor and plateau least affected; however, the situation is reversed with respect to the effects of acid deposition.

6. Future work will involve classification of communities, estimation of biodiversity indices, and modelling the effects of management, climate and pollution changes on biodiversity.

15.1 Introduction

Agriculture is the dominant land use in Scotland, but farming strategies vary widely, from intensive arable cropping to extensive rough grazing on open moorland. Concern has been expressed over perceived changes in the diversity of (semi-) natural flora and fauna on agricultural land, but the causes of these changes are not precisely known, nor the relative importance of the probable causes in different circumstances.

The Institute of Terrestrial Ecology (ITE), together with the Macaulay Land

Use Research Institute (MLURI), Wye College and the British Trust for Ornithology (BTO), are therefore undertaking a study, funded by the Scottish Office, of biodiversity on Scottish agricultural land. The overall objective is to assess the relative importance of agricultural practice, pollution, and climate change in determining biodiversity on agricultural land in Scotland.

15.2 Methods

The general approach is to integrate existing databases, and analyse their interrelations by a variety of methods, including geographical information systems (GIS), statistical modelling and the use of expert knowledge. Data have been collated and examined on land cover and land use, agricultural inputs and changes, pollution loadings, climate change scenarios, and distributions of plant and animal species. These data are at a variety of scales and resolutions, from quadrats of 4 m^2 to 20 km × 20 km Ordnance Survey (OS) grid squares, and from fully quantitative data to simple presence-absence scores.

As an aid to analyses at these diverse scales, we have used a general stratification reflecting the main types or intensities of agricultural land use in Scotland. This distinguishes five strata:

1. *arable*, cropped land, but not including sown grasslands;
2. *inbye*, generally more intensively managed grasslands, usually in an enclosed or lowland situation, with cattle and/or sheep;
3. *moor*, upland rough grazings, moor and bog, typically sheep-grazed;
4. *plateau*, mainly hill ground, where grazing by domestic stock is limited or prevented by cold, wind-exposure or similar climatic extremes; and
5. *non-agricultural* squares where >75% of the ground is not in agricultural use, mainly urban areas or large forest plantations, but also including large inland lochs.

The strata were defined mainly by the land cover composition of 1 km OS grid squares. First, each of the land cover types from the ITE Countryside Survey 1990 (CS90) (Barr *et al.*, 1993) data set was assigned to a stratum. A series of regressions were calculated relating these CS90 land cover types to those used in the MLURI Land Cover of Scotland 1988 (LCS88) data set (Macaulay Land Use Research Institute, 1993). The proportions of CS90 land cover types typical of each stratum were then estimated from the regressions for every 1 km square in Scotland. The land cover data were combined with location and altitude, and squares assigned to the strata according to these combined criteria. The main stratification was at a resolution of 1 km^2 but has also been adapted to all the other scales (both coarser and finer resolution) used in the study.

Although the strata give a single broad category for each 1 km square in Scotland, most squares contain land cover types characteristic of more than one stratum. For example, an 'arable' square might contain some grassland (inbye), woodland (non-agricultural) and perhaps heath or bog (moor). Only in the moor and plateau strata is there a significant proportion of squares with 100% land cover characteristic of the stratum.

PLATE 13 MARINE

*Reefs in Loch Creran formed by Serpulid worms (right - with tentacles extended):
a rare habitat with a rich associated fauna (Photos: S. Scott).*

*The free living form of knotted wrack (Ascophyllum nodosum ecad.
mackaii): the main world populations are in Scotland's sheltered
sea lochs (Photo: S. Scott).*

*The fan mussel, Atrina fragilis, attractive to
collectors and being considered for legal protection (Photo: S.
Scott).*

PLATE 14 SEACLIFFS

A group of razorbills (Alca torda) (Photo: J. Swale).

Dropping-splashed ledges of a seabird colony, Isle of May National Nature Reserve (Photo: L. Gill).

Dickie's bladder-fern (Cystopteris dickieana), confined to a few sea-caves on the north-east coast (Photo: M.B. Usher).

The strata were therefore characterised in terms of their land cover composition. Agricultural inputs, pollution loads, probable climate change, and biological indicators in various plant and animal groups, were then related to the strata.

15.3 Results

15.3.1 The strata

Table 15.1 shows the areas of the strata and the percentage of Scotland that each occupies. The moor stratum clearly dominates, with more than half of all squares, and covers nearly all the north and west of the country. The inbye stratum is concentrated towards the south, but with substantial areas also in Caithness and Orkney, in marginal ground in Aberdeenshire and Speyside, and scattered through the main arable areas. The machairs of the Western Isles are also included in this stratum.

The other strata are relatively small. The arable stratum is mostly confined to a wide irregular belt running down the eastern lowland fringe of mainland Scotland. The plateau stratum lies almost entirely north of the central belt and is quite fragmented. However, although it covers only 6% of Scotland, it represents over 95% of this type of land in Britain.

15.3.2 Land cover composition of the strata

Two data sets, LCS88 and the ITE Land Cover Map of Britain (LCM) (Barr *et al.*, 1993; Fuller *et al.*, 1994) were used to assess land cover composition. They differed in source data, dates, spatial resolution and the land cover classes used, so the proportional composition of the strata was calculated from both data sets. Land cover classes were grouped so as to enable interpretive comparisons between the two.

Whichever data set was used, all strata were clearly dominated by the classes characteristic of the stratum, namely cropped ground in the arable stratum, 'improved' or 'intensive/lowland' grass in inbye, heath/moor/bog classes in moor and plateau, and wooded and urban ground in the non-agricultural stratum.

Table 15.1 Scottish agricultural strata: frequencies and percentage of total.

Stratum	Number of 1 km squares	Percentage of total
Arable	9,393	11
Inbye	16,714	2
Moor	46,451	54
Plateau	4,786	6
Non-agricultural	6,962	8

Nevertheless, all strata contained appreciable amounts of nearly all land cover categories, except for the plateau, which was almost entirely heath/moor/bog.

The main differences between LCS88 and LCM were that LCM composition included a higher proportion of grasslands, and lower proportion of wood-land/forest, especially in the arable and moor strata. The composition of the arable stratum was less dominated by cropped land in LCM than in LCS88. These differences are partly attributable to the differences in spatial resolution, with LCM picking out small distinct areas within what would be a single LCS88 map unit. Generally, LCS88 land cover classes show closer association with the strata, but LCM classes may in some cases provide a truer picture of actual land cover composition.

15.3.3   *Agricultural management, pollution and climate change*

The assessment of management changes was hampered by a lack of comprehensive data that could be related to the strata. However, some broad conclusions could be drawn.

Management practices had changed more in arable areas than in moor, where there are more environmental and economic constraints. Intensification, once begun, was more likely to continue in the high-change areas, where farmers tended to be younger. In more upland areas (mainly moor and plateau strata) there was less change and some extensification.

Arable farms had a relatively low 'environmental stock' index (a measure based on the areas of semi-natural vegetation, extensive grass and broadleaf woodland), but the index had risen between 1984 and 1990, mainly owing to conversion from cropped land to grass. In contrast, inbye and moor, with a higher starting index, had changed relatively little, with gains balancing losses, despite increased fertiliser use and stocking rates in inbye.

Generally, there was a trend to less frequent reseeding of temporary grassland and an increase in the amount of old or permanent pasture. This trend was seen in all strata, but especially in arable and moor. Sheep numbers had increased and cattle decreased. In the arable stratum there had been a large increase in the proportion of winter-sown crops. Most other changes (e.g. in fertiliser or pesticide and herbicide use) were related to these.

Excesses of sulphur and acid deposition over critical loads for soils (Hornung *et al.*, 1995) were compared with the relative proportions of the strata in 20 km × 20 km grid squares. Critical loads tended to be higher, and deposition lower, as the proportion of the arable stratum increased and that of moor and plateau decreased. There were no clear trends in the inbye stratum, partly because of its intermediate nature, partly because of its varied composition (see Section 15.3.2).

Climate change predictions implied a general increase in oceanicity across Scotland, particularly in the eastern and southern parts of the country, where the arable and inbye strata are concentrated. The regions most 'at risk' from climate change included a large proportion of the most important areas of semi-natural ecosystems in Scotland, for example the eastern and central Cairngorms, and several regions declared as Environmentally Sensitive Areas.

15.3.4   Distribution of plant and animal species and communities

These analyses are still at an early stage. Nevertheless, sets of species characteristic of individual strata have been tentatively identified. Some examples of these are arable weeds such as *Papaver* spp. in arable, bog plants such as *Pinguicula* or *Drosera* spp. in moor, or subalpine species like *Sibbaldia procumbens* or *Luzula arcuata* in plateau. Similarly, in diurnal insects, *Pieris rapae* is a species characteristic of the arable stratum, and *Erebia epiphron* of plateau. In many cases the geographical distribution of 'characteristic' species, and species richness of characteristic species groups, closely follows that of the associated stratum. Generally, the results indicate that the strata are not only indicators of agricultural land use and land cover, but also (particularly the moor and plateau strata) valid biogeographical units.

15.4   Discussion

15.4.1   Stratification of Scottish agricultural land

The results so far suggest that the strata are a reasonably accurate reflection of the pattern of land use and land cover in Scotland, and can form a valid base for analyses of how management, pollution and climate change affect biodiversity. The land cover composition of the strata is variable, but each is dominated by its characteristic land cover. Comparisons with similar land cover statistics from England indicates a greater proportion of more diverse land cover types in Scotland than in England, possibly associated with less intensive land management. The analyses of the biological data will show whether this greater diversity of land cover and land use is reflected in greater biodiversity.

15.4.2   Relative susceptibility to change of the strata

Overall, the arable stratum is likely to be most affected by management and climate change, and the moor and plateau least affected by these factors. However, land holdings in moor or plateau can be very large, and management changes in just a few of these holdings could therefore have a significant effect on the overall Scottish landscape. Conversely, the moor and plateau are likely to be most affected by acid deposition and the arable least. The inbye stratum holds a pivotal position, not strongly linked to effects of any one factor, but probably significantly affected by all of them.

15.4.3   Geographical variation

As well as the general associations of the strata with land cover and with the three main factors influencing them (management, pollution, climate change), there are also important exceptions or variants linked with geographical position. Although some are only small, unrelated areas, these 'geographical anomalies' may sometimes involve several different factors within relatively large, coherent regions. One example is northeast Scotland (mainly Banff, Buchan and Aberdeenshire), where critical loads of acidity for soils are exceptionally low in much of the arable and inbye strata. Similarly, the moor and plateau strata show more likelihood of significant climate change here than elsewhere in Scotland. In all the strata within this region, the patterns of occurrence of characteristic species assemblages are to some degree different from the corresponding patterns outside the region.

In regions like this, the combination of deviations from the overall averages or trends may be of sufficient importance to merit special investigation. That would be beyond the scope of the present study, but such 'anomalous' areas must still be taken into account in the final interpretation of the results.

15.5   Future work

This will involve: (i) classification of plant communities and animal assemblages in relation to the strata; (ii) estimation of appropriate biodiversity indices; and (iii) modelling the associations between biodiversity and three factors affecting it, namely agriculture, pollution and climate change. From these models, we will describe the relative importance of these three factors in determining biodiversity on Scottish agricultural land.

Acknowledgements

M. Lobley and C. Potter (Wye College) and R. Crabtree (MLURI) carried out the analyses of socio-economic and farm management factors, J. Hall and H. Dyke (ITE) the main analyses of pollution loadings, and J. Watkins (ITE) provided the climate classifications for the analyses of climate change.

References

Barr, C. J., Bunce, R. G. H., Clarke, R. T., Fuller, R. M., Furse, M. T., Gillespie, M. K., Groom, G. B., Hallam, C. J., Hornung, M., Howard, D. C. and Ness, M. J. 1993. *Countryside Survey 1990: Main Report. (Countryside 1990.* Vol. 2.) London, Department of the Environment.

Fuller, R. M., Groom, G. B. and Jones, A. R. 1994. The Land Cover Map of Great Britain: an automated classification of Landsat Thematic Mapper data. *Photogrammetric Engineering and Remote Sensing,* **60,** 553–562.

Hornung, M., Bull, K. R., Cresser, M., Hall, J. R., Langan, S. J., Loveland, P. and Smith, C. 1995. An empirical map of critical loads of acidity for soils in Great Britain. *Environmental Pollution,* **90,** 301–310.

Macaulay Land Use Research Institute 1993. *The Land Cover of Scotland 1988: Final Report.* Aberdeen, Macaulay Land Use Research Institute.

16 SPECIES BIODIVERSITY AND CONSERVATION VALUE IN AGRICULTURE: GROUND BEETLES AS A CASE STUDY

G. N. Foster, D. I. McCracken, S. Blake and I. Ribera

Summary

1. There is a need to recognise the role of agriculture in sustaining species biodiversity in Scotland.

2. The value of ground beetles (Coleoptera, Carabidae) as an invertebrate group suitable for site evaluation is discussed.

3. Analysis of large data sets available for ground beetles in Scotland indicate that species richness and species quality (as measured by known rarity status) are higher on land subject to conventional agricultural practice than on undisturbed land. The reasons for these findings are discussed.

16.1   Introduction

The Government Panel on Sustainable Development (1996) reported that

> Agriculture is the main land use in the UK, covering over 80% of the total area. It has a major influence on habitats and species throughout the country. Agriculture has created many of the habitats on which wildlife depends – hedges, moorland, grassland, arable crops – and, in the virtually inconceivable situation of no agriculture, we would certainly have a less diverse countryside with a restricted range of species compatible with areas of scrub/woodland/forest, hard development and roadside.

A parallel view by a conservationist might run

> The most intractable feature of assigning conservation value to species in Scotland is that the species of pristine habitats still dominate over much of the area, whereas pioneer species of newly created habitats are often rare (Foster, 1994).

We here apply simple rarity criteria to data for an important group of insects in order to assess the contribution of agriculture to Scotland's biodiversity. We conclude that, far from threatening Scotland's biodiversity, conventional farming practices play a significant role in maintaining it.

16.2   The choice of ground beetles (Carabidae)

Ground beetles were chosen as the basis of our study because of several features.

1. They occupy all terrestrial habitats at all altitudes, including areas devoid of vegetation.
2. They include a mixture of highly mobile species that take readily to flight and many flightless species. Thus they include species indicative of relict, undisturbed habitats and pioneer species.
3. There are enough species (*c.* 190 spp. in Scotland) to provide interest, but not so many that taxonomic problems, such as recognition of new species, will interfere with an inventory.
4. Their distributions are well known because of the operation of an effective recording scheme (Luff, 1982, 1997).
5. Their habitats and life-cycle traits have been the subject of much research.
6. They are the principal ground predators in crops, attacking many soil and foliar pests, but they include some phytophagous species, in particular seed-eaters.
7. They are themselves prey items for birds and small mammals.
8. Many ground beetles have aesthetic appeal, being photogenic.
9. A standardised approach to survey is possible, using pitfall traps. Traps are often criticised for being selective, but the critics cannot offer effective alternatives.

This range of points in favour of ground beetles demonstrate their value as indicator taxa for extrapolation to invertebrates in general (Pearson, 1994).

16.3   Species rarity scores

Within any region, species can be ranked from rarest to commonest using a count such as the number of 10 km × 10 km squares of the National Grid that they are known to occupy. They can be divided into categories of abundance in which the least occurrence doubles from one record to records for 128 or more 10 km squares. When this is done for ground beetles in Scotland (Table 16.1), about 30 species occur in each of the four categories of greatest incidence. Summations of

Table 16.1   Numbers of ground beetle species in each category of rarity in Scotland (starting with 1, for the most common species).

Rarity score	No. species	Rarity score	No. species
1	32	16	17
2	31	32	18
4	31	64	18
8	34	128	3

species scores should not be used as indicators of site quality because they vary with the intensity of recording (Ball, 1986; Foster, 1991). One approach is to divide the total by the number of species detected to produce a Species Quality Score (SQS; also Mean Quality Score, MQS). The usefulness of the SQS has been demonstrated for several invertebrate groups (Eyre, 1996). The SQS also provides a measure of the contribution of a site to the biodiversity of the area for which the original individual species scores were calculated. This approach is important when attempting to explain to a lay person the relative importance of sites for a speciose group of invertebrates that lack common names.

An alternative approach to SQS is to examine the proportions of species within each rarity category for each site. This approach, demonstrated in this chapter, identifies the components of SQS as values independent of each other.

16.4   The data sets

The national recording scheme for ground beetles has been in operation since 1971 (Luff, 1982, 1997). Rarity scores have been assigned to ground beetles in Scotland (Eyre, 1993), the commonest species being *Nebria brevicollis*, known from two hundred and four 10 km × 10 km squares. Eight site data sets were chosen for analysis in this study, based on a range of attributes including geography, land use and intensity of recording (Table 16.2). Disturbance at each site was rated on the basis of the level of agricultural activity, in particular tillage, cutting for silage and intense grazing. The Caerlaverock site was an enclosed part of the National Nature Reserve merseland not subject to grazing. Mersehead Farm was regarded as a site subject to disturbance because its arable areas and neighbouring dune land were surveyed in the first year of its conversion to a reserve of the Royal Society for the Protection of Birds. The Loch Fleet catchment was regarded as an area of low disturbance because, although parts of the moorland had been limed, it was not subject to tillage, apart from in the course of partial afforestation, and grazing, except by occasional use by feral goats. The ground beetle fauna of those areas that had been limed was marked only by an abundance of *Trechus obtusus* (Foster *et al.*, 1995).

16.5   Results

Species richness was highest in the more disturbed sites (Table 16.2). The distribution of these species between the rarity classes, grouped for convenience in Figure 16.1 into five, clearly shifts from dominance by common species in the less disturbed sites to dominance by rarer species in the more disturbed sites. This shift is not based solely on addition of rarer species in the disturbed sites but also on a loss of species in the most common category. The number of species in the categories with greater rarity scores increases greatly, with the farmland and moorland near Fowlis Wester, between Crieff and Perth, having thirteen of the species regarded as most rare within Scotland. Kirkton Farm supports twelve such species, mainly found in areas of close-cropped, spring-flushed upland pasture.

Table 16.2   Data for sites with ground beetle records.

Site	Source	Type of land cover and habitat	Agricultural usage	No. of species
Loch Fleet, Galloway	Foster *et al.*, 1995	moorland and forestry, with some liming	nil	33
Caerlaverock, Dumfries	Blake *et al.*, 1996	merseland above high tide	nil	47
Insh Marshes, Speyside	Baines, 1992	floodplain fen, mire and moorland	low	42
Mersehead, Dalbeattie	S. Blake, unpublished	coastal dunes, pasture and arable land	moderate	64
Kirkton Farm, Tyndrum	D. I. McCracken *et al.*, unpublished	pasture and upland sheepwalk	high	60
Easter Howgate, Penicuik	G. N. Foster *et al.*, unpublished	arable converted to permanent pasture with experimental sowing, cutting and grazing measures	high	42
SAC Crichton Royal Farm, Dumfries	Blake *et al.*, 1996	intensively grazed, improved grassland, some converted to meadows	high	62
Fowlis Wester, Perthshire	D. I. McCracken *et al.*, unpublished	transect through lowland grass, arable and set-aside to sheepwalk/ grouse moor	high	71

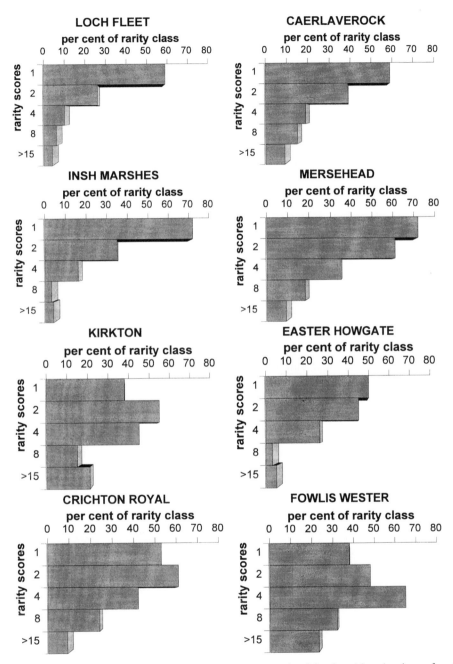

Figure 1.   Percentage occurrences of ground beetle species in each of five Scottish rarity classes for pitfall trap data from eight sites.

16.6   Intuition and counter-intuition

The main conclusion must be that the most disturbed sites not only have more species of ground beetles but also have more of the rarest and fewer of the commoner species. There can be no doubt that disturbed sites have a greater species biodiversity but the shift in species 'quality' is counter-intuitive. Several arguments might be applied.

16.6.1   *Distribution maps are biased towards undisturbed sites*

Some element of bias is inevitable, as those who record wildlife gravitate towards protected, natural or semi-natural areas. Some of the more rarely recorded species are generally regarded as common on a United Kingdom basis, whereas some species in the commonest category, e.g. *Carabus glabratus*, might be regarded as generally rare. This bias is offset by reduced accessibility to upland, undisturbed sites. In any case, the Scottish data set for ground beetles includes many disturbed sites.

16.6.2   *Our perception of what is 'good' is wrong*

Known rarity status and perceived threat are distinct. Some supposedly rare species may be no more than vagrants within Scotland or well-adapted pioneers that can easily survive in disturbed terrain. The loss of many larger species such as *Carabus* spp. from the commonest category in the disturbed sites emphasises how such species should be accorded a higher conservation status on the grounds of habitat fragility; their larval stages last longer and appear to be eliminated by catastrophic events such as tillage (Thiele, 1977). Nevertheless, the primary criterion for consideration of risk must be the rarity of a species. Invertebrate specialists generally avoid assigning too great a priority to apparent changes in distribution; there is a history of rediscovery and recovery associated with insects either considered to be extinct or in rapid decline.

16.6.3   *The recorded rarer species have no conservation value*

This is not true. Upland pastures that are intensively grazed retain many rare and attractive beetles, e.g. *Carabus arvensis*, *C. nitens* and *Elaphrus lapponicus*. Lowland sites retain species such as *Harpalus quadripunctatus*, a species of well-drained areas, classified as a Nationally Scarce species in Great Britain (Ball, 1986).

16.6.4   *The unrecorded rarer species have a greater conservation value than the recorded rarer species*

This does not appear to be true either. The species thus far missing from our inventory of farmed land in Scotland are largely coastal or riparian, or are vagrant species of no conservation value. Some are also found in wetlands, and these include the species most adept at avoiding being trapped.

16.6.5   Scores reflect the distribution of arable land

This appears to be true. Arable land now (Scottish Office Agriculture, Environment and Fisheries Department, 1995) accounts for only 13% of the land mass of Scotland. If species existed that occupied every grid square in Scotland, then any species wholly dependent on arable land would never score less than 4 points. However, because ubiquists do not exist, some species characteristic of arable land, *Trechus quadristriatus* and *Agonum dorsale*, which are also found more rarely in grassland, in fact score 2 points within the present national database.

16.6.6   Ground beetles are exceptional, and provide misleading information about biodiversity and the importance of agriculture

It is fatuous to debate the relative importance of faunal groups. Intuitive, anecdotal and short-term survey results may be applicable for species with a high conservation profile, but data for the invertebrate groups with less aesthetic appeal can be analysed on a wholly objective basis. We are, however, restricted to those few invertebrate groups for which high-quality recording data exist.

16.7   Conclusion

Disturbance often results in an increase in species richness and abundance (Petraitis *et al.*, 1989). Diversity of landform and usage also sustains more species. Agricultural activity can result in an increase in species richness if it diversifies usage, creating a mosaic of habitat types, although intense activity can lead to a uniformity of habitat and a loss of species. Thus the prevailing opinion among conservationists is that *modern* agriculture is non-sustainable in that excessive inputs reduce species diversity and abundance, whereas a return to *traditional* agriculture would bring back a more desirable level of biodiversity. This is an incorrect application of the hump-backed model predicting that species number first increases then declines with increasing disturbance or stress (Grime, 1973), the peak of diversity being achieved by traditional agriculture and the declining tail by intensification. Our results indicate that Scotland's biodiversity for some ground-living insects is high when subject to *conventional*, mixed farming practice. Tracts of land such as those occupied by the farms near Fowlis Wester are species-rich because of the close proximity of a range of disturbed, semi-disturbed and relatively stable sites. We have evidence that intensification within grassland results in loss of diversity (Blake *et al.*, 1996); it remains to be established whether intensive arable use has a similar or greater effect in Scotland.

The results presented here are not in any way final. It is reasonable to suppose that mobile species associated with arable land do not have the same conservation value as sedentary species currently scoring the same number of points. Down-weighting based on pioneer traits must be applied rigorously to the whole distribution data set. This has been done for aquatic Coleoptera in Ireland (Foster *et al.*, 1992), and a similar system needs to be developed for terrestrial Coleoptera.

Two further questions can only result in speculation.

16.7.1 To what extent have ground beetles been lost as a result of agriculture?

One of the main habitats to be lost as a result of agriculture has been lowland wetland. Soil moisture is the main factor dictating ground beetle assemblages (Luff *et al.,* 1989). However, pockets of wetland survive in most farmland and there is no evidence of species extinction. Regional extinction and reduction in range are more likely to be the result of spread of extensive hard surfaces associated with industry, urbanisation and transport infrastructure.

16.7.2 Where were the species of disturbed land prior to the development of agriculture?

Quaternary studies by Professor R. Coope (1979) and others have identified the species associated with the postglacial climatic optimum, which included a period during which trees were absent from much of Scotland. The ground beetles from this period include many associated with disturbed sites, in particular the species of improved grasslands (e.g. *Nebria brevicollis, Loricera pilicornis, Clivina fossor* (Coope, 1979)). Species now typical of arable soils are less evident in the postglacial fossil record, but *Amara* species are particularly abundant. *Carabus* species, which can be eliminated by tillage or heavy poaching, are poorly represented. Ploughing creates a habitat analogue for at least one ancient habitat.

Acknowledgements

This paper is based on result derived from several projects, most of which have been funded by the Scottish Office, Agriculture, Environment and Fisheries Department. In particular, we acknowledge our colleagues in the SOAEFD-funded project on functional diversity, led by Dr Kevin Murphy of the University of Glasgow. Dr David Beaumont, of the Royal Society for the Protection of Birds, kindly provided access to the data collected by Miriam Baines. The conclusions of this paper are largely derived from national recording scheme data collected by Dr Martin Luff of the University of Newcastle upon Tyne, and collated by Dr Mick Eyre, of Entomological Monitoring Services, Newcastle-upon-Tyne.

References

Baines, M. 1992. *A beetle survey of Insh Marshes RSPB Nature Reserve.* Unpublished report. Sandy, Royal Society for the Protection of Birds.

Ball, S. G. 1986. *Terrestrial and freshwater invertebrates with Red Data Book, Notable or Habitat Indicator Status.* Invertebrate Site Register No. 66. Peterborough, Nature Conservancy Council.

Blake, S., Foster, G. N., Fisher, G. E. J. and Ligertwood, G. L. 1996. Effects of management practices on the carabid faunas of newly established wildflower meadows in southern Scotland. *Annales Zoologici Fennici,* **33**, 139–147.

Coope, G. R. 1979. The Carabidae of the glacial refuge in the British Isles and their contribution to the post glacial colonization of Scandinavia and the North Atlantic islands. In Erwin, T. L., Ball, G. E. and Whitehead, D. R. (Eds.) *Carabid Beetles. Their Evolution, Natural History, and Classification.* The Hague, Junk, 407–424.

Eyre, M. D. 1993. *The ground beetles (Coleoptera: Carabidae) of Scotland.* Unpublished report to Scottish Natural Heritage. Newcastle-upon-Tyne, Entomological Monitoring Services.

Eyre, M. D. (Ed.) 1996. *Environmental Monitoring, Surveillance and Conservation using Invertebrates.* Newcastle-upon-Tyne, EMS Publications.

Foster, G. N. 1991. Conserving insects of aquatic and wetland habitats, with special reference to beetles. In Collins, N. M. and Thomas, J. D. (Eds.) *The Conservation of Insects and their Habitats*. London, Academic Press, 237–262.

Foster, G. N. 1994. *Biodiversity Inventory for Scotland: Aquatic Coleoptera*. Scottish Natural Heritage Review No. 26.

Foster, G. N., Nelson, B. H., Bilton, D. T., Lott, D. A., Merritt, R., Weyl, R. S. and Eyre, M. D. 1992. A classification and evaluation of Irish water beetle assemblages. *Aquatic Conservation: Marine and Freshwater Ecosystems*, **2**, 185–208.

Foster, G. N., Eyre, M. D., Luff, M. L. and Rushton, S. P. 1995. Responses of some terrestrial invertebrates to the catchment treatments intended to restore the trout fishery at Loch Fleet. *Chemistry and Ecology*, **9**, 217–229.

Government Panel on Sustainable Development (1996). *Agriculture and Biodiversity*. London, HMSO.

Grime, J. P. 1973. Control of species density in herbaceous vegetation. *Journal of Environmental Management*, **1**, 151–167.

Luff, M. L. 1982. *Preliminary Atlas of British Carabidae (Coleoptera)*. Huntingdon, Biological Records Centre.

Luff, M. L. 1997. *Provisional Atlas of the Coleoptera: Carabidae (ground beetles) of Britain and Ireland*. Huntingdon, Institute of Terrestrial Ecology.

Luff, M. L., Eyre, M. D. and Rushton, S. P. 1989. Classification and ordination of habitats of ground beetles (Coleoptera: Carabidae) in north-east England. *Journal of Biogeography*, **16**, 121–130.

Pearson, D. L. 1994. Selecting indicator taxa for the quantitative measurement of biodiversity. *Philosophical Transactions of the Royal Society of London*, B **345**, 75–79.

Petraitis, P. S., Latham, R. E. and Niesenbaum, R. A. 1989. The maintenance of species diversity of disturbance. *Quarterly Review of Biology*, **64**, 393–418.

Scottish Office Agriculture, Environment and Fisheries Department 1996. *Agriculture Facts and Figures Scottish Agriculture 1995*. Edinburgh, The Scottish Office.

Thiele, H. U. 1977. *Carabid Beetles in their Environment*. Berlin, Springer-Verlag.

PART FOUR
TOWARDS A STRATEGY FOR BIODIVERSITY CONSERVATION: PROCEEDINGS OF WORKSHOPS

PART FOUR

TOWARDS A STRATEGY FOR BIODIVERSITY CONSERVATION: PROCEEDINGS OF WORKSHOPS

If strategies for the conservation of biodiversity are to be effective, it is essential that they have a strong sense of ownership by those who are meant to take forward their constituent parts. The UK Biodiversity Action Plan (Anon., 1994) and the report of its Steering Group (Anon., 1995) have set firm guidance and direction on many of the issues that need to be addressed. The subsequent establishment of United Kingdom and country steering groups (see Foreword, this volume) enables a significant element of participation by the key partners in the process. Inevitably, however, these groups may still seem remote from many of those directly engaged in action to document, record, promote, conserve, and even enhance, our biodiversity.

Accordingly, a number of workshops were included in the conference programme to enable all delegates to make a direct contribution to thinking in this field. The five topics for discussion – inventory, environmental reporting, research, restoration, and public awareness – were selected to address major themes of any biodiversity strategy. Each was chaired by an acknowledged specialist in the field with a brief to examine specific key areas. This section of the book includes their accounts of the proceedings of the five different workshops. Needless to say, with limited time and over 250 participants in all, no workshop could be fully comprehensive. Each report, therefore, reflects the various perceptions, interests and aspirations of the participants. It is nonetheless noteworthy that many common and linking themes emerge.

For example, the workshop on inventory (Shaw) emphasised the need for coordinated management of data, whether for 'all-taxa' biodiversity inventories, biological recording or nomenclatural standards, also a central theme of discussions on environmental reporting (Mackey). Both workshops also recognised the need for additional effort on poorly known or poorly recorded taxa, a point also identified in the workshop on research (Vickery and Newton). The need for the enthusiasm and interests of amateurs to be trained, directed and supported, and for actions to be tailored to meet the needs of a wide variety of end users, were also themes common to these three workshops. Vickery and Newton also note the extent to which positive action for biodiversity depends upon the results of research and how little is known of the basic ecology of many of our priority species. For research to become more useful, researchers need to determine the processes underlying the trends established by monitoring, to adopt a more predictive approach, and to address the impacts of management practice on biodiversity. Continuing this theme, Housden, reporting on the discussions of the workshop on restoration, noted the importance of science, especially inventory, monitoring and research, in planning, recording and undertaking restoration projects. Disseminating the results of such projects, either through the literature or by demonstration sites, is equally important to enable a body of experience and knowledge to develop. Although priority should be given to maintaining existing habitats *in situ*, the opportunities for restoration are significant and range in scale from less than a hectare to full landscapes. Indeed, at the conference a report was presented from a preceding symposium on biodiversity at the landscape scale (Simpson and Dennis, 1996). This highlighted the importance to biodiversity conservation of spatial and temporal dynamics in landscapes, especially in relation to fragmentation, metapopulations and large-scale processes.

All these workshops accepted both that the tasks necessary to conserve biodiversity were resource demanding, whether in grants for research, support for training taxonomists or developing incentives for land managers, and that the receipt of such resources was strongly dependent upon public support. 'Selling biodiversity' to the public, and to decision makers, was the topic for another workshop (Scott). Fundamental to achieving greater public awareness is being able to agree upon a definition of biodiversity that does not exclude the public. The term 'biodiversity' was ultimately seen to be an asset to influence and include all sectors of society. In defining a vision for biodiversity conservation, this workshop recognised the principles of understanding, partnership and commitment that are central to any future strategy.

Although in a book it is impossible to include all of the discussions that took place, it is hoped that the following five accounts of the workshop sessions will highlight the main points of discussion, agreement and disagreement. As editors we hope that these accounts will prove useful in understanding the priorities of people involved in the conservation, management and enhancement of Scotland's biodiversity.

References

Anon. 1994. *Biodiversity: the UK Action Plan.* Cm2428. London, HMSO.

Anon. 1995. *Biodiversity: the UK Steering Group Report.* London, HMSO.

Simpson. I. A. and Dennis, P. (Eds.) 1996. *The Spatial Dynamics of Biodiversity: Towards an Understanding of Spatial Patterns and Processes in the Landscape.* University of Stirling, The International Association for Landscape Ecology (UK Region).

17 INVENTORY

M. R. Shaw

The workshop was asked to: (i) determine whether a complete inventory is a realistic proposition (does it matter if we don't achieve it); (ii) identify the greatest gaps (or priorities) and indicate the scale of effort needed to address these; (iii) determine the obstacles to further inventory work; (iv) identify where future effort is best directed; and (v) determine targets for the future.

It quickly became clear to the 25 or so participants that the term 'inventory' was capable of a much wider span of interpretation than we had time to discuss. The possibilities ranged from an appropriately annotated full list of the species that occur in Scotland, that is an All Taxa Biodiversity Inventory (ATBI), a term not actually used in the workshop, to developing inventory systems for recording and quantifying more complex entities such as communities and habitats. Erupting through all our discussions were keenly felt frustrations that the practical means for progress in recording (or interpreting) our findings are at present badly organised and underdeveloped. For those who are in a position to generate records of occurrence of given groups of taxa, it is often very hard to know what best to do with them, and how to avoid wasting effort. With the exception of a very few well-organised schemes, finding out who is doing what, and what expertise and information exists, is too much a matter of chance. There was a definite consensus that redressing this should be part of a badly needed strategy to review, redirect, and restimulate effort towards a more coherent outcome. It is clear that Scottish Natural Heritage, the National Museums of Scotland, and the Royal Botanic Garden, Edinburgh, could each have important roles in this, although the resource implications are large and, at a time when formal responsibilities seem to grow so paradoxically against the tide of ever-receding budgets, it is difficult to be confident that they could successfully undertake it.

Despite richer possibilities for debate, the workshop focused on the need for an approach suggested by an ATBI, and how that might be taken forward. We confined ourselves to a consideration of species-level taxa, but of course sub-specific or infrasubspecific categories could just as easily be dealt with. Recognising the opportunities of modern technology, it was clear that a database that had the capacity to hold a list of all the species found in Scotland could also incorporate a wealth of additional or qualifying information about each taxon. This could cover distribution, ecological requirements or associates, conservation status, the justification for including the taxon (such as depositories of voucher specimens, nomenclatural status, and any other quality controls that may be appropriate), and more. It doesn't matter that not all of this information will be available at once. The ATBI is not a fixed listing, but rather a framework into which all kinds of information – including the discovery of extra species – is easily fitted. Moreover, it is not a purely electronic device: managing it as a database does not mean that abstracts cannot be issued as hard copy at given publication dates. Indeed that would seem highly desirable, at least for lists of included species, perhaps with codified data relating to their distribution and conservation status, to provide a stimulus for further research and the addition of new knowledge. An ATBI database seemed to many of us to be vital if we are really to improve knowledge of Scottish biodiversity, and hence meet our responsibility to nurture it. However, in recognising this approach as the most logical and cost-effective way to build up, store and interpret knowledge of biodiversity, we were mindful that if this database is established it will carry with it a huge time demand if worthwhile progress (and maintenance) is to be achieved: a cost we could not start to quantify but undoubtedly one that would be substantial. No agency seems to be in a position to take this on without additional funding.

The workshop struggled, at times, to address these ideas with the whole biota in mind. For quite prominent groups of organisms in the visible landscape (or even coastal waters) the idea of 'inventory' in the sense of an ATBI is almost trivial. Knowledge of vertebrates, flowering plants and even some groups of cryptogams and invertebrates has progressed beyond that need, and it is the associated data rather than the species lists that need to be marshalled. In many cases there are already detailed distribution maps for each species, and for the best-known organisms the kind of data capture that may be most relevant would be to do with change over time, either in distribution or in abundance or perhaps, even, genetically. But for less well studied groups we do not yet know what occurs in Scotland: we may know some of the species that do so, but we are certain that there are others here that we do not know about. For some groups – sawflies could be an example – the taxonomy is sufficiently well known, and the competence of potential recorders sufficiently trainable, for the lacuna to be addressed by par- ochial survey and identification. If lots (and lots!) of sawfly enthusiasts spent sufficient time in the field in Scotland we would end up with fairly complete information on the occurrence and distribution of Scottish sawflies. There would probably be a few species new to Britain and maybe even some new to science, but the European literature is in a sufficient state for these to be detected reasonably reliably. However, for other groups there may be greater problems because the taxonomic knowledge of the group is not really yet in place: there may be no reliable keys, the names in current use may have uncertain status, there may be a high proportion of undescribed species, and so on. It is important to appreciate that these problems are on a different scale than that there are simply not enough specialists able to undertake identification. For groups like this, in which the basic taxonomy is still unfolding, it would be strongly counterproductive for the Scottish biota to be approached in isolation. There are some groups of terrestrial invertebrates, more groups of marine invertebrates, and rather a lot of microorganisms at these levels of uncertainty (see Usher, this volume).

One requirement, then, for an ATBI is to find a means of coverage that allows detailed and authoritative information to be collated on a particular species in a poorly known group (say Protozoa) without being constrained by the absence of a complete list of Protozoa into which to plug. The key – and here is the answer to one of the questions we were asked to address – is not to expect 'completeness', but to aspire to it. We should not be at all inhibited by the certainty that we will never arrive at complete knowledge in setting up a system that provides for it. Indeed, we recognise that there will always be change in the biota, with local extinctions and colonisations probably more commonplace than we know. So, in the database that represents the start of the ATBI, the first entry for Protozoa could be simply a list of the higher taxa (perhaps at the family level) and whatever estimation of the number of species, referenced to source, that can be given; or, if available, names (and source reference) of whatever taxa for which we have usable data. That aspires to completion, by providing the framework into which more species can be added and information about them can be given. The essential thing is to give each group of organisms recognition as a part of the overall biota that is our concern.

Another consideration for the ATBI is one of nomenclatural standard. We all recognised a very strong need for the biota of the British Isles to be treated as a whole in this respect. Scotland's (as England's, or Wales') ATBI should therefore be an aspect of a nomenclatural database that covers the whole of the British Isles. A bid to the Millennium Commission for a 'National Biodiversity Network' (see Mackey, this volume) includes the establishment of just such an ATBI nomenclatural database for the British Isles. It is very much to be hoped that that particular project will proceed, whether or not the whole bid is successful. Even if an overall ATBI cannot be established, there is a good chance that certain groups will be dealt with in this way. The case for insects, with some possible functions of such a database, has been outlined elsewhere (Shaw, 1996).

The above discussion touched upon many of the other questions that the workshop was asked to address without quite crystallising, so in the closing few minutes they were each addressed directly. Considering the greatest gaps (or priorities) provoked some scorn for the over-partial views of specialists and arguments for greater consideration for 'users'. There were pleas that the parts of the biota that were of greatest interest to most people, however soft in scientific terms that interest may be, should be addressed first; but also that scientific cases should be heard in selecting groups of organisms that may be good 'indicators'. However, few in the workshop went so far as to regard 'proxy' taxa as an adequate survey base-line. The value of surveying organisms across a span of taxa in the community or habitat dimension was also briefly discussed. We felt that the scale of effort needed to address any of this was unanswerable in such broad terms: it is an important question, but needs to be related to much smaller targets and units, the identification and prioritisation of which requires further and better-informed discussion.

The obstacles to progress were various. Personally, I believe that one of them is lack of a clear will to adopt a sufficiently simple common target: until we are able to articulate clearly what we believe is required, winning funding to do it will be impossible. Even (or perhaps especially!) at the close, the group had not come to a unified view, although we had had an informative discussion towards that end and there seemed a basis for continuing. Indeed, various contributors stressed the strong need not to let the momentum die there, and urged that we go on discussing and evolving a strategy. Other obstacles were (overwhelmingly) related to cost: finding out and collating what we need to know about the organisms that occur in Scotland in the detail that will allow us to approach biodiversity conservation on any but the narrowest (or the most optimistic, proxy-based) fronts cannot be meaningfully started without extra resources. In the current insecure funding environment, long-term commitments to develop and maintain large databases are extremely difficult to make. Further obstacles, or at least matters that need discussion, relate to the validation of records, and also the shortage of expertise: how do we regenerate and stimulate enough taxonomic interest, knowledge and training, both among amateurs and among trained biology graduates (when University courses no longer do so)? Providing suitable courses for enthusiasts didn't seem to a majority to be the solution: the inertia seems to run rather deeper and the absence of career opportunity in taxonomy has certainly not helped the supply side. The workshop paid far less attention to the need for survey, and how to promote and manage that, than several of the participants would have liked.

In identifying our best future effort and first targets, the group did more or less all agree that trying to establish the taxonomic database framework, and trying to put in as much data for the 'easy' groups as we can manage, would be good first steps. A lot of these data already exist: their entry into the database would perhaps provide momentum. We also need to find effective ways to reduce duplication of effort and to increase coordination and stimulation, perhaps by publicising the clear intention to establish an ATBI, and enabling people to feed into it. However, as the workshop drew to a close, we also felt a strong need for more discussion on 'inventory' itself, which we all regarded as a crucial biodiversity issue, however we chose to think of it. As we rose, a key participant had the last word: 'actually, I think this workshop should have been talking about inventories of habitats'. Well, yes, that too needs to be addressed.

Acknowledgements

I am grateful to participants Keith Bland and Andy Whittington for commenting on a draft.

References

Shaw, M. R. (1996). What about the British insect fauna? *Antenna*, **20**, 16–19.

18 ENVIRONMENTAL REPORTING

E. C. Mackey

The conference workshop on Environmental Reporting was asked to address ways of taking forward the recommendations of the Steering Group of the Biodiversity Action Plan.

18.1 Background

Conservation practice and policy should be based upon a sound knowledge base. The UK Biodiversity Action Plan (Anon., 1994), as the national strategy for the conservation of biological diversity and the sustainable use of biological resources, advanced that view. It went on to state that:

> Although we are fortunate in the UK that relatively large amounts of data are collected on biodiversity, much of this is not readily available in a form that assists decisions on the management of species populations or on the direction of land-use change.

Furthermore, there are important information gaps.

Within its remit, the Biodiversity Steering Group (Anon., 1995) was charged with identifying a means of:

> Improving the accessibility and co-ordination of existing biological datasets, to provide common standards for future recording and to examine the feasibility in due course of a single UK Biota Database.

The subgroup on information and data showed that some important groups of species are not covered by rigorous forms of surveillance and monitoring. In addition, the costs of formal monitoring may be high, and lengthy periods of monitoring are often required before results can feed into action. As a result, greater consideration of priorities and cost-effectiveness is required, including sampling and the identification of groups of indicator species.

The Steering Group advised that structured monitoring for environmental reporting requires the establishment of baselines; regular and systematic recording to detect change or progress towards targets; and the investigation of reasons for change, particularly undesirable change, in order to inform action. A need to focus more clearly on priority species and habitats must not be at the expense of the need also for broad-based surveillance of trends. They recommended building on best practice in data collection and data management, while developing biodiversity data standards, information standards and technology standards. Local and national requirements were identified.

Local data are needed to inform the production of Local Biodiversity Action Plans. Local data centres, which serve the locality and which provide information needed for the national picture, were felt to lie at the centre of an effective system.

> However, we start from a dangerous decline in the resourcing of this work at both local and national levels. The local records centre network is not uniformly resourced, and at present centres are closing or barely ticking over (Anon., 1995).

The geographical coverage of Local Records Centres in Scotland (Figure 18.1) shows a clustering of centres about the main population centres, which are predominantly in the Central Belt.

At the national level, it was suggested that a network of custodian organisations should cooperate to maintain and develop data sets that adhere to standards and protocols. A regularly updated directory of data sets and other key information, which is made widely available, can enhance the utilisation of existing data and reduce duplication and redundancy in data collection.

Government (Anon., 1996) expressed a need for biodiversity information 'to inform policy and decision making, monitor progress against targets, and to meet the needs of the Biodiversity Convention'. Building on the work of the Coordinating Commission for Biological Recording (Burnett *et al.*, 1995), the development of a data catalogue and standard terms list is to be an early task of the Biodiversity Information Service in the Joint Nature Conservation Committee (JNCC), with an initial emphasis on species and habitats covered in the Steering Group report. The JNCC is also charged with leadership in promoting the UK Biodiversity Database.

18.2   Workshop discussion

The following points, under six workshop headings, summarise the discussion.

18.2.1   Information needs

1. Reporting on species and habitat action plan targets must be a priority.
2. The components of the national programme relevant to Scotland need to be defined in order to develop a monitoring and reporting programme appropriate to our needs.
3. Local targets also need to be identified in order to translate the Scottish programme into action on the ground.
4. Monitoring is necessary for scheduled species.
5. Non-scheduled species need to be kept under review in order to detect trends.
6. Under-recorded terrestrial (including freshwater) taxa – notably fungi, lower plants and invertebrates – and the marine environment require *additional* inventory and monitoring effort.

18.2.2   The relative value of different data collection methods

1. Common standards do not imply uniform methods. Methods are necessarily varied, for instance in relation to land cover monitoring, based on 'fitness for purpose'.
2. Monitoring methods must be capable of detecting change in extent and abundance.

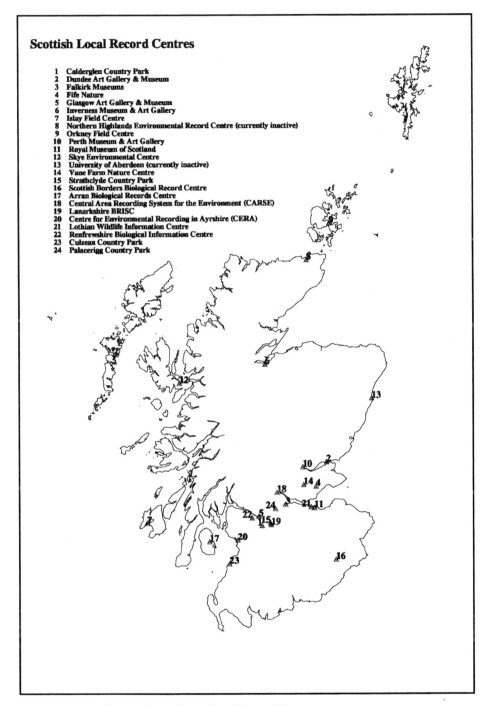

Scottish Local Record Centres

1   Calderglen Country Park
2   Dundee Art Gallery & Museum
3   Falkirk Museums
4   Fife Nature
5   Glasgow Art Gallery & Museum
6   Inverness Museum & Art Gallery
7   Islay Field Centre
8   Northern Highlands Environmental Record Centre (currently inactive)
9   Orkney Field Centre
10 Perth Museum & Art Gallery
11 Royal Museum of Scotland
12 Skye Environmental Centre
13 University of Aberdeen (currently inactive)
14 Vane Farm Nature Centre
15 Strathclyde Country Park
16 Scottish Borders Biological Record Centre
17 Arran Biological Records Centre
18 Central Area Recording System for the Environment (CARSE)
19 Lanarkshire BRISC
20 Centre for Environmental Recording in Ayrshire (CERA)
21 Lothian Wildlife Information Centre
22 Renfrewshire Biological Information Centre
23 Culzean Country Park
24 Palacerigg Country Park

Figure 18.1   Distribution of Scottish Local Record Centres.

3. Monitoring methods should be capable of detecting changes in condition, especially in designated areas.
4. Species monitoring methods will vary according to species scarcity and knowledge of their distribution. Commonly occurring species can be detected by random sampling; less common species require a more systematic census grid-square approach; and rare species require studies to be directed at their known locations (Figure 18.2).
5. Indicator species should be identified for monitoring and summarising information on the state of the environment.
6. The enthusiasm of amateurs, upon whom volunteer effort depends, must be nurtured.

18.2.3 Data collection and transfer

1. There is a need for improved data access at local and national levels.
2. A national data catalogue is a priority need.
3. Ground-studies have not always adhered to the established good practice of securing access permission and explaining survey purpose. There is a need to develop a more broadly based consensus on data collection and data provision, which should seek to facilitate biological recording while safeguarding the confidentiality of site-specific information. A code of conduct for the collection and use of biological information might take the form of a concordat between

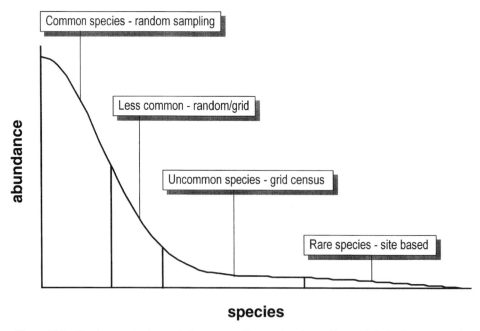

Figure 18.2   Species monitoring techniques according to abundance. From C. Sydes (pers. comm.).

representatives of the biological recording community on the one hand, and representatives of the landowning community on the other.

4. Data collection should not be driven solely by scientific interests, but should be informed by the information needs of the main data users at the local and national levels.

18.2.4   *Obstacles to such a network*

1. The resources available to biological recording are insufficient for adequate geographical and taxonomic coverage.
2. Safeguards are required to guard against the possible abuse of site-specific information, which may adversely affect land values or land management.
3. A general lack of feedback to land managers from the outcomes of studies should be remedied.
4. Complex legislation and data charging may frustrate the aims of freedom of access to environmental information.

18.2.5   *Ideal model for data collection and reporting*

1. Good liaison needs to be established between agencies, voluntary bodies and individual recorders.
2. Meta-data must be an integral requirement of any data set.
3. The greatest possible freedom of access to environmental information, within defined guidelines on data security and confidentiality, is desirable.
4. Guidance is required on priorities for data collection.

18.2.6   *What can practically be achieved?*

1. The information and data initiatives advocated by the Biodiversity Steering Group must be advanced.
2. Standards and guidelines for data collection and use should be developed.
3. Fundamental to the development and maintenance of local records centres, Local Authority interests and requirements need to be considered.
4. Priority must be given to the Biodiversity Action Plan targets.
5. A more balanced monitoring programme is required for species and habitats of the terrestrial and marine environments, for which new work and new funding are required.

References

Anon. 1994. *Biodiversity: the UK Action Plan.* Cm 2428. London, HMSO.

Anon. 1995. *Biodiversity: the UK Steering Group Report.* London, HMSO.

Anon. 1996. *Government Response to the UK Steering Group Report on Biodiversity.* Cm 3260. London, HMSO.

Burnett, J., Copp, C. and Harding, P. 1995. *Biological Recording in the United Kingdom: Present Practice and Future Development.* Middlesex, Department of the Environment.

19 PRIORITIES FOR RESEARCH ON BIODIVERSITY IN SCOTLAND

J. A. Vickery and A. C. Newton

A recent text on biodiversity starts with the sentence: 'Bandwagon, buzz-word, growth industry, global resource, issue and phenomenon ...' (Gaston, 1996). This illustrates the extent to which interpretations of the term 'biodiversity' differ across and within sectors of society. Given that biodiversity represents the 'total variety of life on earth including all genes, species and ecosystems and the ecological processes of which they are a part' (International Council for Bird Preservation, 1992) it is hardly suprising that a tremendous breadth of research is relevant to its study. As a consequence, very different research priorities will be set by scientists, conservationists, economists, sociologists and others interested in the area. Whereas scientists may be concerned with testing theories about diversity, conservationists will tend to be more concerned with applied research directed at solving practical problems, and economists might be more interested in how biodiversity might be valued in financial terms. With such diversity of interest, how may research needs be identified and prioritised?

During the conference workshop, it was suggested that lessons can perhaps be learnt from the USA, where the Biodiversity Uncertainties and Research Needs (BURN) project was developed to identify the primary information needs of decision makers about biodiversity. This information was used to engage scientists in designing policy-relevant research through a series of workshops (Smythe, 1996). The results of this project indicated a need for information about biodiversity that integrates social and biological objectives, and identified the importance of multidisciplinary and participatory research, meta-analysis of existing data, consideration of multiple (hierarchical) scales, and research that is problem- and region-specific. The project also identified six priority research areas that should be pursued to meet the information needs of decision makers (Smythe, 1996): (i) characterisation of biodiversity; (ii) environmental valuation; (iii) sustainable management; (iv) information management strategies; (v) governance and stewardship issues; and (vi) communication and outreach.

In Scotland, a forum for dialogue about biodiversity issues is provided by the Scottish Biodiversity Group, formed in 1995 (see Anon., 1996; Kerr and Bain, this volume). The Group has the primary role of implementing action plans for priority species and habitats identified in *Biodiversity: the UK Steering Group Report* (Anon., 1995). This lists a total of 116 species and 14 habitats, of which 48 species and 12 habitats (the exceptions are lowland heathland and chalk rivers) occur in and are native to Scotland. Research is identified as a requirement to meet the objectives of all the species and habitat action plans. Furthermore, many of the steps outlined in the remaining categories of action depend on output from current or future research. For example, *c.* 74% of 'species advisory' actions, which often require details of aspects such as habitat requirements, depend on the outcome of future research.

Within 'future research and monitoring' for species, action required can be classified into six broad categories (Table 19.1). Classified in this way, the most commonly required research for species action plans is ecological (91.7% of species) (Table 19.1). For half of the species more research is required to identify or quantify threats (50%) and to design conservation strategies (47.9%) with taxonomic, genetic and monitoring-related research being required for relatively few species (16.7% of species). For habitats in Scotland, research required is largely directed at habitat restoration (92% of habitats). Although this analysis is biased

Table 19.1   Number and nature of 'future research and monitoring' actions required under Action Plans (Anon., 1995) for those species and habitats ocurring in Scotland.

	Category[a]					
	Taxonomic	*Genetic*	*Threat*	*Ecology*	*Monitor*	*Active*
Mammals (*n* = 6)	1	0	3	6	5	2
Birds (*n* = 6)	1	1	5	6	0	4
Reptiles and amphibians (*n* = 2)	0	1	1	2	1	0
Fish (*n* = 3)	2	2	3	3	0	1
Insects (*n* = 3)	0	0	2	3	0	2
Other invertebrates (*n* = 5)	0	1	1	5	0	5
Flowering plants[b] (*n* = 7)	1	1	3	6	0	3
Ferns (*n* = 2)	2	2	0	2	0	1
Fungi (*n* = 1)	0	0	1	0	0	0
Lichens (*n* = 5)	0	0	4	5	1	2
Mosses (*n* = 2)	1	0	0	2	0	1
Liverworts (*n* = 3)	0	0	1	3	0	1
Stoneworts (*n* = 1)	0	0	0	1	1	1
Habitats (*n* = 12)	na	na	4	8	0	11
Total (species only)						
Number (*n* = 48)	**8**	**8**	**24**	**44**	**8**	**23**
Percentage	**16.7**	**16.7**	**50.0**	**91.7**	**16.7**	**47.9**

[a]Action points were grouped according to the following categories: *Taxonomic*, clarify taxonomic status (e.g. the pipistrelle bat, *Pipistrellus pipistrellus*); *Genetic*, speciation, genetic variation within and/or between populations (e.g. speciation between allis shad, *Alosa alosa,* and twaite shad, *A. fallax*); *Threats*, identify or quantify nature of threats (e.g. the harmful effects of sulphur dioxide pollution on orange-fruited elm lichen, *Caloplaca luteoalba*); *Ecology*, basic ecological studies (e.g. identifying habitat requirements of medicinal leech, *Hirudo medicinalis*); *Monitor*, develop monitoring and survey methods, where survey methods need to be developed, not simply where surveys are required (e.g. for the great crested newt, *Triturus cristatus*); *Active*, reintroduction, translocation, captive breeding (e.g. methods of artificial rearing and translocation techniques for the narrow-headed wood ant, *Formica exsecta*).
[b]Considers three eyebright (*Euphrasia*) taxa under the same Action Plan (i.e. as one species).

towards 'ecological research', which is a very broad category in relation to the others, it indicates that in many cases we simply do not know enough of the basic ecology to meet action plan objectives and targets.

This apparent lack of information obscures the fact that some species have been the subject of intensive previous research. For example, 23 of the 48 Scottish species are listed in the Steering Group Report (Anon.,1995) as each requiring six or more 'future research and monitoring' actions, despite the fact that current knowledge about these species varies widely in extent. This is indicated by the number of publications on the international science citation index (accessed via BIDS) in the past ten years (1986–1996), which range from 79, 19 and 54 for the water vole (*Arvicola terrestris*), skylark (*Alauda arvensis*) and grey partridge (*Perdix perdix*) respectively, to 0–2 for species such as the narrow-mouthed whorl snail (*Vertigo angustior*), the river jelly lichen (*Collema dichotomum*) and the Killarney fern (*Trichomanes speciosum*).

Discussion during the workshop focused on the extent to which the Steering Group Report (Anon., 1995) provides a guideline for future research on biodiversity in Scotland. The published action plans are clearly biased in favour of certain groups: a relatively high number of birds and mammals are included (Table 19.1) considering the total number of species in these groups (see Usher, this volume). This perhaps suggests a need for research to be concentrated on those groups of organisms that currently receive relatively little emphasis in the action plans. If additional research were focused on the conservation status of under-represented groups such as fungi, for example, many additional species would undoubtedly qualify under the published criteria for the production of an action plan. To a large extent these biases reflect the interests of the various organisations that were involved in drafting the plans, and therefore arguably the public at large as well. As development of action plans is intended to be an ongoing process (Anon., 1995, 1996), it is to be hoped that such biases will be remedied in future.

Despite its limitations, the UK Biodiversity Action Plan and its subsequent Steering Group Report (Anon., 1994, 1995) is a landmark for conservation in Britain. The approach described is the most logical for directing immediate 'fire-fighting' conservation action. However, it was noted at the workshop that the Steering Group Report does not provide a comprehensive assessment of the research and action required to ensure that biodiversity, in its fullest sense, is adequately conserved. As noted elsewhere in this volume, the concept of biodiversity embraces all native species and habitats, as well as intraspecific variation. For commitments such as the Biodiversity Convention to be adequately met, all species and habitats – not just those considered to be threatened at the present time – need to be maintained. The conference workshop therefore also focused on the question of what research is needed, in addition to that listed in the Steering Group Report, to enable all native Scottish biodiversity to be conserved. Some of the key points of discussion and principal conclusions follow.

For ecological research to be most effective, it should be based solidly on *predictive theory*. Much research relating to biodiversity is necessarily adaptive or applied in nature. However, there is a need both in applied and fundamental research for a strong theoretical basis. Such an approach not only increases the precision of research findings, but provides a framework on which generalisations may be based. Theoretical and empirical relationships that can provide a basis for rigorous hypothesis testing are already available, such as species-area curves (Peters, 1991) and body size relationships (Siemann *et al.*, 1996).

The development of useful theories requires an understanding of the *key processes influencing biodiversity*. Attention to processes rather than states (such as current distribution patterns) leads to a recognition of the importance of dynamics in ecological systems. For example, in terms of regulation of species number within a given area, the key processes are speciation, migration and extinction (Rosenzweig, 1995). Understanding the process of extinction is clearly of vital importance for the maintenance of threatened taxa, and for indicating which species may become threatened in the future. For many groups of organisms, rather little is known about the factors influencing their mobility within landscapes, and the habitat features required to maintain evolutionarily viable populations. Given the possibility of future climate change, the provision of appropriate habitat to enable migration and adaptation of species is of paramount importance if they are to be maintained in the long term. At present, understanding of the operation of such processes at the landscape scale is particularly lacking (e.g. the dynamics of metapopulations).

The *impact of management* on key ecological processes needs to be elucidated. As the conservation of biodiversity will not be achieved solely by designation of protected areas (Cassells, 1995), the development of land use systems that are compatible with conservation objectives is of paramount importance. This is particularly relevant to land uses geared to intensive production, such as agriculture and plantation forestry. The development of such approaches is essential if management of natural resources is to be genuinely sustainable. Research is required that will enable the impacts of different management options on biodiversity to be predicted. This may be achieved by a combination of field experimentation, monitoring and modelling approaches. Current research initiatives, such as the Biodiversity Research Programme of the Forestry Commission (see Newton and Humphrey, this volume), should ultimately provide managers with practical guidelines on how biodiversity may be most effectively conserved and assessed, perhaps through the development of suitable indicators. There is a clear need to focus research more closely to the needs of conservation practitioners, and to ensure that the results of research reach those who can implement them practically.

Research is needed that meets the needs of policy and society. It is essential to remember that research costs money. Who will provide the financial support for the research required? There is perhaps a lack of appreciation by society in general of the importance of research into biodiversity. An example is provided by the recent injection of massive financial resources into ecological restoration projects funded by the Millennium Commission, which specifically avoids funding research. Why is this? Is it not appreciated that without a solid foundation of research, such ambitious projects run the risk of failure? Or do such decisions betray a perception that scientific research does not meet the needs of society in general? If this is the case, not only do researchers need to focus on problems which have been identified by society rather than by themselves, but scientists need to convince the public why their research is important. Only with the sympathy of the public can researchers hope to lobby successfully for the financial resources they will require if they are to provide the information needed for effective conservation of biodiversity.

References

Anon. 1994. *Biodiversity: the UK Action Plan.* Cm 2428. London, HMSO.

Anon. 1995. *Biodiversity: the UK Steering Group Report.* Vol. 2, *Action Plans.* London, HMSO.

Anon. 1996. *Government Response to the UK Steering Group Report on Biodiversity.* Cm3260. London, HMSO.

Cassells, D. S. 1995. Considerations for effective international cooperation in tropical forest conservation and management. In Sandbukt, O. (Ed.) *Management of Tropical Forests: Towards an Integrated Perspective.* Norway, Centre for Development and the Environment, University of Oslo, 357–375.

Gaston, K. J. (Ed.) 1996. *Biodiversity. A Biology of Numbers and Difference.* Oxford, Blackwell Science.

International Council for Bird Preservation 1992. *Putting Biodiversity on the Map: Priority Areas for Global Conservation.* Cambridge, ICBP (Birdlife International).

Peters, R. H. 1991. *A Critique for Ecology.* Cambridge, Cambridge University Press.

Rosenzweig, M. L. 1995. *Species Diversity in Space and Time.* Cambridge, Cambridge University Press.

Siemann, E., Tilman, D. and Haarstad, J. 1996. Insect species diversity, abundance and body size relationships. *Nature,* **380**, 705–706

Smythe, K. 1996. BURN: Biodiversity Uncertainties and Research Needs. *The Globe,* **30**, 7–8.

PLATE 15 URBAN

Children enjoying a school nature area, Cambuslang (Photo: L. Gill).

Two examples of recent speciation associated with anthropogenic habitats: Young's helleborine (Epipactis youngiana) and Welsh groundsel (Senecio cambrensis) (Photos: M.B. Usher).

Red admiral (Vanessa atalanta), a typical summer visitor to urban gardens (Photos: M.B. Usher).

PLATE 16 AGRICULTURE

A mixed farmland landscape with arable land, pasture and small farm woodlands, Fife (Photo: L. Gill).

Two species that have suffered declines as a result of changing agricultural practice: cornflower (Centaurea cyaneus), an arable weed, and song thrush (Turdus philomelos) (Photos: M.B. Usher).

20　RESTORATION

S. Housden

The workshop aimed to address the following points: (i) the scope and priorities for habitat restoration in Scotland; (ii)how restoration may become accepted as an integral part of land use policy and practice; and (iii) priorities for the future.

20.1　Introduction and principles

Habitat restoration is a key action for delivering our obligations to enhance and conserve biodiversity. *Biodiversity: the UK Steering Group Report* (Anon., 1995) ident-ifies priorities for such work. Although the workshop focused on habitats, it was agreed that the quality of habitat restored must be such that it meets the needs of priority species that depend on that habitat.

In certain cases, it may be desirable to attempt to rehabilitate or even re-create a habitat, especially when it is threatened in conservation terms or it supports rare species. However, the workshop recognised that a restored habitat is usually second best to an example that has not been degraded and that priority should, therefore, be given to preventing loss or degradation in the first place. Often, it is not possible to re-create in full the biodiversity of a habitat that has developed over a long period of time. The external factors that moulded that habitat and the assemblage of species it supports will have changed over time: the initial melting pot and recipe often no longer exist.

In such cases, only the known constituent elements can be introduced or created on site. For example, the re-creation of a reedbed or wetland on land that is now arable farmland would involve several stages. The hydrological regime would be managed to retain high water levels, and the desired botanical communities would be introduced and managed appropriately. With good management and an element of luck, the system would begin to be colonised by some of the more mobile wetland species such as birds, invertebrates and fish. In a relatively short period of time the area could superficially resemble a natural wetland. However, when compared with a wetland that had developed over hundreds or thousands of years, it might lack many specialist species and features.

Many habitats that are stable and have developed over a long time support species that do not have the ability to survive outwith these conditions. Such species are very poor colonisers of newly created and apparently suitable sites. In such cases greater colonisation could be achieved by re-creating habitat adjacent to an existing relict site, or on a former site where relict populations may still be present and can recolonise suitable areas.

Even with close proximity to existing habitat, and the re-introduction of known characteristic species, the end result will still be of different character to the original habitat, although in time these differences could diminish. The workshop accepted that it was biologically more valuable and economically much more effective to maintain, manage and conserve existing habitats of high biodiversity interest than to attempt to re-create them. By the same logic, it is also more effective to rehabilitate a degraded habitat *in situ* than to attempt the creation of that habitat from a 'sterile' site.

The workshop came to the following conclusions.

1. Geology, climate and other 'immovable' factors must be considered when selecting areas/habitats for re-creation. It is no use seeking to create, say chalk grassland in acid conditions.
2. Areas where the habitat (and typical species) still occur should be targeted for expansion, or linking up, of existing pockets.
3. The cost of such actions on any scale is considerable and usually outside and beyond the capabilities of private owners and managers. The voluntary conservation bodies and Scottish Natural Heritage (SNH), therefore, have an important role to play in such action.
4. Quality is important. Creating or re-creating areas inappropriately so they cannot support species of conservation concern is a wasted effort.
5. The science underpinning such work is still rudimentary for some habitats. Any schemes undertaken must, therefore, be carefully monitored so the lessons learned can be more widely applied.

20.2   Prioritisation of habitats for restoration and re-creation

Given that there are relatively few habitats that lend themselves to successful creation, or restoration to a near-pristine condition, and that the costs are high, it is essential to prioritise action to achieve the best possible results. Owing to recent loss of habitat, particularly in the past 50 years, many species assemblages and populations of dependent species are now at very low levels and continue to decline. This can lead to local and national extinctions and loss of biodiversity. It is valid, therefore, to attempt to restore and re-create habitats to a state where they will sustain populations of particularly threatened species or species assemblages.

Identifying and prioritising suitable habitats for restoration work could be done by using the following criteria.

1, The habitat is identified as a priority in *Biodiversity: the UK Steering Group Report* (Anon., 1995) and also supports priority species identified from the Biodiversity Action Plan process.
2. It is physically possible to undertake the work.
3. The site is appropriate for restoration and once supported the target habitat (correct soil and climatic conditions).
4. The factors which have caused the loss or decline are understood.
5. The skills, finance and knowledge exist to underpin the work.

Using these criteria, the workshop felt that there are considerable opportunities for successful habitat restoration and re-creation in Scotland at both a large and a small scale.

Some of the most important and suitable habitats are as follows.

Caledonian pinewood.

Wetlands and river systems.

Lowland peat habitats, such as raised bogs.

Upland native woodland and scrub.

Blanket mire.

Maritime and serpentine heath.

Heather moorland (targeted to benefit priority open ground species).

20.3 The scope for habitat restoration in Scotland

In terms of scale, Scotland has the potential to support extensive whole-landscape or ecosystem restoration over thousands or tens of thousands of hectares. Caledonian pine woodland is a good example of an ecosystem that was much more extensive in former years. Where relict sites are targeted and regeneration schemes designed around these, very large areas can be considered for restoration. Such an ecosystem could also include a variety of constituent habitats in its mosaic and form a significant conservation and socio-economic resource.

The scale of restoration has to be appropriate to achieve real species and habitat gains. Large-scale schemes are preferable but can be complex and costly, requiring long-term management and resource input before they are self-sustaining. Targets need to be challenging but realistic and achievable.

In many situations it is valid to restore habitats on a much smaller scale. For example, raised mires, wet grassland or water meadows and poor fens can all be quite localised, covering tens or hundreds of hectares. Restoration in these cases will generally have a large and positive impact on the local biodiversity of an area. Further attractions of small-scale restoration are the lower costs and resources involved. The methods are usually known and technically available and are highly cost-effective. In addition, there are existing examples of such restoration projects with detailed documentation, from which a great deal can be learned. Many of the statutory and non-statutory nature conservation bodies have undertaken restoration management on land that they manage.

Even small schemes will require a degree of ongoing management to maintain optimum states. This input can be quite considerable over time and until Government policies (and incentives) are adequate most action is likely to continue to be undertaken by statutory and non-statutory nature conservation bodies such as Scottish Natural Heritage, the Royal Society for The Protection of Birds, the Scottish Wildlife Trust and the National Trust for Scotland.

20.4 Research, survey and monitoring

For any proposed restoration scheme, no matter how large or small, consideration must be given to preliminary research and ongoing survey and monitoring once restoration is underway. The preliminary research is vital in establishing the suitability of a site and the ecological achievability of the scheme.

Initially, this may take the form of desk studies to collate existing information on the site history, the past and current distribution and extent of the target habitat(s) and species and their current conservation status. Preliminary field research could include detailed topographic and hydrological studies, soil studies, vegetation mapping and biological surveys to establish the existence of relict populations on site or on adjacent sites.

Gathering detailed background information can take time and usually needs to be done with a degree of pragmatism. It takes knowledge and skilled judgement to decide which pieces of information are crucial to determine whether restoration can be achieved. A decision will also have to be taken on the time available to do this background research. This can be a frustrating time while data is gathered and no visible change occurs on the ground. However, overlooking or neglecting to properly resource this preliminary research can result in costly failure at a later stage.

At this stage, baseline surveys should be undertaken to document the existing state of the habitat. A range of indicators should be taken, from simple fixed-point photography to detailed botanical mapping. Again, this is a vital part of any restoration project. Only by undertaking thorough and targeted baseline surveys can rates of change and success or failure be measured in future years.

Once the major restoration works have been completed, ongoing monitoring of key indicators should take place at regular intervals. This monitoring supplies the 'health-check' required to assess what ongoing management is needed. Such monitoring will also measure the success of the whole scheme.

The benefits of undertaking research, survey and monitoring are many. The feasibility of the restoration scheme can be assessed, its success judged and the published results can be used by others to design their own restoration schemes. In this manner, experience and expertise is spread as widely as possible in the conservation community. Demonstration sites with well documented management histories (including costs), have a key role in influencing further successful habitat restoration projects.

20.5   Looking to the future

The costs of habitat or ecosystem restoration will continue to rise owing to a number of factors. For instance, habitats will continue to degrade if the current *status quo* is maintained and so will be more expensive to restore in the long run. Examples of this include continued overgrazing of woodland, or former woodland sites, and continued damage to wetlands and peatlands through existing drainage schemes and other agricultural operations.

The view of the workshop was that it is important, therefore, to effectively target resources towards achievable restoration of habitats and ecosystems of highest priority for conservation and enhancement of biodiversity. Larger and more widespread restoration is desirable but the costs should not be underestimated. To this end, Government could extend the use of agri-environment measures, and Forestry Commission or SNH grants, to focus attention on to those key habitats that would benefit most from restoration action in the short term. If such grants worked in an integrated fashion, land managers would not receive competing signals. In this way restoration may become attractive to landowners, resulting in more extensive schemes.

Targets need to have achievable timescales. For example, the extension of native pinewood to the natural treeline at Abernethy Forest may take 100–150 years. Restoration of large areas of wet grassland could take as little as five years, given ideal conditions and resources. The regular monitoring of ongoing projects and regular appraisal of priorities is paramount to ensuring continued success and cost-effective targeting of effort.

As more restoration takes place and lessons are learned, the scale and complexity of restoration should increase (given adequate funding). The benefits of restoration should become clear and demonstrable for biodiversity and socio-economic reasons. It should not be forgotten that such management can create employment opportunities, which may benefit rural communities. The latter should not be neglected because without the willingness and support of local communities, habitat restoration will never succeed on a large scale. The RSPB's Abernethy Forest, which employs 11 staff directly and uses local goods and contractors to help manage the restoration of native pinewood habitat, provides significant economic benefits to the local community (Royal Society for the Protection of Birds, 1997).

20.6   Conclusion

In the medium term, habitat restoration should become accepted as an integral part of land use policy and practice. This can be championed by the implementation of biodiversity action plans at local and national levels and the provision of economic support for biodiversity priorities, through legislation and policy initiatives underpinned by appropriate financial incentives. Such measures are essential if the interest and involvement of a wider audience, particularly private landowners, industry and local communities, is to be achieved.

Building on past and current achievements we now need to think of taking on the large and difficult challenges: for example, restoration of the blanket bogs of the Flow Country, restoration of once extensive wetland ecosystems, restoration of whole catchments and rehabilitation of overgrazed upland tracts. All are enormous projects with vast potential to benefit biodiversity in Scotland.

References

Anon. 1995. *Biodiversity: the UK Steering Group Report*. London, HMSO.
Royal Society for the Protection of Birds 1997. *Working with Nature in Britain: Case Studies of Nature Conservation, Employment and Local Economies*. Sandy, Royal Society for the Protection of Birds.

21 AWARENESS AND EDUCATION: SELLING BIODIVERSITY

M. Scott

With such a wide-ranging subject, it is inevitable that this paper must be a personal reflection on a very challenging and stimulating workshop. Nevertheless, I hope it fairly represents the main views that were put forward by the participants and the broad areas of consensus that emerged.

The particular objectives set for the workshop were as follows.

1. To determine how to disseminate the science and knowledge of biodiversity to as wide an audience as possible.
2. To determine the effectiveness of current mechanisms and programmes, and identify whether there were any gaps.
3. To list the key steps for the government, communities, key sectors and education.
4. To identify the most appropriate methods and organisations to target different audiences.
5. To determine our long-term vision, and consider how it might be achieved.

It was noted that this corresponded closely to the remit for the Public Awareness Subgroup of the Scottish Biodiversity Group, convened by the Scottish Office, and that it was unlikely that the workshop would be able to cover such a wide field in 90 minutes (as indeed proved to be the case, and as this chapter will reflect). Nevertheless it was valuable to have the Chairperson on the Public Awareness Subgroup as a member of this workshop.

21.1 A question of terminology

To launch the workshop and stimulate discussion, the role of devil's advocate was conveniently filled by two papers in a recent issue of *Ecos*, the magazine of the British Association of Nature Conservationists (BANC), which had the cover title *Bio-whatever*. In the first paper (Pollock, 1996), John Pollock, who carried out research on public attitudes to the term 'biodiversity' for the government's Biodiversity Steering Group, expressed a view that is often offered more widely. He suggested that the term 'biodiversity' is, by its very nature, unlikely to win public support. He wrote:

> A major problem with the Graeco-Roman hexasyllabic neologism 'biodiversity' is that it smacks, precisely, of the kind of enclosed and elitist language guaranteed to elicit widespread incomprehension and ennui.

Pollock went on to suggest that 'biodiversity' is in fact largely synonymous with the older, and 'more meaningful', concept of 'Nature'. This led the workshop directly into an attempt to define biodiversity.

The term 'biodiversity' was coined by Walter G. Rosen in 1985 for a major conference held the following year in Washington, DC, entitled the 'National Forum on BioDiversity' (Wilson, 1988). However, the term was largely popularised (without the intrusive capital D) by E. O. Wilson, who is often wrongly stated to be the author of the neologism. It came to prominence at the United Nations Conference on Environment and Development (UNCED), often known as the Earth Summit, in Rio de Janeiro in 1992, at which the British Prime Minister was one of 160 signatories of the Convention on Biological Diversity.

The definition used in this convention – and therefore widely quoted in government publications – is a rather inaccessible one. It states:

> 'Biological diversity' means the variability among living organisms from all sources, including, *inter alia*, terrestrial, marine and other aquatic ecosystems and the ecological complexes of which they are part; this includes diversity within species, between species and of ecosystems.

This definition is rendered into slightly simpler language in the definitive *Global Biodiversity Assessment* (Heywood, 1995), which also adds some further explanation of the scope of the term:

> Biodiversity means the variability among living organisms from all sources and the ecological systems of which they are part; this includes diversity within species, between species, and of ecosystems. Were life to occur on other planets, or living organisms to be rescued from fossils preserved millions of years ago, the concept could include these as well. It can be partitioned, so that we can talk of the biodiversity of a country, of an area, or of an ecosystem, of a group of organisms, or within a single species. (Bisby, F. A. in Heywood, 1995, p.27.)

The introduction to Heywood offers a more succinct definition:

> Biodiversity is defined as the total diversity and variability of living things and of the systems of which they are part. (Heywood, V. H. and Baste, I. in Heywood, 1995, p.9.)

Perhaps the most user-friendly definition of all is one that the present author has attributed to Heywood (Scott, 1996), but whose original source he cannot now reconfirm. It states:

> Biodiversity is the totality of the world's living things, including their genetic make-up and the communities they form.

Using these definitions, the workshop went on to consider whether 'biodiversity' was indeed a pseudo-scientific synonym for 'Nature'. It was evident that the word 'Nature' could be substituted in any of the definitions above, and they continue to make sense, so the terms must be closely parallel. However, the workshop agreed that there were significant differences in the meanings of the two words. 'Biodiversity' is a more objective concept, freed from the value-judgment that is implicit in the more philosophical concept of Nature. It emphasises the spectrum of the natural world, in which a medicinal leech (*Hirudo medicinalis*), for example, is an component of equal importance to a golden eagle (*Aquila chrysaetos*) or a giant panda (*Ailuropoda melanoleuca*).

It was agreed that there was certainly an element of repackaging an old concept in the term 'biodiversity', but it was felt that this new terminology also symbolised a new, more determined approach to the conservation of what we now call biodiversity. The term precisely pinpointed the objectives that a broad community of politicians, scientists and conservationists were setting out to achieve, and, as such, the workshop felt it should not be seen as a liability in achieving these goals, but as an asset to be used more widely.

21.2   Nature for nerds?

The second, 'devil's advocate', quote from *Ecos* came from a paper by Paul Evans, chair of BANC (Evans, 1996), who wrote:

> Nature will not be confined to the abstractions, theories, data sets, computer models and species lists of the nerds – and neither should we.

The symposium to date had indeed concentrated mainly on species lists and data sets, and certainly would not have served, of itself, to win public support for biodiversity conservation, but that was not its purpose nor its intended audience. The workshop agreed that the concept of 'biodiversity' had been a powerful tool in winning the support and commitment of government, academics and 'conservation bureaucrats', precisely because it was apparently objective and lacking in value-judgement.

It was noted that policy-makers appeared rather uncomfortable with the emotional response to Nature that is the prime motivation for many non-governmental conservationists (and, privately, also for many individuals in the statutory conservation sector). The data-set approach of biodiversity therefore had made more acceptable many of the arguments that long have been put forward as the main justification for nature conservation, helping to forge the new-found determination to address the conservation of the whole range of natural heterogeneity.

However, it was noted that this objective, cataloguing approach – the often-quoted concept of biodiversity as the 'living bank balance' – did tend to lead to a concentration on species diversity (which is relatively easy to assess), at the expense of the genetic and ecosystem components of biodiversity (which cannot be measured comprehensively by any accepted methodology). This overemphasis on one component of biodiversity is seen in many published papers and had been evident already in the conference (and thus also in this volume; see, for example, Usher; Young and Rotheray; etc).

In focusing in on this one component, something of the inspirational power of the term biodiversity – evident, for example, in books by E. O. Wilson, such as *The Diversity of Life* (Wilson, 1992) – had been lost. The 'diversity' in biodiversity means variety, and variety in this context should imply constant surprises, the thrill of the unexpected, the wonder of extremes and the variability of biological adaptation.

Above all, diversity should be exciting. Perhaps this thrill of discovery has indeed been lost a little in what Evans (1996) called 'the data-sets... of the nerds'. Certainly the workshop agreed that the dynamic, exciting nature of biodiversity needs to be further emphasised to ensure that scientific credibility is not maintained at the expense of public commitment.

21.3   Valuing biodiversity

The workshop went on to consider why biodiversity mattered to the participants (as relatively informed practitioners in the field), and, by extension, why it should matter to a wider constituency of the public. Many of the points made emphasised the fundamental importance of biodiversity. It is 'the putty that holds together the mosaic of life', according to one participant. It is the raw material of evolution, it is a fundamental part of our heritage, and it is what we want to see in our countryside, according to others.

The point was emphasised that we depend on biodiversity for survival, while another participant suggested that biodiversity represents our past and is the key to our future. It was seen as a critical test of sustainability. Indeed the two terms can be regarded as opposite sides of the same coin: by definition, any development which risks reducing biodiversity cannot be sustainable, while biodiversity is itself a resource that, with careful management, can be utilised sustainably to our mutual benefit. Above all, it was agreed that biodiversity was a unifying concept, which brought together many disciplines into a new consensus and concerted action.

This led into discussion about how to win over a broader audience to the concept of biodiversity. Among the points raised were that any publicity should emphasise the positive action that members of the public themselves can play. This should start locally and focus first on practical, small-scale measures. It might be necessary to target a series of different messages at different audiences, tailored to their own understanding and aspirations. A first step should be to target the 'policy cascaders' (i.e. the influential individuals whose views have the most effect on the policies adopted by individuals and organisations at lower levels down the management chain), but the workshop members also emphasised that those promoting biodiversity should be prepared to listen, especially to young people who have a clear concept of the importance of biodiversity, even if they did not necessarily recognise the term itself.

21.4   A vision for biodiversity

It was abundantly clear that all of the workshop group were themselves highly committed to the concept of biodiversity, or they would not have chosen to participate in this workshop. The final task for the group therefore was to work towards a long-term vision for biodiversity. There was not time to fully refine and perfect this vision, but the following are the key points that were discussed as elements of an ultimate vision.

1. There should be widespread respect for living things and natural habitats.
2. The natural heritage should be recognised as a key component of the quality of life for everyone.
3. Both policy decisions by society and all personal lifestyle choices by individuals should be based on a sound understanding of biodiversity.
4. All actions, by society or the individual, should take biodiversity (the health of the natural world) into account, just as they already take personal health into account.
5. The target of our policy objectives should be a net gain in the range, numbers and ecological health of species and habitats.
6. To this end, it is vital that the targets in the government's *Biodiversity Action Plan* (and new targets set by the various focus groups now at work) should be achieved, and resources must be provided to ensure that they can be achieved.
7. To achieve these objectives, there must be a partnership of all sectors working to ensure that biodiversity is delivered.
8. Personal and corporate purchasing decisions should be informed by an awareness of the origins of consumed resources and there should be a preference, wherever practicable, for products of local origins.
9. Every individual should feel he or she has something to contribute to biodiversity.
10. Every individual should feel he or she has something to gain from biodiversity.
11. Notwithstanding all that is stated above, the future vision should be constantly refined in the light of experience and the perceptions of the day, not set in stone by us at the end of the 20th century.

In the light of this germinal vision, the conference (and this volume) seem very small steps in a momentous journey, but they are important nonetheless in helping to define the nature of the new understanding, the new partnership and the new commitment that the term 'biodiversity' has come to symbolise.

Acknowledgement

The author thanks all the workshop participants for their stimulating and thoughtful contributions and apologises to those whose comments have been omitted or misrepresented.

References

Evans, P. 1996 Biodiversity: Nature for Nerds. *Ecos,* **17**, 7–12.
Heywood, V. H. (Ed.) 1995. *Global Biodiversity Assessment.* Cambridge, UNEP/Cambridge University Press.
Pollock, J. 1996. Negative science, positive science: a Jeremiad on 'biodiversion'. *Ecos,* **17**, 2–6.
Scott, M. 1996. Entry on 'Biodiversity'. In *Oxford Children's Encyclopaedia.* Oxford, Oxford University Press.
Wilson, E. O. 1988. *Biodiversity.* Washington D.C., National Academy Press.
Wilson, E. O. 1992. *The Diversity of Life.* Cambridge, Massachusetts, Harvard University Press.

PART FIVE
ACTION FOR BIODIVERSITY

PART FIVE

ACTION FOR BIODIVERSITY

Four major causes of loss of biodiversity have been identified, the so-called Evil Quartet (Diamond 1984): habitat loss and degradation, overkilling of plants and animals, introduction of alien species, and secondary effects of extinction. To reduce these losses we have to know when and how to act, develop and implement appropriate practical and political conservation strategies, and identify research needs to devise such strategies where knowledge is lacking. Whereas the preceding four sections have considered biodiversity largely from a biologist's point of view, this section considers it from the view of policy makers and practitioners. Conservation action can be classified into three broad categories: conservation within protected areas, conservation beyond reserves (wider countryside measures, pollution control, etc.), and *ex situ* conservation (zoological and botanical collections, captive breeding programmes, etc.). All three must be deployed in a complementary and effective way to conserve Scottish biodiversity.

The chapter by Fleming outlines the complex legal mechanisms designed to conserve biodiversity in Britain and attempts to assess the effectiveness with which they have been applied in a Scottish context. The measures have been relatively successful in identifying areas of special conservation significance and/or the species most at risk from direct human threat. For example, within Scotland the biogeographic zones with the greatest percentage cover of Sites of Special Scientific Interest (SSSIs) are two montane ones, the areas that are also of most value, in a Scottish context, for rare plants. However, identifying benefits arising from legal protection is far from straightforward. Although deductions based on population trends of species afforded differing degrees of such protection are problematical, the analyses in this chapter do not provide convincing evidence of such benefits. Similarly, changes in distribution of species in three taxonomic groups – dragonflies, liverworts and scarce vascular plants – suggest that overall occurrence on Sites of Special Scientific Interest (SSSIs) has made little difference to their persistence.

Gaining legal protection is not a conservation achievement in its own right. To be effective it must be actively enforced as an integral part of the sort of wider conservation measures outlined in the chapter by Kerr and Bain. The most recent and comprehensive attempt to address the conservation of biodiversity in Britain is the UK's Biodiversity Action Plan (Anon., 1994, 1995), detailing specific action plans for 116 species and 14 habitats and foreshadowing plans for many more. Legal protection and site safeguard are only two of a wide range of actions, which include advisory work, research, monitoring, publicity and communication to raise public awareness of the issues. Wider countryside measures are becoming increasingly important for conservation in Scotland. The partial reform of the Common Agricultural Policy in 1992 provided important incentives for farmers to integrate farming and conservation through the agri-environment scheme. Environmentally Sensitive Areas now cover more than 25% of the land surface in Scotland and the Millennium Forest initiative aims to promote the establishment and spread of native woodlands. Databases and information sources underpin effective action and new initiatives by Government agencies are intended to coordinate approaches to data holding. Concerted action at the local level to promote environmental education should help disseminate this information and raise public awareness.

In his summing up of the conference Professor Aubrey Manning reminded us of the fact that, although Scotland may be relatively sparsely populated, having 10% of the UK's human population but 25% of its land area, ever-increasing human population size is the single biggest threat to the world's biodiversity. Manning, as a scientist and conservationist of international standing, offers some personal thoughts on the 'hows' and 'whys' of 'ensuring Scotland has a healthy environment rich in wildlife and sustainably supporting human communities'. The shifting emphasis and broadening scope of modern conservation is based on the realisation that 'wildlife has more chance allied to humanity than divorced from it' (Western, 1989). The motivation and support for such an alliance must come from society as a whole (see also Scott, this volume). Convincing the public at large of the vital need and enormous benefits of conserving biodiversity is a major educational task; it is hoped that this book might provide a step towards achieving this objective.

References

Anon. 1994. *Biodiversity: the UK Action Plan*. Cm2428. London, HMSO.

Anon. 1995. *Biodiversity: the UK Steering Group Report*. London, HMSO.

Diamond, J. 1984. Normal extinctions of isolated populations. In Nitecki, M. H. (Ed.) *Extinctions*. Chicago, University of Chicago Press, 191–246.

Western, D. 1989. Why manage nature? In Western, D. and Pearl, M. (Eds.) *Conservation for the Twenty-first Century*. Oxford, Oxford University Press, 133–137.

22   PROTECTING BIODIVERSITY: MECHANISMS AND THEIR EFFECTIVENESS

L. V. Fleming

Summary

1. Key legislation and international obligations for the protection of biodiversity are outlined.

2. Numbers of protected taxa have increased significantly since 1954 as a result of quinquennial reviews and international obligations, yet it is difficult to assess unequivocally the benefits of such protection. Many protected species continue to decline because they are affected by factors that protection cannot remedy.

3. The effectiveness of SSSIs in the safeguard and representation of three taxonomic groups, namely dragonflies, liverworts and scarce plants, is analysed. The majority of species are associated with SSSIs more than would be expected by chance alone. Dragonflies have a significantly greater preference index for SSSIs than liverworts or scarce plants.

4. Species strongly associated with SSSIs have a significant part of their British range in Scotland. In the case of scarce plants these species tend to be montane species. When changes in distribution of the species groups are analysed, a strong association with SSSIs does not appear to have any positive influence on their persistence.

5. The relative effectiveness of site and species conservation is discussed by reference to examples from the Solway of natterjack toads and barnacle geese.

22.1   Introduction

Laws to protect selected species have been enacted over the centuries though the purpose and rationale for doing so has changed markedly through time. Statutory site protection for the purpose of nature conservation is a more recent phenomenon, postdating the Second World War yet with its forerunners in early nature reserves. However, the number, scope and complexity of these laws have increased significantly. This paper considers the application of these mechanisms in Scotland and attempts to gain some measure of their relative effectiveness at conserving the species, rather than habitat, element of biodiversity. A comprehensive inventory or explanation of wildlife law in Britain is beyond the scope of this chapter.

22.2 Protection of species

22.2.1 *Legislation*

Laws for the express purpose of the conservation of a group of organisms *for their own sake,* first arose with the Bird Protection Acts of the second half of the 19th century. Not until the Protection of Birds Act 1954, however, was there a uniform application of protection across a taxonomic group. All birds were given a measure of protection apart from those pest or quarry species exempted from protection; the most threatened species were given enhanced protection. Subsequently, similar protection was applied to some other animals and selected plants in the Conservation of Wild Creatures and Wild Plants Act 1975. Here, an alternative approach was taken in that only species included in relevant schedules received a measure of protection, though all plants were protected from unauthorised uprooting. This dichotomy of approach continues through domestic and international legislation.

A significant development in recent decades has been the enactment of several international treaties, conventions and agreements to which Britain is party, and which require that protection be given to species or elements of their habitat, such as breeding or resting sites. In many cases, these have been developed further by Directives of the European Community (EC) which, to take effect, have to be translated into domestic legislation. The Wildlife and Countryside Act 1981 (WCA), for example, arose largely as a result of the need to meet the requirements of the EC *Directive on the conservation of wild birds* (the 'Birds Directive') and the 'Bern' *Convention on the Conservation of European Wildlife and Natural Habitats.* This Act superseded earlier legislation and also included enhanced measures for the protection of sites for nature conservation. The WCA gives protection to a range of taxa, some listed in schedules, and covers offences from the taking, sale, deliberate disturbance, killing and injuring of protected species, to damaging or obstructing their places of shelter and protection. Some species now have dual protection because of regulations arising from the EC *Directive on the conservation of natural habitats and of wild fauna and flora* (the 'Habitats Directive').

22.2.2 *Numbers of protected species*

The number of taxa in Scotland given some form of legal protection has increased significantly since 1954 (Figure 22.1). In part this results from the international obligations outlined above; indeed, 30% of all species currently receiving some legal protection do so, in part at least, for these reasons. Equally, since 1975 there has been a statutory requirement for the conservation agencies to review and recommend, at five-yearly intervals, changes to the schedules of protected animals

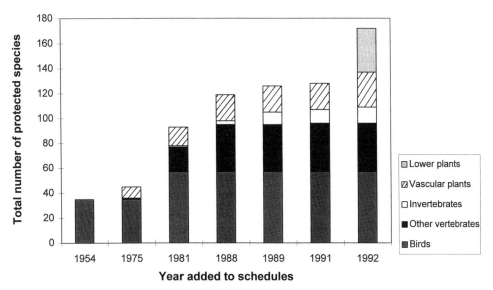

Figure 22.1. Cumulative numbers through time of species given legal protection in Scotland.

and plants (but not those of birds). These quinquennial reviews have also resulted in further increases in protected species (Figure 22.1). Although the number of protected bird species has remained unchanged since 1981, there have been gradual increases in the numbers of protected species of plants, other vertebrates and invertebrates. Until the addition in 1992 of a large group of hitherto neglected lower plants (lichens, bryophytes and charophytes), the rate of increase in numbers appeared to have slowed. This trend is unlikely to change. Despite the shift towards a greater representation of other taxonomic groups on schedules (Figure 22.1), the 'popular' groups of birds, mammals and vascular plants still account for 72% of all protected species. Clearly, the application of protection to species groups has not been applied *pro rata* to the size of the group in question (Usher, this volume). Only 1% of Scotland's 1,459 lichen species receive special protection compared with 28% of birds and 100% of the native herpetofauna.

Although the quinquennial reviews may remove species from, as well as add species to, schedules, since 1981 no species has lost protection entirely. Only one species, the chequered skipper butterfly (*Carterocephalus palaemon*), has been given reduced protection. In this case, a decline to extinction in England largely coincided with the discovery of many new populations in western Scotland (Ravenscroft, 1995).

It should not be surprising that the number of protected species has increased. The criteria defined in the WCA for inclusion on a schedule require either an international obligation, or that a species is 'in danger of extinction' or is 'likely to become so endangered unless conservation measures are taken'. Increasing knowledge of species, their distribution and the threats to them, means that more species are likely to be found to qualify for protection than are likely to recover to the point where protection is no longer required. However, adding species to schedules has not been based upon extinction risk alone; to have done so would result in schedules substantially longer than those we have today. A key determining factor used in quinquennial reviews is to question whether a species will actually benefit from the act of listing on a schedule. If legal protection will have no influence on the factors contributing to the risk of extinction, then there is little to be gained from such protection. Although this is not a requirement in current legislation, it was explicitly stated in the Conservation of Wild Creatures and Wild Plants Act 1975 that a species must be threatened by 'any action designated as an offence under this Act' to be considered for scheduling. This largely explains the bias towards 'charismatic' species in the schedules. Most of the species present in Scotland are not threatened by the kind of intentional acts that might apply to a persecuted predator or a species attractive to collectors.

22.2.3   *Effectiveness of protection*

Given that some species have been protected now for over 40 years, can we assess the effectiveness of these measures? In Table 22.1, population trends in protected species are identified based upon the most recent available assessments (Anon., 1995 and unpublished data), although these do not necessarily cover the same time periods. Surprisingly, even among the vertebrates, trends in populations, range or distribution are often poorly known. Indeed, the majority of protected species fall into this category. Only among birds and vascular plants do a significant number show evidence of increase or, at least, of stability.

At first sight, this does not provide convincing evidence of the benefits of legal protection. However, scheduled species, by definition, are likely from the outset to be at some risk of extinction. Equally, a number of factors may influence the effectiveness of such protection. For some species, especially those added recently to the schedules, protection may not have been in place long enough for it to take effect. Alternatively, and perhaps most critically, many species are also affected by factors such as isolation and habitat fragmentation, small population size, stoch-

Table 22.1   Population trends by taxonomic group in species given legal protection in Scotland.

Taxonomic group	Trend			
	Increase	*Stable*	*Decline*	*Unknown*
Birds	13	20	14	7
Other vertebrates	2	3	5	34
Invertebrates	0	4	6	1
Vascular plants	1	13	6	5
Lower plants	0	11	8	17
Total	**16**	**51**	**39**	**64**

astic events and genetic drift, about which legal protection can do nothing. Even where protection might have a significant effect, there may be practical difficulties of enforcement. Most wildlife crime is likely to be committed in remote country well away from potential witnesses.

By contrast, even when a protected species shows signs of recovery, it may be difficult to attribute this unequivocally to the effects of protection alone. For example, otter (*Lutra lutra*) and peregrine (*Falco peregrinus*) populations in Scotland have now regained much of their former range and abundance following a period of decline and subsequent recovery. Although protection, and the consequent abandonment of organised hunting and a reduction in persecution, undoubtedly played a part in this recovery, both may have benefited equally, or even more, from reductions in the use of pesticides (Crick and Ratcliffe, 1995; Harris *et al.*, 1995; Jefferies, 1996). Similarly, pine martens (*Martes martes*) began a recovery long before they gained legal protection in 1988. This may be attributed as much to the spread of afforestation and diminished numbers of gamekeepers as to protection alone (Balharry, 1992; Harris *et al.*, 1995). Some of these species may, for the foreseeable future, be dependent on protection even after they have achieved a favourable conservation status. Removing protection following recovery may simply make the species vulnerable again to factors that caused its decline originally.

Gaining legal protection is not a conservation achievement in its own right, although it may raise the public profile of the species in question. If protection is to be effective it needs to be applied proactively and enforced. As lists of protected species become ever longer, and laws more complex, it is unreasonable to expect that lay people (including enforcement agencies), or even competent naturalists, will be able to identify most or all protected species, especially in the more difficult groups of invertebrates and lower plants. Scheduling then may become counterproductive (Klemm and Shine, 1993). If we want land managers and others to avoid accidental damage to protected species on their land, contact needs to be established to inform them of the occurrence of such species, their legal responsibilities and to initiate positive dialogue over their conservation. Such contact needs to be balanced by a degree of confidentiality if species may be put at risk of persecution or damage by releasing information on their whereabouts. Whether to release such information, or not, is a dilemma facing conservation managers internationally (Klemm, 1990).

22.3   Release of species to the wild

The spread of alien species is considered to be one of the greatest risks to biodiversity conservation world-wide, second only to habitat destruction. The WCA prohibits the release into the wild of any non-native animal. Exotic animals already established in the wild and selected invasive plants are listed on Schedule 9 of the Act, which also outlaws their further deliberate release or introduction to the wild. Yet when population trends of the 17 non-native species in Scotland on this schedule are examined, most (9) are increasing, trends for six are insufficiently known and two species are either stable or have declined. In this case, it is questionable whether the legislation can do anything other than, at best, slow the rate of human-induced spread. At worst, this legislation is entirely ineffectual. Already, exotic plants account for up to 40% (Sydes, this volume) of the Scottish flora and 18% of the non-marine vertebrate fauna (Anon., 1995). Although few of these cause significant problems, those that do so may be ecologically damaging and costly to control.

Schedule 9 is also used to regulate releases of some native species. For example, barn owls (*Tyto alba*) were added following concern over a spate of poorly planned *ad hoc* introductions (Mead, 1992). Such releases may obscure real trends in the natural population, interfere with official recovery programmes, or raise significant welfare concerns. The release of other native organisms may also be damaging to biodiversity. Current legislation does not prohibit the movement of animals indigenous to any part of Britain beyond their native range. Hedgehogs (*Erinaceus europaeus*) or foxes (*Vulpes vulpes*) for example, may be moved with impunity to offshore islands where they may have never occurred naturally and where they may be damaging to colonies of ground-nesting seabirds. Recent regulation to control the release of genetically modified organisms represents an example of legislation that seeks, amongst other things, to protect the genetic element of biodiversity. Judging its effectiveness is likely to be extremely problematic.

22.4   Site protection

22.4.1   *Designation and selection of Sites of Special Scientific Interest*

The protection of sites complements that of species. Here, only those protected areas notified as Sites of Special Scientific Interest (SSSI) under the WCA will be considered. SSSIs were first created under the National Parks and Access to the Countryside Act 1949, their title reflecting the strong academic force behind the early conservation movement (Nature Conservancy Council, 1984). Other areas protected for their biodiversity interest include National Nature Reserves (NNR), and, at the international level, Special Protection Areas under the EC Birds Directive, Special Areas of Conservation under the EC Habitats Directive, and 'Ramsar' sites designated under the Ramsar Convention on Wetlands of International Importance. These all form part of a network of protected areas throughout Scotland. The designations are not mutually exclusive; indeed a single site may conceivably be all of these, but it is the mechanism of the SSSI that forms the domestic framework for site protection.

There are currently some 1,400 SSSIs in Scotland, covering 11% of the land area. Of these, 1,059 are notified for their biological or mixed (rather than earth science alone) interest and cover 10.5% of Scotland's land area. SSSIs are selected according to guidelines (Nature Conservancy Council, 1989; Joint Nature Conservation Committee, 1992), which cover various habitats and species groups, yet the majority of biological SSSIs are selected for their habitat interest (Blackstock *et al.*, 1996). A site may qualify on several grounds, however, and there may be additional interest on sites apart from the qualifying features. On the species side, the majority of biological sites in Scotland qualify for notification on the basis of their bird (16% of total), vascular plant (12%) or invertebrate (9%) interest, consistent with results from elsewhere in Britain (Blackstock *et al.*, 1996). Some groups, such as lower plants, are hardly used so far for site selection (3% of sites) despite Scotland's international importance for these taxa. Apparently, only two sites are notified for their herpetofaunal interest. It is a basic assumption of the selection process that the representation of less well known or surveyed groups, especially lower plants, will have to rely on their 'chance occurrence in sites selected on other grounds' (NCC, 1989). However, this assumption remains to be critically tested.

22.4.2   Effectiveness of site protection

Typically, one measure of the success (or failure) of the SSSI series is the extent and incidence of damage (see, for example, Chambers, 1996). For instance, in 1994–5, 21 Scottish SSSIs suffered long- or short-term damage representing less than 0.01% of the total area. This approach is more a measure of adherence to the law, or the extent of monitoring for compliance, than any measure of biological quality. Accordingly, this section attempts to consider how effective the SSSI series of protected areas has been in the conservation of selected species groups.

Roy *et al.* (in press) took distribution data for three taxonomic groups – dragonflies, liverworts and scarce plants – and, using a Geographical Information System, overlaid their occurrence on digitised boundaries of SSSIs. All three species groups had recent atlases (Merritt *et al.*, 1996; Hill *et al.*, 1991; Stewart *et al.*, 1994, respectively), had good geographical cover of records with good spatial resolution (to 100 m or 1 km) and were covered by SSSI selection guidelines (Nature Conservancy Council, 1989; Joint Nature Conservation Committee, 1992). For each of these species a preference index was calculated, to give an indication of how strongly associated with SSSI they were, and the proportion of their overall British range in Scotland was determined. Results for all groups showed that over 76% of species had a preference index greater than zero. In other words, they occurred more frequently on SSSIs than might be expected by chance alone (Roy *et al.*, in press). However, these results have to be interpreted with some caution. The naturalists that supply most of these records may target their effort selectively on SSSI and may thus make them appear to be richer than the undesignated countryside around them.

There were, however, differences between species groups, with dragonflies having a significantly greater mean preference index (93.5) than scarce plants (41.8) or liverworts (30.1). In all three cases, those species most strongly associated with SSSIs had a greater part of their British range within Scotland, in the case of scarce plants significantly so (Roy *et al.*, in press). Those scarce plants most strongly associated with SSSIs are, almost without exception, montane species, typically, but not exclusively, calcicoles (Table 22.2). By contrast, those that are poorly associated with SSSIs are predominantly lowland, often at the northern limit of their range in Scotland, and include species of arable, coastal and disturbed habitats

Table 22.2   Examples of scarce plants either most strongly or most weakly, when ranked in order of preference index, associated with SSSIs in Scotland (derived from Roy *et al.*, in press).

Strongly associated with SSSIs	*Weakly associated with SSSIs*
Mountain avens	Northern knotgrass
Dryas octopetala	*Polygonum boreale*
Rock sedge	Spignel
Carex rupestris	*Meum athamanticum*
Alpine bearberry	Wall whitlow-grass
Arctostaphylos alpinus	*Draba muralis*
Net-leaved willow	Creeping lady's tresses
Salix reticulata	*Goodyera repens*
Two-flowered rush	Eyebright
Juncus biglumis	*Euphrasia rostkoviana* var. *rostkoviana*
Alpine saxifrage	Lady's mantle
Saxifraga nivalis	*Alchemilla glomerulans*
Downy willow	Cornflower
Salix lapponum	*Centaurea cyanus*
Dwarf birch	Northern hawk's-beard
Betula nana	*Crepis mollis*
Hair sedge	Dense-flowered fumitory
Carex capillaris	*Fumaria densifolia*
Chestnut sedge	Greater broomrape
Juncus castaneus	*Orobanche rapum-genistae*

as well as some pinewood and grassland species. Neither dragonflies nor liverworts show such a marked contrast in habitat preferences.

To gain some measure of the trends of these species on and off SSSIs, an arbitrary sample of those species most strongly, or most weakly, associated with SSSIs, when ranked in order of preference index, was selected (Table 22.3). For each of these species, an estimate of their change in distribution, pre- and post-1970, was made based on their recorded occurrence in 10 km squares held at the Biological Records Centre (Dring, 1994). Although a relatively crude measure, this approach gives some indication of the changing fortunes of these species. In liverworts and dragonflies, both groups had been recorded from many more squares after 1970 than before, the results suggesting a greater increase in species most poorly represented on SSSIs (Table 22.3). However, in these two groups the results seem to be more an artefact of increased recording effort after 1970, to enable the production of atlases, than to any real change in distribution.

By contrast, vascular plants have a much longer history of thorough recording and the results were less likely to be biased by recording effort. In this instance, those scarce plants most poorly associated with SSSIs had apparently declined to a greater degree than their strongly associated counterparts (Table 22.3), suggesting that SSSIs may have had a positive effect on the persistence of these species.

Table 22.3   Mean preference index (see text for explanation), proportion of range in Scotland, and mean change in distribution (post–1970) in samples of species from three taxonomic groups, either strongly or weakly associated with SSSIs in Scotland when ranked in order of preference index (derived in part from Roy *et al.*, in press).

Taxonomic group (total no. of spp. in group)	Association with SSSIs (sample size)	Mean preference index	Mean % range in Scotland	Index of mean change in distribution[a]
Liverworts	Strong (15)	49	78	7.7
(297)	Weak (15)	−2	37	10.2
Scarce plants	Strong (15)	122	96	0.8
(136)	Weak (15)	−2	43	0.5
Dragonflies	Strong (10)	63	42	3.7
(26)	Weak (10)	9	28	4.8

[a] Based on numbers of occupied 10km squares before and after 1970, where 1.0 equals no change.

However, when the entire scarce plants dataset was analysed, there was no significant relationship between preference index and change in distribution. Overall then, it seems that at this scale of analysis occurrence on SSSIs has made little difference to the persistence of any of these groups.

However, these results can be affected by many factors in addition to recorder effort. For example, the reference date for assessing changes in distribution, 1970, pre-dates by over 10 years the enhanced protection given to SSSIs under the WCA. In addition, SSSIs are not distributed evenly over Scotland. When the proportion of different biogeographical zones (Usher and Balharry, 1996) occupied by SSSIs is considered (Table 22.4), the two montane zones have the greatest percentage cover of SSSIs. The largest zone, of the central and southern lowlands, has the greatest number of SSSIs but the lowest percentage cover. This demonstrates the general pattern of few but large SSSIs in the uplands and many small sites in the lowlands (Blackstock *et al.*, 1996).

In general, then, any upland species has a better chance of occurring in an SSSI than has a lowland species, further complicating any interpretation of the success of SSSIs in conserving species. Yet this analysis does suggest that, for scarce plants at least, SSSI selection has achieved the aim of identifying those areas with features that are most special for nature conservation in a Scottish context. For vascular plants this is Scotland's montane flora (Sydes, this volume) rather than, for example, its arable weed flora. Nevertheless, we must recognise that there are gaps in species groups represented in the SSSI series. Whether this represents a limitation of the SSSI approach, or whether other mechanisms are best suited to their conservation, needs to be established.

22.5   Site and species protection

It is difficult to demonstrate unequivocal benefits of either site or species protection and there have been few studies attempting to assess the comparative effectiveness of each. It is, however, reasonable to ask how these two mechanisms

Table 22.4   Distribution of SSSI between different biogeographical zones (see Usher and Balharry, 1996; Plate 2) in Scotland (from Holbrook *et al.*, in press).

Biogeographic zone	Area of zone (km²)	Area of SSSI (km²)	% of zone as SSSI	Number of SSSI[a]
Western Highlands	5,056	1,388	27	94
Cairngorm	3,650	949	26	66
Barra and Tiree	243	49	20	35
Western Isles (N) & North Mainland	12,649	1,910	15	256
Western Isles (South)	1,522	218	14	79
Western Highland Fringe	10,915	1,314	12	267
Northern Isles	990	95	10	76
Argyll and the Inner Hebrides	6,890	618	9	228
Grampian Fringe & Southern Uplands	14,975	754	5	338
East Coast	4,316	174	4	135
Galloway Coast	617	19	3	32
Central and Southern Lowlands	16,532	381	2	553
Total Scotland	**78,353**	**7,869**	**11**	**2,159**

[a] Numbers of SSSIs are not mutually exclusive between biogeographic zones.

compare. Two separate studies from the Scottish Solway are considered in this regard.

The Svalbard population of barnacle geese (*Branta leucopsis*) wintering on the Solway was reduced to about 500 birds by 1945. After protection against shooting in Britain in 1954 and the establishment of Caerlaverock NNR in 1958, the population subsequently increased. Thereafter, further protection on the Norwegian breeding grounds and the establishment of a feeding refuge on the Solway contributed to further growth (Owen, 1980) such that, by 1980, the population numbered an estimated 8,000 individuals, increasing to *c.*13,000 by 1996. In this instance, there has been a clear and measurable response to protection measures, in which both site and species protection, in Britain and in Norway, have acted in synergy.

However, only 10 km from Caerlaverock, an isolated population of natterjack toads (*Bufo calamita*) has been in a long-term decline since the early part of this century (Fleming *et al.*, 1996). This decline has continued despite protection given to the species from 1975 (subsequently strengthened in 1981 and 1994), and to some of its habitat by notification as an SSSI from 1973 (and again in 1988). The causes of this decline have largely involved habitat destruction, by agricultural improvement and drainage, outside the SSSI. Species protection was ineffectual because, in most cases, land managers had not been made aware that they had natterjack toads on their property and so could not take measures to comply with the law. A wider national study found that site protection was much more effective than species protection in countering threats to natterjacks (Banks *et al.*, 1994).

There are few other studies comparing site and species protection. Warren (1994), for example, found that marsh fritillary butterflies (*Eurodryas aurinia*) in Britain were declining as much on protected sites, including nature reserves, as outside them. It is evident that the relative merits of site or species protection in the conservation of selected species have to be assessed individually.

22.6   Conclusion

There is now, in Britain, a complex series of legal mechanisms aimed at conserving elements of our biodiversity. These measures have largely been successful in identifying those areas of special significance for biodiversity and those species most at risk from direct human threat, yet it is far from straightforward to identify benefits arising from this protection. Often this is because site and species protection cannot always address those factors that detrimentally affect the conservation status of a species. In particular, these mechanisms may be most effective in countering acute threats that might result in direct destruction, but they are likely to be much less effective at countering chronic threats operating at population level (Sydes, this volume). A species may have legal protection and occur on protected areas, even NNRs, but unless appropriate management or intervention is known and practised, protection from acute change alone may not be sufficient to maintain long-term viability.

Unlike legislation in some other countries, such as the Endangered Species Act in the USA, there are no requirements attached to protection in Britain that require the production and implementation of positive action plans for species. Protection will rarely be sufficient in its own right to guarantee the recovery of most species, though there may be some spectacularly successful examples (such as barnacle geese). Ultimately, protection is just one of many mechanisms that may need to be deployed to effectively conserve biodiversity.

Acknowledgements

I am grateful to an anonymous referee, Professor M. B. Usher, Dr A. C. Newton, Dr D. Phillips and M. Shewry for their helpful comments. The views expressed here are my own and not those of SNH.

References

Anon. 1995. *The Natural Heritage of Scotland: an Overview*. Perth, Scottish Natural Heritage.

Balharry, E. 1992. Pine martens in Scotland. *British Wildlife*, **4**, 35–41.

Banks B., Beebee T. J. C. and Cooke, A. S. 1994. Conservation of the natterjack toad *Bufo calamita* in Britain over the period 1970–1990 in relation to site protection and other factors. *Biological Conservation*, **67**, 111–118.

Blackstock, T. H., Stevens, D. P. and Howe, E. A. 1996. Biological components of Sites of Special Scientific Interest in Wales. *Biodiversity and Conservation*, **5**, 897–920.

Chambers, D. 1996. Extent, distribution and safeguard of habitats within biological SSSI in Scotland. Scottish Natural Heritage Information and Advisory Note No. 30.

Crick, H. Q. P. and Ratcliffe, D. A. 1995. The peregrine falcon *Falco peregrinus* population of the United Kingdom in 1991. *Bird Study*, **42**, 1–19.

Dring, J. C. M. 1994. *Support for the Biological Records Centre 1993/94: First Annual Report*. Part 3, *Summaries of Species Occurrence*. Joint Nature Conservation Committee Report No. 187. Peterborough, Joint Nature Conservation Committee.

Fleming, L. V., Mearns, B. and Race, D. 1996. Long term decline and potential for recovery in a small, isolated population of natterjack toads *Bufo calamita*. *Herpetological Journal*, **6**, 119–124.

Harris, S. Morris, P., Wray, S. and Yalden, D. 1995. *A Review of British Mammals: Population Estimates and Conservation Status of British Mammals other than Cetaceans*. Peterborough, Joint Nature Conservation Committee.

Hill, M. O., Preston, C. D. and Smith, A. J. E. 1991. *Atlas of the Bryophytes of Britain and Ireland*. Vol. 1, *Liverworts*. Colchester, Harley Books.

Holbrook, J., Gardner, S. and Shewry, M. In press. A profile of Sites of Special Scientific Interest within Scottish biogeographical zones. Scottish Natural Heritage Research, Survey and Monitoring Report.

Jefferies, D. 1996. Decline and recovery of the otter – a personal account. *British Wildlife*, **7**, 353–364.

Joint Nature Conservation Committee 1992. Guidelines for the selection of biological SSSIs: non-vascular plants. Peterborough, Joint Nature Conservation Committee.

Klemm, C. de 1990. Wild plant conservation and the law. IUCN Environmental Policy and Law paper No. 24. Cambridge, International Union for the Conservation of Nature and Natural Resources.

Klemm, C. de and Shine, C. 1993. Biological diversity conservation and the law. IUCN Environmental Policy and Law Paper No. 29. Cambridge, International Union for the Conservation of Nature and Natural Resources.

Mead, C. 1992. Barn owl releases – why the new controls are so necessary. *British Wildlife*, **3**, 262–265.

Merritt, R., Moore, N. W. and Eversham, B. C. 1996. *Atlas of the Dragonflies of Britain and Ireland*. ITE research publication No. 9. London, HMSO.

Nature Conservancy Council 1984. *Nature Conservation in Great Britain*. Peterborough, Nature Conservancy Council.

Nature Conservancy Council 1989. *Guidelines for Selection of Biological SSSIs*. Peterborough, Nature Conservancy Council.

Owen, M. F. 1980. *Wild Geese of the World*. London, Batsford.

Ravenscroft, N. O. M. 1995. The conservation of *Carterocephalus palaemon* in Scotland. In Pullin, A. S. (Ed.) *Ecology and Conservation of Butterflies*. London, Chapman and Hall, 165–179.

Roy, D. B., Eversham, B. C., Preston, C. D. and Wright, S. M. In press. Biodiversity of protected areas: a pilot study. Scottish Natural Heritage Research, Survey and Monitoring Report No. 89.

Stewart, A., Pearman, D. A. and Preston, C. D. 1994. Scarce plants in Britain. Peterborough, Joint Nature Conservation Committee.

Usher, M. B. and Balharry, D. 1996. *Biogeographical Zonation of Scotland*. Perth, Scottish Natural Heritage.

Warren, M. S. 1994. The UK status and suspected metapopulation structure of a threatened European butterfly, the marsh fritillary *Eurodryas aurinia*. *Biological Conservation*, **67**, 239–249.

23 PERSPECTIVES ON CURRENT ACTION FOR BIODIVERSITY CONSERVATION IN SCOTLAND

A. J. Kerr and C. Bain

Summary

1. This chapter traces the milestones in biodiversity conservation from 1972 to 1996. The Earth Summit at Rio in 1992 provided a huge stimulus for action by a wide spectrum of government agencies, voluntary bodies and the private sector.

2. Traditionally, biodiversity conservation has relied on the protection of species and sites, although such approaches can only safeguard a few species and a small proportion of the planet's land and water resources.

3. The integration of biodiversity conservation into the wider aspects of resource use will require greater accessibility of information and a raising of public awareness of biodiversity issues.

4. There are many challenges for the future. New ways of working, new partnerships, even new perspectives, will be required if the challenges of the Earth Summit are to be met.

23.1 Introduction

Biodiversity is such an all-encompassing concept that it is possible to give only some perspectives on what is, or purports to be, current action for biodiversity conservation in Scotland. A recent review for the UK gives a useful summary (Hill *et al.*, 1996). This chapter focuses on some significant actions, but there are so many activities that only a few representative schemes are highlighted and this review does not aim to be entirely comprehensive. Actions that have received impetus from the current popularity of biodiversity are commented on, but that does not imply that other work is regarded as being of less significance.

The Earth Summit at Rio de Janeiro in 1992 was the culmination of a long process of making the public and politicians aware of the fundamental importance of ecology and the economic significance of biodiversity. Some of the milestones in this process are identified in Table 23.1. This is the context for the initiatives stimulated by the Rio conference, and in particular by the requirements of the Convention on Biological Diversity (United Nations, 1992). The UK Biodiversity Action Plan (UK BAP) was produced as a result of a commitment given by the Prime Minister at Rio and is the key document (Anon., 1994) in setting out the Government's agenda for action. The voluntary sector were quick to appreciate the political impetus that was under way and seized the opportunity to help set a challenging agenda for Government (Wynne *et al.*, 1993).

Site protection is one of the traditional forms of action for biodiversity conservation (Bishop *et al.*, 1997). Representative areas covering the range of ecosystems in a country need protection from major modification by mankind, but only a small part of the Earth's surface, or of any individual country, is ever likely to be designated for protection. The use of all land, fresh water and sea should be sustainable in the yield of renewable resources if the chance of survival of biodiversity is to be maximised. Sustainable use of resources requires fundamental ecological knowledge about species, their distribution, abundance and population

Table 23.1   Milestones before and after the Earth Summit in Rio de Janeiro in 1992.

Date	Milestone	Comments
1972	A Blueprint for Survival	A seminal work, which generated considerable intellectual debate and research world-wide (Goldsmith *et al.*, 1972)
1980	The World Conservation Strategy	Governments and NGOs agreed on the major action needed at the international level (IUCN *et al.*, 1980)
1983	The Conservation and Development Programme for the UK	A major piece of constructive thinking about how to respond in the UK (Anon., 1983)
1987	Our Common Future	The Brundlandt Report, which gave political credence to what had gone before (World Commission on Environment and Development, 1987)
1991	Caring for the Earth	Seeks to establish a new ethic for sustainable living (IUCN, 1991)
1992	Global Biodiversity Strategy	Produced to help set the agenda and win minds for the Rio Summit (IUCN, 1992)
1992	Convention on Biological Diversity	Signed by 157 Heads of Government at the Rio Summit and quickly ratified by many countries
1993	Biodiversity Challenge	The NGOs set out an Agenda for the UK Government (Wynne *et al.*, 1993)
1994	Biodiversity: the UK Action Plan	The UK Government reviewed previous work, set 59 objectives and established a committee structure to develop thinking on priorities for action (Anon., 1994)
1995	Biodiversity: the UK Steering Group Report	Thinking was revealed in Vol. 1; habitat and species action plans formed Vol. 2 (Anon., 1995a, b)
1996	Government Response to the UK Steering Group Report	Acceptance of the Report with significant caveats about funding (Anon., 1996a)

dynamics, and an understanding of how ecosystems function. It is these ecological processes that have increasingly been referred to as 'life-support systems' (Kapoor-Vijay, 1993).

Information must be captured and stored in a consistent manner and be readily accessible to those responsible for taking decisions about the use of natural resources. Because of the magnitude of the task of bringing interested parties together and reaching agreement, discussion on this topic continues. Ensuring a rational approach to biodiversity requires that the most important decision-makers and politicians, at both local and national levels, are well informed. They must be a prime target for any campaign aiming to increase awareness of the importance of biodiversity conservation. Some of the current initiatives have therefore explored approaches to influencing them directly, or to influencing them through public opinion.

23.2 Conserving biodiversity: a planned approach

The UK BAP (Anon., 1994) and Steering Group Report (Anon., 1995a) set out challenging targets in the field of site and species protection. Work on species has received a major boost. Action plans for individual species and habitats were the major thrust of the voluntary sector's challenge to the Government (Wynne *et al.*, 1995). Key elements of this approach are the setting of species and habitat priorities and the production of action plans with clear objectives and targets. The UK Government adopted the approach on the recommendation of the UK Steering Group (Anon., 1995a,b). National species and habitat action plans will guide the activities of all sectors. The plans contain objectives such as maintaining or increasing the range and/or numbers of particular species and have targets for population size or area of habitat to be conserved in given time periods (generally by 2000 and 2010). Shortage of funding produces a need to prioritise in terms of both the selection of species and the choice of plans to implement. In relation to the former, the Steering Group report sets out clear criteria; in relation to the latter, there are no obvious criteria, although pragmatic approaches, depending on the aims of different organisations, seem to have been adopted. The process for prioritising species applies the new International Union for the Conservation of Nature and Natural Resources (IUCN) approach for the production of Red Lists for Species (IUCN, 1994; Palmer *et al.*, 1997). Application of clearly defined, and internationally acceptable, criteria makes the process of species selection more objective. However, given the discrepancies in the amount of information available, there will always be an element of subjectivity. For example, we have a lot of information on the populations of birds and mammals (see, for example, Racey, this volume), a reasonable amount on populations of vascular plants, butterflies and dragonflies, but a dearth of information about the majority of marine populations (Matthews *et al.*, this volume), terrestrial invertebrates (Young and Rotheray, this volume) and cryptogamic plants (Watling, this volume). The selection of species for the plans published in Anon. (1995b) does, however, represent our best attempts at objectivity on the basis of our current knowledge.

The action plans also take into consideration such factors as the need to influence government policies, legislation, research needs, education requirements and the public relations aspects of the work. The plans do not focus solely on site-related management (Anon., 1995b), although this is recognised as being of prime importance. It could be said, therefore, that the plans are more holistic than many previous plans, recognising that success requires a partnership of the biological and social sciences, a partnership of the thinkers and the practitioners, and a partnership of the professionals and the public. This emergence of partnerships for preparing and implementing species and habitat plans is a relatively new phenomenon. For example, in relation to the corncrake (*Crex crex*), a globally threatened species, the Royal Society for the Protection of Birds (RSPB) have established links with Scottish Natural Heritage (SNH) to run an incentive scheme developed in collaboration with the Scottish Crofters' Union. The latter are essential partners because they are the representative body of the main managers (crofters) of the land used by the corncrake in Scotland. Action in the past few years has led to the first increase in the corncrake population, which went through a massive decline at the end of the 19th century and had remained at a low and still declining level ever since (Crockford *et al.*, 1995). The initiative has also benefited from advisory action by the Scottish Office and the Scottish Agricultural Colleges. On reflection, success could only have been achieved by the involvement of such a range of partners: the land owners and land managers, Government and its statutory agencies, and the voluntary sector. It establishes a model for other partnerships for the conservation of other highly endangered species.

Lowland raised bogs provide an example of a habitat facing threats from a number of sources and also requiring joint action. Plantlife took the lead on developing thinking on the issues of commercial peat extraction, landfill and agricultural activities (Plantlife, 1992). Together with the RSPB they developed their initial analysis of the problems into a series of strategies (Stoneman, 1997). The Scottish Wildlife Trust played a very significant role in Scotland and were able to obtain funding from the European Union and so make the Scottish experience available to other countries. This initiative was successful in identifying positive solutions, which achieved protection for the bogs, provided gardeners with alternatives to peat, assisted in waste management and secured jobs in the horticultural industry.

23.3 Site and species protection

Scottish Natural Heritage has been required by Government to give high priority to a programme of work relating to the Birds Directive (European Commission, 1979) and the Habitats Directive (European Commission, 1992). The work initially entails the selection of a series of sites, at the UK level, for the protection of species listed in Annex 1 of the Birds Directive, and of habitats and species listed on Annexes 1 and 2, respectively, of the Habitats Directive. At a European level, these have been agreed to be species and habitats of priority, but they might not have been either Government's or SNH's priorities in Scotland. For example, the stocks of Atlantic salmon (*Salmo salar*) have declined throughout much of the European Union, with the major stocks now located in the UK and Ireland. Albeit of priority at the European level, this is not a species that would have been accorded much priority in Scotland.

Work on the two European Directives has therefore forced a reorientation in SNH's own internal priorities. As a result, the preparation of management plans for all biological Sites of Special Scientific Interest (SSSIs) has progressed more slowly than would have been the case. However, the Habitats Directive has stimulated one entirely new area of work, that concerning the marine environment, where difficulties of survey work and the complexities of legislation mean that progress has been extremely slow. Extensive debate and consultation are now taking place on how to make effective provision for protected areas in the marine environment (Anon., 1996c). The voluntary sector continues to push for action in the marine environment (Gubbay, 1997), but the nature and direction of that action remains unclear. What is again certain is that progress will only be made by partnership approaches, rather than by prescription, and these increasingly are being formed for the implementation of management actions on sites designated for marine species and habitats under the Habitats Directive.

The protection of species by legal means has benefited greatly from closer working between an array of statutory and voluntary agencies that form the Partnership for Action against Wildlife Crime. This is a unique grouping in Europe and has published a handbook of practical guidelines for staff involved in preventing or investigating crimes against protected species of flora and fauna (Taylor, 1996).

23.4   Use of the land and sea

Despite the fact that agricultural and forestry operations are usually exempt, the town and country planning system is a key mechanism for controlling development and integrating biodiversity conservation in land use planning. Local and Subject Plans are of significance in that they set out the likely acceptability to local authorities of different activities within an area. This can have major consequences for the ease or otherwise of maintaining natural heritage assets. The Scottish Office have completed a review of local planning in Scotland; it is also jointly funding work with the RSPB on the production of local biodiversity action plans for four local authorities: Argyll and Bute, Orkney, South Lanarkshire and North East Scotland (Aberdeen City, Aberdeenshire and Moray). Other areas carrying out work are West Lothian, Glasgow, Edinburgh and Fife. This could be a really significant step forward if it leads to greater integration of biodiversity conservation in the land-use planning system. The key to this is probably resources, in terms of both people and funds, rather than knowledge of how to undertake the process. Significant increases in funding will be required to extend these pilot studies to the whole of Scotland, but the demand for resources will not end there. National and local government will then need to invest in the actions prescribed by their

Local Biodiversity Action Plans; will politicians at all levels be sufficiently forward-looking to recognise the strategic importance of this work?

In the maritime environment, the voluntary bodies have shown a considerable capacity to lead on various aspects and to be creative and innovative about new approaches to old problems (see, for example, Gubbay, 1997). The Scottish Office has published a discussion paper looking at the resource, the pressures on it and the planning framework within which decisions are made (Anon., 1996b). SNH has a major initiative called 'Focus on Firths'; this will lead first to the preparation of Management Strategies and then Plans for the major estuaries in Scotland (initially the Moray Firth, the Solway Firth and the Firth of Forth, but also to incorporate the Firths of Clyde and Tay and linking to a similar project for The Minch). This entails the voluntary approach of encouraging all interested parties to discuss the issues in a constructive and collaborative manner. The aim is to promote integrated management of the natural resources of the estuary and hence the conservation of biodiversity. Similar projects involve the Wild Rivers Initiative of the World Wide Fund for Nature, which encourages a holistic approach to river management (Gilvear *et al.*, 1995) and the need for better control of water resources, which has also been set out by the Biodiversity Challenge Group (Hunt, 1996).

A step or two forward was achieved through the partial reform of the Common Agricultural Policy in 1992. This introduced greater flexibility, providing incentives to encourage farmers to operate in an environmentally friendly way while removing a number of other incentives that had led to adverse affects on biodiversity. Environmentally Sensitive Areas (ESAs) are intended to help to conserve areas of high landscape, wildlife and historic values, which are vulnerable to changes in farming practice. There are now ten ESAs in Scotland, covering more than 25% of the land surface. They depend on farmers voluntarily agreeing to use specified practices that protect or enhance the biodiversity interest, and in return the farmers receive payments based on the area of land involved. Although the schemes are generally regarded as having been successful, many people had hoped for greater uptake by farmers. The downside of the voluntary principle is that incentive schemes need to be very attractive in order to ensure the desired endpoint. A recent analysis of the situation in Europe concludes that, to maintain biodiversity, there is scope for continued incentives aimed at the retention of land under low-intensity farming (Baldock *et al.*, 1996). A new Scottish Countryside Premium Scheme, launched in 1997, subsumes some previous agri-environment schemes and affords to the remainder of the countryside opportunities that are currently available only on land within an ESA. This could potentially be of enormous benefit to biodiversity conservation. However, the funding available is limited and discretionary and much remains to be done to target the scheme at areas that could benefit most (McCracken and Bignal, 1994; Badger, 1996).

The Millennium Forest Initiative (Millennium Forest Scotland, 1996) succeeded in winning a grant of £5,750,000. A Trust has been established to progress the vision of nurturing the existing native woodlands of Scotland and establishing a linked network through the creation of new woodlands of native tree and shrub species. The first tree was planted by the Secretary of State for Scotland in February 1996. This is a multifaceted scheme. The funding is being used to assist a variety of organisations to realise their own ambitions in respect of creating and maintaining woodlands into the next millennium. The World Wide Fund for Nature is concentrating on riverine woods, the Forest Authority has a scheme for hazel woods in Argyll and a partnership is involved in restoration of native woodlands in Tayside. The Native Pinewoods Scheme run by the Forestry Authority (Forestry Authority, 1995) continues to promote conservation. This takes the form of enhanced rates of grant aid to the managers of woodland which is agreed to be of native origin, but it also enables managers to extend existing remnants by using appropriate regeneration techniques.

23.5 Data and information sources

Information is at the heart of being able to take effective action. The UK BAP stresses how fortunate we are in the UK in terms of the existing institutions and the strength of the voluntary sector. This applies no less to the Scottish scene, although in a UK context the lower human population density in Scotland means that the amount of biological recording tends to be less intense.

SNH has been active in the field of assembling a comprehensive picture of the biodiversity of Scotland and in creating a summary of the existing databases. A first attempt at defining biodiversity 'hotspots' has been completed (Usher, this volume). Another requirement of the UK BAP already met is the publication of biogeographic zones for Scotland (Usher and Balharry, 1996). The database of countryside surveys in Scotland has been improved through work on the National Countryside Monitoring Scheme. This is based on analysis of air photo cover from the 1940s, 1970s and 1980s, recording changes in the distribution and extent of habitats and linear features between these periods (Tudor *et al.*, 1994). Similarly, the Land Cover of Scotland, also included as an objective in the UK BAP, has been progressed (Macaulay Land Use Research Institute, 1993), as has the Countryside Survey 1990 and the Countryside Information System. An overview of the natural heritage in Scotland has been produced (Mackey, 1995).

A preoccupation of all concerned with the creation and maintenance of databases that are related to biodiversity has been the development of a system that would enable both national oversight and local use (Mackey, this volume). This subject has been reviewed by the Coordinating Commission for Biological Recording under the Chairmanship of Sir John Burnett (Coordinating Commission for Biological Recording, 1995). Current action is two-fold. First, the country agencies (SNH, English Nature and the Countryside Council for Wales) are coordinating their approaches to data holding under the auspices of the Biodiversity Information Service at the Joint Nature Conservation Committee, through which they part-fund the work of the Biological Records Centre. This has entailed working out protocols for the development and validation of current and future data sets, and the development and promotion of the *Recorder* database as a standard tool for amateur and professional recorders covering all species groups. Second, a wide spectrum of interests came together and submitted a bid for funding to the Millennium Commission for £30,000,000. The bid was unsuccessful but the idea of creating a national infrastructure in conjunction with a network of local records centres, all of which would be accessible to the general public, is now generally accepted, well thought out and being pursued by other means.

23.6   Education and the raising of awareness

This sector has seen an impressive surge of activity since the Rio Conference. As acknowledged in the UK BAP, Scotland is in a leading position. The publication of a major environmental education report (Anon., 1993) set an agenda that encouraged all relevant organisations and all sectors of education – formal, informal and training – to implement appropriate environmental education programmes. Encouraging the adoption of the recommendations in the report is one of the objectives of the UK BAP. *Learning for Life* was subsequently adopted by the Government as the basis for its policies in environmental education in Scotland (Anon., 1995c).

There has also been considerable effort by the new local councils to develop programmes of action under Local Agenda 21. SNH launched its Environmental Education Initiative with a budget of £1,300,000 over three years (Scottish Natural Heritage, 1995). It comprised programmes aimed at key target groups. For the public it aimed to encourage communities and individuals to take action to benefit the environment. Projects included the Environmental 'Community Chest', a box of materials and a users' handbook delivered through the community education service. This was subsequently expanded to a 'Sea Chest', aimed at raising awareness of the maritime environment, and a 'Tree Trunk', with similar aims for forests and woodlands.

Public awareness is a key area in which voluntary bodies are active. Most of the literature and media material produced by these bodies is geared towards changing peoples' attitudes and encouraging a general understanding of, and support for, the importance of conserving biodiversity. It also aims to assist in the direct conservation of particular species and habitats. The process of reorienting programmes and publications to account for the often subtle differences between a traditional view and a more holistic view, reflecting biodiversity and its link to sustainability, is ongoing. The greater challenge lies in gaining an understanding of biodiversity, and its importance to both the general public and professional and vocational groups. The Scottish Wildlife Trust has launched an initiative entitled the 'Web of Life', which is a useful exemplar aimed at the general public. It is proving both popular and effective. For senior school and university students there is a series of publications produced jointly by the RSPB and SNH. Both of these initiatives aim to show the importance of biodiversity and how the individual has responsibility and can take action to help to conserve it. The key to much of this work is engendering that sense of personal responsibility for biodiversity.

Table 23.2   Some current initiatives for raising awareness about biodiversity.

Title of initiative	Lead organisation	Audience
Biodiversity Data Support Sheets	World Wide Fund for Nature	Secondary schools
Flowers of the Forest	Plantlife	General public
Focus on Firths	SNH	All
Golf's Natural Heritage	SNH	Golfers
Grounds for Learning	Scottish Environmental Education Council, SNH	Schools
Nature's Prize	Shell UK, Grampian TV, SNH	All
Ocean Watch	Scottish Museums Council, Crown Estate, SNH	General public
Plant for Wildlife	SNH	Gardeners
Scottish Rare Plants Trail	Royal Botanic Garden Edinburgh	General public
Seasearch	SNH	Scuba divers
Seas for Life	RSPB	General public
Web of Life	Scottish Wildlife Trust	Amateur naturalists
Wild Rivers Initiative	World Wide Fund for Nature	River managers

Some of the earlier efforts have failed because they lacked an ability to engender this sense of responsibility; we must ensure that the current and future efforts recognise this essential ingredient. In a sense, to extend the partnership theme further, educational activities need to foster the partnership between the individual taking action for himself or herself, the organisation that can support these individual actions, and nature, which will benefit from these actions (Scott, this volume).

Some examples of the many other initiatives in progress are presented in Table 23.2.

23.7   Priorities for future action

A global view of priorities has been set out by IUCN (Holdgate, 1996). The top priorities for all involved are to be clear about the objectives that must be attained, to have realistic but nevertheless challenging timetables within which to achieve them, and to use reliable criteria for measuring the degree of success. The approach stemming from the UK Biodiversity Action Plan is laudable and must be pursued. In particular, the integration of effort across Government Departments is a key requirement for success. Conserving biodiversity has also been accepted as a key test of sustainable development; hence, biodiversity considerations must be an integral part of the development process. Recent changes in evaluating projects under various European Union funding mechanisms in Scotland have been beneficial, but further refinement is necessary if the maximum use is to be made of this external source of funding. Key sectors for attention will be agriculture, forestry, fisheries and transport since all have major implications for important species and habitats. The voluntary conservation bodies, supported by many tens of thousands of members, have a unique role in advising on this process and monitoring the impacts of society's actions for biodiversity maintenance and enhancement. They also have a proven capacity to build partnerships across a broad range of activities at both local and national levels.

One of the significant benefits already derived from the Government's initiatives on biodiversity has been the building of partnerships. A challenge for the voluntary sector is to encourage greater financial support from industry and commerce; a challenge for the public sector is to know how and when to use such support. The concept of 'champions' for individual species and habitats is in its infancy. Plantlife has received funding from ICI for work on the sticky catchfly (*Lychnis viscaria*) in a partnership involving the Royal Botanic Garden Edinburgh, Scottish Wildlife Trust, Historic Scotland and Scottish Natural Heritage. ICI has also agreed to champion the action plan for the pearl-bordered fritillary (*Boloria euphrosyne*) by Butterfly Conservation in the UK, as well as that for the large blue (*Maculinea arion*) in England. Persuading sponsors to support less glamorous species is a real challenge for the future.

By 2004 all biological SSSIs will have agreed management statements. Implementing the actions identified will be a challenging and expensive task. There is therefore scope for new initiatives that create jobs in the countryside, but to do this we need to find ways to generate the necessary funding. New forms of partnerships and trusts based on the activities of Local Enterprise Companies is one model for attracting funding. How many other models are there, and have we yet gained any experience of harnessing such resources? Implementation of species action plans – written for the UK and involving both statutory and voluntary sectors – will require new forms of working to achieve the plan's objectives and to monitor progress. Making plans work, rather than just writing them, will be the new challenge!

The inclusion of biodiversity conservation as a consideration in all planning and development activities is an obvious goal. Local Biodiversity Action Plans may pave the way for such an approach. Action at the local level will be encouraged by plans that assist in creating greater awareness and involvement. However, it is this latter task that is so important in changing the view of both officials and elected representatives, and perhaps it could be suggested that, until they take a longer-term view that recognises the strategic importance of biodiversity, little success will be achieved. To be sustainable, development must protect or enhance the biological capital of the area, and must not lead to its degradation.

Data on biodiversity will remain a prime requirement. There is now an acceptance that more emphasis needs to be given to training in taxonomy (Shaw, this volume). The national database and the local networks envisaged in the Steering Group Report (Anon., 1995a, b) need to be created. The Biological Recording in Scotland Campaign acts as a focus for this, but needs greater support, greater acceptance in the public, private and voluntary sectors, and perhaps greater vision. Finding ways to facilitate a greater monitoring effort in Scotland is one of the urgent challenges for both the statutory and voluntary sectors. It will require the involvement of academic and research institutions as well as volunteers.

To keep the significance of biodiversity conservation in the forefront of thinking – of both politicians and the general public – considerable ingenuity will be required. Much remains to be done in Scotland to convince politicians of the fundamental economic, social and biological significance of our natural assets. A new generation of children will expect to be able to explore facts and issues on interactive CD-ROM; this will be expensive to create but will have a commercial value. There is a huge amount of work needed to raise awareness amongst all sections of Scottish society.

Finally, new designations and the implementation of action plans for species and habitats are worthwhile objectives. However, these are futile in the absence of a sound approach to the use and management of the countryside, in the absence of public and political support, or in the absence of an adequate system of data storage and management. This means that it is more important than ever that those fortunate enough to lead in this field have a broad vision. They will need to engage all sorts of specialists, but equally, they must encourage specialists to see their activities set in the larger frame. The main question about priorities for future action is whether we dare to think large enough to turn into reality a dream of a Scotland renowned for its biodiversity.

Acknowledgements

Many colleagues in SNH and RSPB have assisted in the completion of this manuscript. Special thanks are due to Professor Michael B. Usher for his help. The views expressed are none the less those of the authors and not of their organisations. Assistance with word processing has been diligently and effectively provided by Jane Lawson and Jo Newman. The back numbers of the monthly digest of environmental news, *Scenes,* provided a very useful *aide-memoire* in reviewing recent action. Our sincere thanks go to all of those involved.

References

Anon. 1983. *The Conservation and Development Programme for the UK.* London, Kogan Page.
Anon. 1993. *Learning for Life. A National Strategy for Environmental Education in Scotland.* Edinburgh, The Scottish Office.
Anon. 1994. *Biodiversity: the UK Action Plan.* Cm2428. London, HMSO.
Anon. 1995a. *Biodiversity: the UK Steering Group Report.* Vol. 1, *Meeting the Rio Challenge.* London, HMSO.
Anon. 1995b. *Biodiversity: the UK Steering Group Report.* Vol. 2, *Action Plans.* London, HMSO.
Anon. 1995c. *A Scottish Strategy for Environmental Education.* Edinburgh, Scottish Office.
Anon. 1996a. *Government Response to the UK Steering Group Report.* Cm3260. London, HMSO.
Anon. 1996b. *Scotland's Coasts: a Discussion Paper.* Edinburgh, HMSO.
Anon. 1996c. *Scotland's Seas and the Habitats Directive.* Edinburgh, The Scottish Office.
Badger, R. J. 1996. *Wildlife and Agriculture in Scotland: a Secure Future.* Edinburgh, Royal Society for the Protection of Birds.
Baldock, D., Beaufoy, G., Brouwer, F. and Godeschalk, F. 1996. *Farming at the Margins: Abandonment or Redeployment of Agricultural Land in Europe.* London, Institute for European Environmental Policy.

Bishop, K., Phillips, A., and Warren, L. M. 1997. Protected areas for the future: models from the past. *Journal of Environmental Planning and Management,* **40**, 81–110.

Coordinating Commission for Biological Recording 1995. *Biological Recording in the United Kingdom: Present Practice and Future Development.* London, HMSO.

Crockford, N., Newbery, P., Watts, O., Williams, G. and Wynde, R. 1995. Conservation action for priority birds species in the UK. *RSPB Conservation Review,* **9**, 7–13.

European Commission 1979. *Directive on the Conservation of Wild Birds* (79/409/EEC). Brussels, European Commission.

European Commission 1992. *Directive on the Conservation of Natural Habitats and of Wild fauna and Flora* (92/43/EEC). Brussels, European Commission.

Forestry Authority 1995. *Grant Aid for Native Pinewoods under the Woodlands Grant Scheme (WGS).* Edinburgh, Forestry Authority.

Gilvear, D., Handley, N., Maitland, P. and Peterken, G. 1995. *Wild Rivers: Phase 1, Technical Paper.* Aberfeldy, WWF Scotland.

Goldsmith, E., Allen, R., Allaby, M., Davoll, J. and Lawrence, S. (Eds.). 1972. A blueprint for survival. *The Ecologist,* **2**, 1–43.

Gubbay, S. 1997. *Scottish Environment Audits. Paper 1: the Marine Environment.* Perth, Scottish Wildlife and Countryside Link.

Hill, D., Treweek, J., Yates, T. and Pienkowski, M. 1996. *Actions for Biodiversity in the UK.* London, British Ecological Society.

Holdgate, M. 1996. *From Care to Action.* London, Earthscan Publications.

Hunt, I. D. 1996. *High and Dry.* Sandy, Biodiversity Challenge Group.

IUCN, UNEP and WWF 1980. *World Conservation Strategy.* Gland, International Union for the Conservation of Nature and National Resources.

IUCN 1991. *Caring for the Earth: a Strategy for Sustainable Living.* Gland, International Union for the Conservation of Nature and National Resources.

IUCN 1992. *Global Biodiversity Strategy: A Policy Makers Guide.* Gland, International Union for the Conservation of Nature and National Resources.

IUCN 1994. *IUCN Red List Categories.* Gland, International Union for the Conservation of Nature and National Resources.

Kapoor-Vijay, P. 1993. The Trinidad Report: identification of key species for conservation and socio-economic development. In Kapoor-Vijay, P. and Usher, M. B. (Eds.) *Identification of Key Species for Conservation and Socio-economic Developments: Proceedings of a Workshop.* London, Commonwealth Secretariat, 1–20.

Macaulay Land Use Research Institute 1993. *The Land Cover of Scotland 1988.* Aberdeen, Macaulay Land Use Research Institute.

Mackey, E. C. (Ed.) 1995. *The Natural Heritage of Scotland: an Overview.* Perth, Scottish Natural Heritage.

McCracken, D. I. and Bignal, E. M. (Eds.) 1994. *Farming on the Edge: the Nature of Traditional Farmland in Europe.* Peterborough, Joint Nature Conservation Committee.

Millennium Forest Scotland 1996. *Project Update.* Glasgow, Millennium Forest Scotland Trust.

Palmer, M. A., Hodgetts, N. G., Ing, B., Stewart, N. F. and Wigginton, M. J. 1997. The application to the British Flora of the World Conservation Union's Revised Red List Criteria and the Significance of Red Lists for Species Conservation. *Biological Conservation,* **82**, 219–226.

Plantlife 1992. *Commission of Inquiry into Peat and Peatlands: Commissioners Report.* London, Plantlife.

Scottish Natural Heritage 1995. *Learning to Live with our Natural Heritage.* Perth, Scottish Natural Heritage.

Stoneman, R. E. 1997. *The Scottish Raised Bog Conservation Strategy.* In Parkyn, L., Stoneman, R. E. and Ingram, H. A. P. (Eds.) *Conserving Peatlands.* Wallingford, CAB International, 424–432.

Taylor, M. B. (Ed.) 1996. *Wildlife Crime: A Practitioner's Guide to Law Enforcement in the Conservation and Protection of Wildlife in the United Kingdom.* London, Department of the Environment.

Tudor, G. J., Mackey, E. C. and Underwood, F. M. 1994. *The National Countryside Monitoring Scheme: the Changing Face of Scotland: 1940s to 1970s.* Perth, Scottish Natural Heritage.

United Nations 1992. *Convention on Biological Diversity.* New York, United Nations.

Usher, M. B. and Balharry, D. 1996. *Biogeographical Zonation of Scotland.* Perth, Scottish Natural Heritage.

World Commission on Environment and Development. 1987. *Our Common Future.* Oxford, Oxford University Press.

Wynne, G., Avery, M., Campbell, L., Gubbay, S., Hawkswell, S., Juniper, T., King, M., Newbery, P., Smart, J., Steel, C., Stones, C., Stubbs, A., Taylor, J., Tydeman, C. and Wynde, R. 1993. *Biodiversity Challenge: an Agenda for Conservation in the UK,* 1st Edn. Sandy, Royal Society for the Protection of Birds.

Wynne, G., Avery, M., Campbell, L., Gubbay, S., Hawkswell, S., Juniper, T., King, M., Newbery, P., Smart, J., Steel, C., Stones, C., Stubbs, A., Taylor, J., Tydeman, C. and Wynde, R. 1995. *Biodiversity Challenge: an Agenda for Conservation in the UK,* 2nd Edn. Sandy, Royal Society for the Protection of Birds.

24 BIODIVERSITY CONSERVATION IN SCOTLAND: PERSONAL REFLECTIONS

A. Manning

Summary

1. A personal view is presented of the challenges facing us in the conservation of biodiversity in Scotland.

2. Scotland should not be seen in isolation from pressures in the rest of the world. Despite Scotland's relatively low population density, human population growth is still the greatest threat to the conservation of biodiversity.

3. How we conserve biodiversity, and why we should do so, are both addressed. The need to engage and convince the general public of the importance and relevance of conserving biodiversity is essential.

24.1 Introduction

In this concluding chapter, I do not intend to provide a summary of the findings of this volume or, indeed, of the conference itself. Contributors have come from suitably diverse backgrounds and each of us has brought our own set of priorities to the fore. I can try to pull out some common themes, or at least what I would take to be some of the underlying assumptions behind what has been presented earlier, but also, of course, I shall bring my own set of priorities and prejudices.

I am an ethologist by training and most certainly not an ecologist. However, I do have some background in the matter of conservation, which has been an obsession of mine for many years. All will recognise two clichés from our field of concern, 'Think globally, act locally' and 'Only one earth'. They catch nicely a feeling I have had throughout as I listened to speakers at the conference. One senses the uniqueness of Scotland and its astonishing beauty of life and form, although my rush of love is always tempered with anxiety. It is so obviously fragile and, in some regards, so obviously threatened. Then again we all recognise that uniqueness is relative. Scotland can never be isolated and accordingly my first thought is global and relates our situation to that of the planet as a whole.

Some years ago, a well-known religious leader set out some directives under the heading '*Veritatis splendor*'. At the time many of us felt that he was being selective with the truth whose splendour he celebrated. Table 24.1 represents my selection, which I believe has more relevance to our concerns here and to the future pattern of life on our planet. These figures are, of course, approximations derived from annual estimates for the most part. Nevertheless they are real and err on the side of caution. Unless we address some of these issues within the next decade or so, then our Scottish actions on biodiversity are in danger of being reduced to trivial tinkerings on the margin: another example of deckchair-shifting on the Titanic.

'Sustainability' sits alongside 'biodiversity' as a currently fashionable term, much paraded by politicians when talking about plans for development. Some argue that the term is meaningless and that our attempts to define it are both feeble and inconsistent. Yet whatever sustainability is or could be, it most certainly is *not* to be seen in the data in Table 24.1. The most crucial factor is the increase in human population. Upon this all the other factors ultimately depend, because here we are charting the resource demands laid on the planet. These demands are best assessed by numbers multiplied by resource consumption *per capita*, which immediately reveals to us that 'the population problem' is not just 'out there' in southeast Asia or sub-Sahelian Africa. It is also here in the rich north where surely any informed biologist would agree (as did the group attending the Institute of Biology's sym-

Table 24.1 · Veritatis splendor: a biologist's selection, erring on the safe side. Modified from an original idea by Professor R. W. Short.

VERITATIS SPLENDOR?
A day in the life of our world towards the end of the millennium

385,000 births

135,000 deaths
(40,000 of infants under 4 years of age)

The population will increase by 250,000
(more than 2 extra per second)[a]

Over 300 square kilometres of tropical forest will be cleared[b]

Over 50 species of living organism will become extinct[b]

Fossil fuel (oil, gas, coal) equivalent to over 25 million tons of oil will be consumed[c]

[a] Derived from United Nations (1996).
[b] Derived from Wilson (1992). In 1989 he calculated 390 km^2 d^{-1} cleared with rates of clearance likely to rise. He calculated extinction at 74 species d^{-1}, *for rain forest habitats alone*.
[c] Derived from the GlobEcho model of the Resources Use Institute, Edinburgh. See King and Slesser (1994).

posium over 30 years ago (Taylor,1970)) that, along with the rest of Europe, Britain is already grossly overpopulated in relation to its land and its resource consumption. Population control alone will not solve our problems, but again biology teaches us that actions that are not sufficient to achieve a result may nevertheless be absolutely essential. The most optimistic forecasts of the United Nations (UN) suggest that, barring disasters, Earth will have to cope with about twice the present number of people by the middle of the next century. Apart from any other considerations, it will be hard to maintain biodiversity in the face of this voracious monoculture.

I make no excuse for that diversion onto global territory. There *is* only one earth and population growth will soon affect us all directly. Consider the attention we conservationists pay to the agricultural 'set-aside' schemes and what they can do to support biodiversity. These are all predicated on the existence of embarrassing food surpluses in Europe. How long are those going to persist into the next millennium? We already import huge amounts of food, especially for feeding livestock. During the Second World War we 'dug for victory' and grew vegetables on every available space; even the quadrangles and courts of Oxbridge were not always immune! I can envisage a near future in which once again we have to hammer productive ground to its very limits with so-called marginal land coming under increasing pressure.

However, one has to be a short-term optimist; how else can we be motivated into action and persuasion? Apart from the fates of species such as the great auk (*Pinguinus impennis*) and the thylacine (*Thylacinus cynocephalus*), nothing is yet totally irreversible if we make rational decisions. Speaker after speaker at this meeting has outlined constructive action that can be taken now.

So, I return to our first duty, to act locally to put our own house in order. We start with certain advantages, for Scotland – numerically at least – has a better balance than the rest of the UK: 10% of the population and 25% of the area. Much of that area is designated as 'least favoured' and that has protected it from some of the worst excesses of human development. Cynics might add that the climate and the midge have done the rest!

Several contributors have grappled with definitions of biodiversity (see, for example, Scott, this volume). I appreciated Professor Michael Usher's frank recognition that Scottish Natural Heritage (SNH) could not realistically address the conservation of the Protista, bacteria, viruses and the like! None of us can; we lack even the scales to comprehend their diversity properly, nor do we have any real idea of what their conservation means, let alone how to bring it about.

Of course, their fate is inextricably tied up with the rest of us metazoans, although many of them will be much more resilient than we are. With our familiar anthropocentric viewpoint we can readily put the issue to one side for the moment and accept that we know pretty well what we mean when we define our aims in relation to biodiversity as 'ensuring Scotland has a healthy environment rich in wildlife and sustainably supporting human communities', or something along these lines. Two main questions then arise: (i) how? and (ii) why?

24.2   How?

The question of how we conserve biodiversity most closely engages our professional expertise as biologists, for the first essential is to know what is there now and also what was there in the past. Because it immediately involves specialist knowledge, it is difficult to regard 'biodiversity' as a research field in its own right; rather, it is a topic that brings together specialists around a common aim, which is certainly not based just in biology. This volume contains contributions from just such a group of distinguished specialists, some dealing with particular taxonomic groups, others at the level of ecosystems. At all stages we will need a combination of the particular and the holistic. I am very glad to note contributors point out that survey work is a continuous process. Long-term monitoring of flora and fauna using consistent sampling techniques is essential if we are to extract signal from noise in what is going on in our environment (see Mackey, this volume). It is the basis for any management procedure and will, for example, become a crucial part of national attempts to assess the effects of predicted climate changes. Biologists may have to emphasise again and again to the outside world that with every accumulating year of data, the next year's data become *more* valuable. The three-year research grant round is not the appropriate funding basis for this kind of vital work.

Other contributors emphasise our urgent need to train young biologists in the neglected classical techniques of systematic botany and zoology (see Shaw, this volume). Some groups do attract a great deal of popular attention; there is never a shortage of ardent ornithologists and the British Trust for Ornithology's common bird census shows how they can be employed to yield first-class data on biodiversity changes. The Botanical Society of the British Isles can do likewise. Most invertebrate groups can muster only a fraction of such specialists and most University biology courses are doing little training that would redress the balance. I do believe that there is now the beginning of a recognition that taxonomy is not just for pedants and that apart from its important applications it is also a means to explore important issues in ecology and evolutionary biology. The new techniques of molecular genetics have opened up a new range of possibilities to add to the old skills. However, Dr. Mark Young drew attention to an extraordinary problem with some modern students. Such are the sensibilities about the use of animals in science that he was made to feel diffident about wielding a butterfly net! We shall get nowhere unless we can educate about balance, emphasising the difference between sentimentality and informed concern, which carries with it respect.

A recurring issue of debate concerning the 'how' of biodiversity is whether we manage for species or for habitats. It has cropped up at this conference and many would argue that it is an artificial distinction, for whichever is chosen, it will always benefit the other. In practical terms I suggest that one first concentrates on habitats, beginning from soils and plants, and the more extensive the continuous areas the better. Although luckier than much of England, Scotland still suffers from fragmentation of habitats, reducing population sizes of their component species. For the most part, with a habitat-based approach, the rest will follow, although we have heard at the conference of a number of cases where some specific actions may be able to encourage particular species and increase their success without reducing that of others: water voles and corncrakes, for example (Racey, this volume; Kerr and Bain, this volume). In addition there will be places where we are involved in habitat creation or restoration and sometimes with deliberate re-establishment (a better term than re-introduction) of species that have become extinct in Scotland through human agency. In practice, surely, there is little distinction to be made: whatever we do we are managing habitats. The degree to which we emphasise individual species is likely to be determined as much by public relations as by biology.

Management is a continuous process and it is rarely possible to interfere once only. Humans have so dominated Scotland's environment since the retreat of the ice that now only the coasts and the high mountain tops remain unaltered. To enhance biodiversity we must interfere again and restore some of the original features, most notably with the expansion of large areas of native woodland. There can remain enough of the open hills to satisfy anybody and their biodiversity can also be greatly enhanced, for there is plenty of evidence that they are in a degraded state. We inherit a mess! Large areas are still managed as 'traditional sporting estates' for deer stalking or for grouse as they have for over 150 years (Thompson *et al.*, this volume). The Victorian gamekeepers left good records of their prey and they make fascinating, if heartbreaking, reading now. Lister-Kaye (1994) and Smout (1992) both give accounts of the predatory 'vermin' that Scotland supported then, living alongside fine harvests of deer, red and black grouse To quote from Smout:

> ... on a single estate south of the forests of Gaick and Glen Feshie, well over a thousand kestrels and buzzards, 275 kites, 98 peregrine falcons, 78 merlins, 92 hen harriers, 63 goshawks, 106 owls, 18 ospreys, 42 eagles and sundry other hawks, in only three years, 1837–1840 ... The estate near Glen Feshie reported the destruction of some 650 pine martens, wild cats, polecats, badgers and otters in three years.

Lister-Kaye reports on similar and equally staggering carnage. We may quibble about exact figures and allow for some exaggerations but, none the less, as Smout says,

> At least as significant as the fact of their destruction is what the former volume of predators reveals about the volume of prey species in the Highlands – voles, mice, hares, small birds and so on. They clearly no longer exist in anything like the densities necessary to support such numbers of predators, presumably because of damage done to their habitat by two centuries of modern land use.

Lister-Kaye (1994) is clear that '... a sustainable land ethic ... has to be based upon fundamental ecological principles which permit nature to replenish the exhausted habitats upon which the whole sporting economy is founded: ...'. It is truly ironic that even now the call from the overgrazed, overburnt Scottish uplands is for restoration of the culling of raptors because of low grouse populations! We have a major task of persuasion ahead of us. SNH's practical demonstration of restoration at Creag Meagaidh and the work of the Royal Society for the Protection of Birds at Abernethy are helping to lead the way.

Interference has to be repeated and regular in those transitional habitats, such as woodland clearings, field verges or sand dunes, which we also cherish for their biodiversity. Here, at its best, human activity and biodiversity go together and this is a point which conservationists should sing out loud and clear for public consumption. The two are still so often portrayed as being in inevitable conflict, but we can all recognise the contribution that traditional farming methods make towards biodiversity in the Scottish countryside. Several contributors have suggested various straightforward ways in which the pattern of agricultural and forestry subsidies could be made to work well for the Scottish environment while retaining rural incomes (see Housden, this volume; Kerr and Bain, this volume). The Common Agricultural Policy was set up for admirable social reasons; it has been a disaster for the most part but there are now good grounds for optimism because its reform cannot be delayed much longer. With some sensible biological input and some teeth to ensure cross-compliance between grant and conservation result, we could easily move forward.

As we develop management procedures for Scotland's biodiversity it will behove us to recognise that we face limitations. Firstly, the earth is a dynamic system even without human activities. Inevitably we inherit and, just as inevitably, we shall pass on an unstable situation; species will always wax and wane whatever we do. We have heard at this meeting something of the dynamic nature of the genus *Ranunculus*, and especially the survival of the hybrid Loch Leven spearwort long after one of its parents became extinct (Sydes, this volume). However we try to protect some species, they may evade us by hybridising with other species, the hybrids often flourishing at the expense of the parents! Habitats are bound to change and their inhabitants must change with them. Climate change seems already in train and this will have major effects in Scotland, where many species seem now to be living close to the edge of their range (Watt *et al.*, this volume). It will be vitally important to monitor changes accurately but we must not become biological Canute's courtiers and waste effort trying to hold back the inevitable.

Secondly, we must repeatedly examine our own procedures and ask whether they are themselves sustainable. For example, the essential protection of Scotland's goose populations coincided with an increase in their wintering areas in the cultivation of improved grasslands and autumn-sown arable crops to provide early spring feeding for livestock. Geese have descended on the pastures and crops and the damage they cause has often had to be met by compensation payments to farmers and by the establishment of some reserves specially cultivated for the birds. The result has been a dramatic increase in the goose populations: barnacle geese on Islay have more than quadrupled from 5,000 to almost 25,000 in 30 years, for example. It may be that food resources in the wintering areas are a limiting factor for the recruitment of young birds into the breeding populations in Greenland, Iceland and elsewhere. Whatever the cause of the increase, it will have to be managed. A number of years ago the Nature Conservancy Council for

Scotland, SNH's predecessor, was faced with a rapid increase of herring gulls: up to 15% per annum. Juvenile mortality was being greatly reduced by access to municipal waste tips in winter. An extensive cull had to be carried out on the Isle of May, where the gulls were swamping out tern colonies. This cannot have been an easy decision, but one significant human intervention had to be followed by a second one, certainly required in the interests of biodiversity. Recently the Chief Executive of SNH suggested that the goose situation required close watching and that we should attempt to plan for a sustainable maximum (Crofts, 1995). He was pilloried by some conservationists for hinting at the future control of numbers, a completely unjustified response to a very reasonable suggestion. There are few cases where we can just let things run and expect that non-human processes will lead to an optimum balance. Indeed it may be some measure of our success in promoting biodiversity that such management problems will increase.

24.3 Why?

The question of why we should conserve biodiversity is of particular importance, for it relates our work and endeavours to the aspirations of society as a whole. A glib answer to the question, although none the less true for that, is that we need a healthy environment for our own survival and that biodiversity is often an excellent measure of such health. In the eyes of the public this answer will rarely be sufficient on its own, because many people's idea of a healthy environment is determined as much by television advertisements for breakfast cereal as by biological concern.

At one time, I had fairly regular contact with a reporter on a Scottish tabloid newspaper who was genuinely interested in environmental issues and wanted to help get them across. He would always bring me down to earth with, 'What's in it for my punters?' He intended no condescension then, nor do I now. More and more people are becoming better informed about environmental issues: after all, wildlife programmes compete on level terms with sex and violence for prime-time television. Nevertheless we must acknowledge that we still have a major educational task in order to convert this interest into a concern by the great majority of people for other living things, especially in relation to the environment as a life support system for all of us.

Myers (1996) reviews very clearly what he calls the 'environmental services of biodiversity'. Certain elements are regularly paraded as justifications. First is the 'gene bank' argument. There could be no better measure of how narrowly focused is the popular idea of how 'Nature' supports us. The crucial role that tropical moist forest plays in the regulation of climate, especially patterns of rainfall, is relatively ignored. Instead we hear constantly in the media of how the loss of these forests may deprive us of the cancer cures that might await us in some unexplored plant species! There may indeed be something to this argument and certainly a good deal of serious exploration is being undertaken. The value of ancient breeds of domestic animals is also acknowledged. Molecular genetics may enable us to isolate desirable genes and synthesise their products or transpose them into other plants or animals. The genes could come from the wild just as readily as from domestic stock. Whether this kind of exercise will provide sufficient economic will power to preserve large sources of biodiversity remains to be seen. It would be unwise to rely on it, but we should nevertheless embrace it for now. The end is so important that all arguments must be pressed. There are, after all, many different ways to skin a cat!

A related, severely practical argument for biodiversity concerns the perpetual danger we run with our concentration on a pathetically few plant species for food for ourselves and for livestock. We rely on gigantic monocultures of about 20 species of plant to provide food for about 90% of the human population; three species – maize, rice and wheat – supply over half. We know already of the ceaseless arms race we have to wage with pest organisms to keep this narrow range of food species in a productive state. Common prudence would suggest the value of a major effort to extend our range of possibilities and this will involve good botany as well as good agriculture.

24.4   Conclusions

The 'how' and the 'why' of the conservation of our biodiversity come finally together when we consider how to develop public involvement. One of the most valuable post-Rio policies for Britain has been the development of the local Agenda 21 plans. This should concentrate minds upon acting locally. If we can get real public involvement, the results will eventually show up at the political level and ensure that environmental issues stay high enough on the agenda to be given resources.

We have to accept that, for the present, the public's interest in biodiversity will run along different lines from many of us. Within the Wildlife Trusts we often refer to the 'OOOh Factor!', whereby we acknowledge that orchids, ospreys and otters are good for grabbing the public's attention. Unhesitatingly we should go along with this, because once attention is gained it can be retained and extended if we are any good at our jobs. For this reason, the concentration on species lists in various biodiversity action plans may be very important. My earlier backing for habitats rather than species related to management action on the ground. It will be easier to capture the imagination of the public with individual plants and animals to which a story can be attached, together with a plan for the conservation of each and the monitoring of its progress. How many of us were first lured into biology via the magic of identifying and studying particular groups?

For all its empty hills and glens, Scotland is a highly urbanised country. I suppose about 85% of Scots live in towns or cities, many of them offering too few opportunities for regular contact with the natural world. Surely it is part of everyone's birthright to have such opportunities. In purely practical terms for all of us here, there can be no better opportunity to increase biodiversity than by paying some attention to urban green spaces and their management for wildlife. We shall not be dealing with endangered species for the most part (although all sorts of suprises may await us: a possible new species of orchid turned up on a

Glasgow spoilheap!) but the relative degree of enrichment can be spectacular compared to what we can achieve in more complex habitats. The effect on public opinion may be orders of magnitude higher.

I know we are all totally convinced that conservation is not a luxury, an optional extra we can afford when the economy picks up, but a vital necessity if our species is to have a long-term future. Ultimately it will require some tough decisions about our lifestyle, although I believe that if we succeed then in retrospect they won't seem tough at all. We are still a minority in holding these views, although an increasing one. We are beginning to win the arguments and we have a very positive message to put across. In what we have heard at this conference and the manner in which it has been delivered, it is clear to me how much of our motivation comes from a love of the natural world. In this we are fortunate; let us hold on to it as we try to share it with others. Socrates is quoted as saying, 'Without love there is no wisdom, only learning'. We shall need plenty of wisdom as we work to conserve biodiversity in this beautiful country.

References

Crofts, R. 1995. The environment – who cares? Scottish Natural Heritage Occasional Paper No. 2.

Lister-Kaye, J. 1994. Ill fares the land. Scottish Natural Heritage Occasional Paper No. 3.

Myers, N. 1996. Environmental services of biodiversity. *Proceedings of the National Academy of Sciences, USA,* **93,** 2764–2769.

King, J. and Slesser, M. 1994. The natural philosophy of natural capital: can solar energy substitute? In van den Bergh, C. J. M. and van den Straaten, J. (Eds.) *Towards Sustainable Development: Concepts, Methods And Policy.* Washington D.C., Island Press, 139–163.

Smout, C. 1992. The Highlands and the roots of green consciousness. Scottish Natural Heritage Occasional Paper No. 1.

Taylor, L. R. (Ed.) 1970. *The Optimum Population for Britain.* London & New York, Academic Press.

United Nations. 1996. *World Population Prospects: The 1996 Revision.* New York, United Nations.

Wilson, E. O. 1992. *The Diversity of Life.* Cambridge, Massachusetts, Harvard University Press.

INDEX

INDEX

Printed in Scotland for The Stationery Office Limited. J20864, c8, 10/97, CCN 020249

**Published by the Stationery Office
and available from:**
The Stationery Office Bookshops
71 Lothian Road Edinburgh EH3 9AZ
(counter service only)
59–60 Holborn Viaduct London EC1A 2FD
(temporary location until mid-1998)
Fax 0171-831 1326
68–69 Bull Street Birmingham B4 6AD
0121-236 9696 Fax 0121-236 9699
33 Wine Street Bristol BS1 2BQ
0117-926 4306 Fax 0117-929 4515
9–21 Princess Street Manchester M60 8AS
0161-834 7201 Fax 0161-833 0634
16 Arthur Street Belfast BT1 4GD
01232 238451 Fax 01232 235401
The Stationery Office Oriel Bookshop
The Friary Cardiff CF1 4AA
01222 395548 Fax 01222 384347

**The Stationery Office publications
are also available from:**
The Publications Centre
(mail, telephone and fax orders only)
PO Box 276 London SW8 5DT
General enquiries 0171-873 0011
Telephone orders 0171-873 9090
Fax orders 0171-873 8200
Accredited Agents
(see Yellow Pages)
and through good booksellers